R. D. Driver

Ordinary and Delay Differential Equations

Applied Mathematical Sciences 20

Springer-Verlag New York · Heidelberg · Berlin

R. D. Driver
University of Rhode Island
Department of Mathematics
Kingston, Rhode Island 02881

QA
1
.A647
v. 20

AMS Classifications: 34-01, 34B05, 34D99, 34Kxx

Library of Congress Cataloging in Publication Data

Driver, Rodney David, 1932-
   Ordinary and delay differential equations.

   (Applied mathematical sciences ; v. 20)
   Bibliography: p.
   Includes index.
   1. Differential equations. 2. Differential equa-
tions--Delay equations. I. Title. II. Series.
QA1.A647   vol. 20   [QA372]   510'.8s [515'.352]
                                          76-58452

Printed in the United States.

ISBN 0-387-90231-7   Springer-Verlag New York

ISBN 3-540-90231-7   Springer-Verlag Berlin Heidelberg

To Carole, Dave, Karen, and Bruce

PREFACE

This textbook is designed for the intermediate-level course on ordinary differential equations offered at many universities and colleges. It treats, as standard topics of such a course: existence and uniqueness theory, linear systems, stability theory, and introductory phase-plane analysis of autonomous second order systems.

The unique feature of the book is its further inclusion of a substantial introduction to delay differential equations. Such equations are motivated by problems in control theory, physics, biology, ecology, economics, inventory control, and the theory of nuclear reactors.

The surge of interest in delay differential equations during the past two or three decades is evidenced by thousands of research papers on the subject and about 20 published books devoted in whole or in part to these equations. The books include those of Myškis [1951][*], El'sgol'c [1955] and [1964], Pinney [1958], Krasovskiĭ [1959], Bellman and Cooke [1963], Norkin [1965], Halanay [1966], Oğuztöreli [1966], Lakshmikantham and Leela [1969], Mitropol'skiĭ and Martynjuk [1969], Martynjuk [1971], and Hale [1971], plus a number of symposium and seminar proceedings published in the U.S. and the U.S.S.R.

These books have influenced the present textbook. But none of them can be considered as an introductory text -- with worked examples, problems, and answers or hints for the problems. I believe that the present volume does meet these

---

[*]A name followed by a date in brackets indicates an item listed in the References starting on page 466.

specifications, and I hope that it will facilitate the introduction of delay differential equations in the senior and
beginning-graduate level curricula.

Chapters I and III provide introductory material on
scalar differential equations to make the book self-contained.
This material will probably be familiar to most readers and
can be passed over lightly. However, Chapter II *is* recommended for study as it sets the style of the book in seeking to
answer the uniqueness questions for any differential system
at an early stage.

For some parts of the book, the reader will find it
helpful to have had a semester of advanced calculus, or equivalent exposure to ε-δ arguments. The actual results required from advanced calculus are presented in the Appendices
starting on page 449.

Starred problems are perhaps a little more challeng-
ing than the other problems in the same section. A star on a
section, theorem, or proof indicates material which can be
omitted without loss of continuity.

I owe thanks to many people in the writing of this
book. The first debt is to the students who worked through
several preliminary editions of the manuscript as it developed
in courses taught at the University of Rhode Island since
1970. Their questions, objections, and suggestions have
helped to smooth out many rough spots. Further valuable suggestions have been made by my colleagues and by a number of
reviewers, some of the latter being anonymous to me. Among
the individuals whose contributions I gratefully acknowledge
are Mike Berman, Dean Clark, Kenneth Cooke, William Dandreta,

Chris Doyle, Michael Greenberg, Deb Guha, Jack Hale, Kenneth Hoffman, Deh-phone Hsing, G. E. Ladas, V. Lakshmikantham, Ken Levasseur, Larry Sirovich, Roy Streit, Stephen Travis, and Peter Tsen.

An author quickly discovers the importance of a good typist, and I have been fortunate in having the help of several excellent ones. Patricia Haigh, Pam Paden, and Dale Potter efficiently typed the four preliminary mimeographed editions of this book for classroom use; and Kate MacDougall quickly and accurately prepared the final camera-ready copy for this volume with the help of Eleanor Addision on the figures.

R. D. Driver
Kingston, R. I.
November, 1976

TABLE OF CONTENTS

Chapter I   ELEMENTARY METHODS FOR ORDINARY
                DIFFERENTIAL EQUATIONS OF FIRST ORDER   1

  1.   Examples and classification   2

  2.   Linear equations   7

  3.   Separable equations   19

Chapter II   UNIQUENESS AND LIPSCHITZ CONDITIONS
              FOR ORDINARY DIFFERENTIAL EQUATIONS   26

  4.   First order scalar equations   27

  5.   Systems of equations   40

  6.   Higher order equations   51

  7.   Complex solutions   63

  8.   A valuable lemma   72

  9.   A boundary value problem   78

Chapter III   THE LINEAR EQUATION OF ORDER n   84

  10.   Constant coefficients (the homogeneous case)   85

  11.   Linear independence and Wronskians   98

  12.   Constant coefficients (general solution for
        simple h)   109

  13.   Variation of parameters   124

Chapter IV   LINEAR ORDINARY DIFFERENTIAL SYSTEMS   139

  14.   Some general properties   140

  15.   Constant coefficients   150

  16.   Oscillations and damping in applications   168

  17.   Variation of parameters   180

  18.   Matrix norm   189

*19.   Matrix exponential   197

  20.   Existence of solutions (successive approximations) 212

Page

Chapter V   INTRODUCTION TO DELAY DIFFERENTIAL
                EQUATIONS   225

   21.   Examples and the method of steps   226

   22.   Some distinguishing features and some "wrong"
        questions   245

   23.   Lipschitz condition and uniqueness   255

Chapter VI   EXISTENCE THEORY   268

   24.   Ordinary differential systems   269

   25.   Systems with bounded delays: notation and
        uniqueness   285

   26.   Systems with bounded delays: existence   299

Chapter VII   LINEAR DELAY DIFFERENTIAL SYSTEMS   313

   27.   Superposition   314

   28.   Constant coefficients   319

   29.   Variation of parameters   331

Chapter VIII STABILITY   340

   30.   Definitions and examples   341

   31.   Lyapunov method for uniform stability   352

   32.   Asymptotic stability   362

   33.   Linear and quasi-linear ordinary differential
        systems   374

   34.   Linear and quasi-linear delay differential
        systems   384

Chapter IX   AUTONOMOUS ORDINARY DIFFERENTIAL SYSTEMS   400

   35.   Trajectories and critical points   401

   36.   Linear systems of second order   414

   37.   Critical points of quasi-linear systems of
        second order   424

   38.   Global behavior for some nonlinear examples   431

APPENDICES

   1.   Notation for sets, functions and derivatives   449

Page

APPENDICES

   2. Some theorems from calculus   454

REFERENCES   466

ANSWERS AND HINTS   476

INDEX   498

Chapter I. ELEMENTARY METHODS FOR ORDINARY
            DIFFERENTIAL EQUATIONS OF FIRST ORDER

Whenever one seeks to analyze mathematically a problem arising in the "real" world, the first order of business is the selection of a mathematical "model" or formulation for the problem. That is, one must choose or invent some mathematical problem tᴏ represent the real problem.

The mathematical models chosen are often differential equations, i.e., equations involving the derivative(s) of some unknown function(s). Typical elementary examples have the form

$$x'(t) = f(t,x(t)),$$

where x' is the derivative of the unknown function x, and f is a given continuous function. And the goal is to find the unknown function, or at least to obtain some useful information about it.

In this book we shall discuss problems from physics, engineering, biology, and ecology.

Applied scientists will presumably have a natural interest in such examples. For differential equations have contributed immeasurably to the advancement of physics and engineering; and more recently they are playing a significant role in biology and ecology and even economics. But many mathematicians have great interest in applications too. In fact, real-world-motivated problems in differential equations have inspired some of the most exciting mathematical research.

1. EXAMPLES AND CLASSIFICATION

One of the simplest examples of a differential equation is the standard model for the decay of a radioactive material. If $x(t)$ is the quantity of some radioactive substance present at the instant of time t then it is generally assumed that

$$x'(t) = -ax(t),$$

where a is a positive constant. This equation expresses the assertion that the rate of decay of the material at any instant is proportional to the quantity of material remaining at that instant.

It is easily verified that for any constant c the function defined by

$$x(t) = ce^{-at}$$

is a "solution" of this differential equation. Some members of this infinite family of solutions are indicated in Figure 1. Note that negative values of c are perfectly acceptable

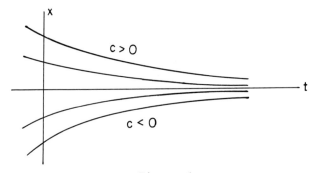

Figure 1

as far as the differential equation is concerned, although they would not be meaningful for our physical problem.

We apparently pulled the solutions, ce^{-at}, out of thin air. So it is natural to ask whether any other type of function, besides these exponentials, could also satisfy this same differential equation. The answer to this question is "No", and we shall prove it in Section 2. Thus the function defined by $x(t) = ce^{-at}$ with unspecified constant c is called the "general solution". Every possible solution can be obtained from this by an appropriate choice of c.

To completely determine a particular solution, some additional data is needed. This often takes the form of an "initial condition", say $x(t_0) = x_0$, the amount of material present at the "initial instant" $t = t_0$. Substituting $t = t_0$ in the general solution gives $x_0 = ce^{-at_0}$ or $c = x_0 e^{at_0}$. Thus we must have

$$x(t) = x_0 e^{-a(t-t_0)}.$$

This represents an exponential time decay from the initial amount.

Real-world problems are often represented by simple differential equations which can be solved exactly, as in the above example. But it should always be recognized that the differential equation itself is at best only an approximate representation of the problem being considered. Approximations might enter because of our imperfect knowledge of the laws governing the behavior of the real world or because of imperfect measurement of input data (such as initial conditions) for the problem. And in many cases we deliberately make approximations in order to reduce the mathematical equation to a tractable form.

For example, the representation of the decay of a radioactive material considered above is certainly not perfect.   If we believe in atoms we must admit that the function x will not even be continuous. And it certainly cannot have a derivative at points of discontinuity. If we overlook this difficulty we should still admit that, for a random collection of atoms, the assertion that the rate of decay is some constant multiple of the number of atoms present is at best a statistical average. Such an approximation would become quite unacceptable if only a few atoms were involved.

In other situations more serious approximations enter.

One of the most pervasive approximating assumptions, often made unconsciously, is that of "simultaneity". One usually assumes that the values of the unknown functions and their derivatives, which occur in differential equations, should all be evaluated at the same instant. For example, in modeling radioactive decay we wrote $x'(t) = -ax(t)$ and not $x'(t) = -ax(t-1)$ or $x'(t) = -ax(t/2)$.

Thus most books on differential equations treat equations such as

$$x'(t) = -ax(t),$$

where a is a given constant, or

$$x''(t) = -\frac{K}{x^2(t)},$$

where K is a given constant, or

$$mx''(t) + bx'(t) + kx(t) = h(t),$$

where m, b, and k are given constants and h is a given function, or

$$\frac{\partial^2 u}{\partial x^2}(x,y) + \frac{\partial^2 u}{\partial y^2}(x,y) = 0,$$

where u is an unknown function of x and y.

Of these four examples, the first three, involving unknown functions of only a single variable, are called <u>ordin-
ary differential equations</u>. The second of these equations
describes the free fall of a mass under the sole influence of
gravity. The third equation covers, for example, the oscillations of a mass on a simple spring.

Differential equations, involving partial derivatives
of unknown functions of two or more variables are called
<u>partial differential equations</u>. The fourth example given
above is a famous partial differential equation of physics --
Laplace's equation. This text does not treat partial differential equations.

It has been so universally assumed that all functions
should be evaluated at the same point that one often omits
the "independent variables" (arguments of the unknown functions) altogether. Thus the above equations are usually
written simply as

$$x' = -ax,$$

$$x'' = -\frac{K}{x^2},$$

$$mx'' + bx' + kx = h(t),$$

and

$$\frac{\partial^2 u}{\partial x^2} + \frac{\partial^2 u}{\partial y^2} = 0.$$

However, since the middle of the 18th century, mathematicians have from time to time considered problems involving

relations between unknown functions and their derivatives evaluated at different arguments. Thus one might encounter an equation such as

$$mx''(t) + bx'(t) + qx'(t-1) + kx(t) = h(t).$$

Since the 1930's and 40's the number of such "delay differential equations" arising in practical applications has escalated rapidly. Chapter V introduces several examples in which delay differential equations arise and begins the study of their solutions.

The first four chapters are devoted to certain ordinary differential equations. In particular, the present chapter gives two elementary but useful methods for solving certain special types of ordinary differential equations. We shall use the accompanying examples as vehicles for introducing the concepts of "uniqueness" and "continuability" of solutions.

Some important terminology is as follows.

The order of a differential equation is the order of the highest derivative involved in that equation. In this chapter we will consider ordinary differential equations of first order. Such equations have the general form $F(t,x(t),x'(t)) = 0$, where F is a given function. But we shall restrict ourselves to those which can be written in the more special form

$$x'(t) = f(t,x(t)),$$

or briefly

$$x' = f(t,x),$$

where f is a given function. Such an equation is often
considered together with an appropriate initial condition,
say

$$x(t_0) = x_0.$$

A _solution_ is any continuous function, on some open
interval, which satisfies the given differential equation together with the initial condition, if any. Throughout this
chapter solutions will be real-valued functions.

We shall say a solution x is _unique_ if every other
solution agrees with x, as far as both are defined.

Certain standard mathematical terms and symbols are
described in Appendix 1 for reference as needed.

2. LINEAR EQUATIONS

The general _linear_ ordinary differential equation of
first order which we will consider is

$$x' + a(t)x = h(t). \tag{1}$$

Here a and h are given functions continuous on some open
interval $J = (\alpha, \beta)$ where $-\infty \le \alpha < \beta \le \infty$. We shall often
seek a solution of (1) which also satisfies the initial con-
dition

$$x(t_0) = x_0 \quad \text{where} \quad t_0 \in J. \tag{2}$$

An equation of the form

$$b(t)x' + c(t)x = g(t),$$

where b, c, and g are given continuous functions, can be

put in the form (1) on an interval J provided $b(t) \neq 0$
for all t in J. On such an interval one merely divides
through by $b(t)$.

Equation (1) becomes particularly simple in the special
case when $a(t) \equiv 0$. Indeed the resulting differential equation,

$$x' = h(t),$$

with initial condition (2) is solved at once using the fundamental theorem of calculus. One has

$$x(t) = x_0 + \int_{t_0}^{t} h(s)ds.$$

A more interesting example of a linear first order
equation was encountered in Section 1. In that example we
had $a(t) = a$, a constant, and $h(t) = 0$. We shall now complete the analysis of that equation.

<u>Example 1.</u> Find all solutions (on an interval J) of the
differential equation

$$x' + ax = 0.$$

We proceed as follows. <u>Assume</u> that some function x
is a solution. Then multiply both sides of the differential
equation by the exponential e^{at}. (The reason for this will
be seen shortly.) It follows that x must satisfy

$$x'e^{at} + axe^{at} = 0,$$

or

$$\frac{d}{dt} [xe^{at}] = 0,$$

and the only functions with zero derivative (on an interval)
are the constant functions. Hence

$$x(t)e^{at} = c$$

for some constant c, or

$$x(t) = ce^{-at}.$$

In other words the only functions which <u>can</u> <u>possibly</u> be solutions of our differential equation are of the form $x(t) = ce^{-at}$ where c is some (real) constant.

But we already agreed, in Section 1, that such exponentials <u>do</u> satisfy the differential equation. Thus the problem is completely solved. For each c, $x(t) = ce^{-at}$ is a solution, and no other type of function can be a solution. Hence we refer to ce^{-at} as the "general solution" of $x' + ax = 0$.

The key trick in the solution of Example 1 was the multiplication by e^{at}. This multiplicative factor yielded the combination

$$x'e^{at} + axe^{at} = \frac{d}{dt}[xe^{at}],$$

which is easily integrated. The multiplier e^{at} is called an "integrating factor".

Students usually ask at this point, "How did you know in advance that multiplication by e^{at} would be helpful?"

We will not give a very satisfactory answer to this question. Let us just say that we have used a special case of a standard procedure for the solution of equations of this type. More generally, we shall always begin the analysis of Equation (1) by multiplying both sides of the equation by the "integrating factor"

$$e^{v(t)} \quad \text{where} \quad v(t) = \int a(t)dt \qquad (3)$$

is any indefinite integral of $a(t)$.

Some mathematician more than 200 years ago discovered that multiplication of Equation (1) by $e^{v(t)}$ gives an integrable combination on the left hand side, as in Example 1. Without worrying about how this trick was first discovered, let us accept it as a method worth using. We now show that it always works.

Justification of the Method. Assume, at first, that (1) has a solution x on some open interval $J_1 \subset J$. We refrain from assuming that a solution exists on the entire given interval J . Let $v(t) = \int a(t)dt$ as in (3), and multiply Equation (1) by $e^{v(t)}$ to obtain

$$x' e^{v(t)} + a(t)x e^{v(t)} = h(t) e^{v(t)}.$$

or

$$\frac{d}{dt} [x(t) e^{v(t)}] = h(t) e^{v(t)} \tag{4}$$

on J_1 . Verify this. Now integrate (4) to find

$$x(t) e^{v(t)} = \int h(t) e^{v(t)} dt + c,$$

or

$$x(t) = c e^{-v(t)} + e^{-v(t)} \int h(t) e^{v(t)} dt, \tag{5}$$

where c is an arbitrary constant.

If, in addition to satisfying (1), x satisfies $x(t_0) = x_0$ for some t_0 in J_1 , we obtain a more explicit form for x as follows. Replace t by s in Equation (4) and then integrate from t_0 to t to find

$$x(t) e^{v(t)} - x_0 e^{v(t_0)} = \int_{t_0}^{t} h(s) e^{v(s)} ds.$$

Thus

$$x(t) = x_0 e^{v(t_0)-v(t)} + e^{-v(t)} \int_{t_0}^{t} h(s) e^{v(s)} ds. \qquad (6)$$

It is easy to verify that (6) does reduce to $x(t_0) = x_0$ when $t = t_0$. However, it should be emphasized that we have <u>not</u> yet shown that either (5) or (6) gives a solution of (1). What we have shown is that <u>if</u> Equation (1) has a solution, <u>then</u> it must have the form (5) or (6).

Note that, for any given value of c, (5) can always be considered as a special case of (6) obtained by an appropriate choice of x_0. So, to complete our analysis, the reader is now asked to show that the function x defined by (6) -- the sole <u>candidate</u> for a solution of (1) and (2) -- actually <u>is a solution</u> of (1) and (2). You can do this either by justifying the reversal of each step in the derivation of (6), or by directly substituting (6) into Equation (1). Furthermore, in this calculation, observe that you can actually take $J_1 = J$.

Thus we conclude that, <u>Equations (1) and (2) have a solution on the entire interval</u> J. <u>This solution is unique and is given by (6)</u>.

It should be noted that, in particular examples, the integrals occurring in Equations (5) and (6) may be difficult to evaluate. They may even be impossible to evaluate in terms of elementary functions. Thus, even though (6) exactly represents the unique solution of Equations (1) and (2), it may not be easy to interpret the result.

As already observed, (5) is implied by (6) if x_0 is considered arbitrary. Nevertheless, Equation (5), which emphasizes the arbitrariness of the constant c in the absence

of any specified initial condition, is the form often referred
to as the "general solution" of (1).

When we described the integrating factor $e^{v(t)}$ in
(3), $v(t)$ was "any" indefinite integral of $a(t)$. Suppose
we had chosen a different $v(t)$. Why would this not affect
the outcome? Problem 1.

In order to solve a specific linear, first order equation one could simply plug the appropriate functions a and
h into (5) or (6). However, it is strongly recommended
that, instead, the reader make a practice of using the method
of the general case -- and not the "formula" for the solution. Then all one must remember is that $\exp \int a(t)dt$ is
an integrating factor for (1).

Example 2. Solve the differential equation

$$t^{-1}x' = 3 - t^{-2}x$$

with the initial condition

$$x(2) = 3.$$

We must first put the differential equation into the form of
(1) by multiplying through by t and moving all terms involving x to the left hand side. This yields

$$x' + \frac{1}{t} x = 3t.$$

Since $a(t) = t^{-1}$, $h(t) = 3t$, and $t_0 = 2$, we should take
$J = (0,\infty)$. Why? Taking $v(t) = \int t^{-1}dt = \ln |t| = \ln t$, we
multiply the differential equation by the integrating factor
$e^{v(t)} = e^{\ln t} = t$ to obtain

$$x't + x = 3t^2.$$

Now, as they should, the two terms on the left hand side represent exactly the derivative of xt. (You should always
verify this step carefully. For if there has been any mistake in the calculation of $e^{v(t)}$ this is your chance to
catch it.) Thus

$$\frac{d}{dt} [xt] = 3t^2.$$

Integrating both sides from $t_0 = 2$ to t we find

$$tx(t) - 2x(2) = \int_2^t 3s^2 ds = t^3 - 8,$$

or, since x(2) = 3,

$$x(t) = t^2 - 2/t \quad \text{for} \quad t > 0.$$

It is no longer essential to verify that this x is a solution of our original problem, because our general analysis
says it must be (if we have made no mistake). However, a
direct substitution of x into the original equations does
provide the ideal check on our work.

Example 3. (A Mixing Problem). A tank contains 100 gallons
of a salt water brine. An inlet pipe discharges into this
tank, at the rate of 5 gallons per minute, brine containing
2 lb. of salt per gal.
At the same time fluid
leaves the tank through
an outlet hole at the
same rate. The mixture
in the tank is continually

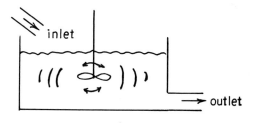

Figure 1

stirred so that the concentration can always be considered
uniform throughout. If initially the mixture contains one
pound of salt per gallon, determine the future concentration
as a function of time.

Let x(t) be the concentration of salt in the tank in
lb/gal at time t minutes after the initial instant t = 0.
Then the total amount of salt in the tank at time t is
100x(t) lb. Salt is being added to the tank at the rate of
5 × 2 = 10 lb/min., and is leaving at the rate of
5x(t) lb/min. Thus

$$100x' = 10 - 5x \quad \text{and} \quad x(0) = 1.$$

To put this differential equation into the form of Equation
(1), we divide through by 100 and get

$$x' + 0.05x = 0.1.$$

Then multiplying by the integrating factor $e^{0.05t}$ we find

$$\frac{d}{dt}[e^{0.05t}x] = 0.1\, e^{0.05t}$$

or, upon integrating,

$$e^{0.05t}x(t) - x(0) = 2(e^{0.05t} - 1).$$

Since x(0) = 1, this gives

$$x(t) = 2 - e^{-0.05t}.$$

This function describes, as would be expected, an increasing
concentration starting with 1 lb/gal at t = 0 and approaching 2 lb/gal as $t \to \infty$.

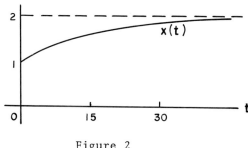

Figure 2

One assumption of the mathematical model in Example 3
appears particularly questionable. How can the brine always
be so perfectly mixed that the liquid leaving the tank has
just the same concentration as does the tank as a whole?
This assumption will be relaxed in a similar example in
Chapter V.

Problems.

1. If in (3) we chose, as the indefinite integral of a(t),
   some $v_1(t) \neq v(t)$, then $v_1(t) = v(t) + c_1$ for some constant c_1. Show that the use of $v_1(t)$ instead of $v(t)$
   always leads to the same form for $x(t)$, namely (5) or
   (6).

2. Show that (6) is equivalent to

   $$x(t) = x_0 e^{-\int_{t_0}^{t} a(s)ds} + \int_{t_0}^{t} h(s) e^{-\int_{s}^{t} a(u)du} ds.$$

3. (a) Find the general solution of

   $$x' + x = \sin t \quad \text{on} \quad J = (-\infty, \infty).$$

   That is, find a function involving an unspecified

constant, c, which satisfies the differential equation and which gives as special cases all solutions
of that equation. Hint. Proceed as in the derivation of (5) and assume the validity of the assertion
of part (c) below.

(b) Verify your solution by substitution. Why can you be
sure that an arbitrary initial condition of the form
$x(t_0) = x_0$ will always uniquely determine c in
your general solution?

(c) Using two integrations by parts, obtain

$$\int e^t \sin t \, dt = \frac{1}{2} e^t (\sin t - \cos t) + c.$$

Consider the equations in Problems 4 through 12 on $(-\infty, \infty)$.
Find a solution for each problem and determine some interval
J on which your solution is valid. Decide whether the solu-
tion is unique there.

4. $x' - x = 1$ with $x(2) = 3$.

5. $x' = (x + 1)\sin t$ with $x(t_0) = x_0$, where t_0 and x_0
are given constants.

6. $x' + 2x = t$.

7. $Li' + Ri = E$ with $i(0) = i_0$, where $L > 0$, $R \geq 0$, E
and i_0 are given constants. This is the equation for
the electrical current $i(t)$ (in amps) flowing in a
series circuit having inductance L (in henries), resistance R (in ohms), and source E (in volts). Figure
3.

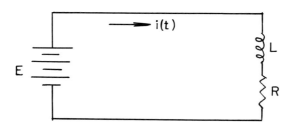

Figure 3

8.   (a) tx' + x = t + 1   with   x(1) = 0.

     (b) tx' + x = t + 1   with   x(-1) = 2.

     (c) Sketch graphs of the solutions of parts (a) and (b)
          and describe the difference between them.

9.   x' + 2tx = 1   with   x(0) = 5.

10. tx' - x = t.

11. (t-1)x' - x = t   with   x(-1) = -2.

12. (cos t)x' + (sin t)x = 1   with   x(0) = 1.

*13. Show that Problem 12 has a solution on   (-π/2,3π/2), but
     that the solution is not unique there.   Why does this not
     contradict the assertions in the text?

14. Assume that the radioactive isotope Strontium 90 decays
     exponentially according to the law   x' = -ax   considered
     in Section 1.   Use the fact that Strontium 90 has a half
     life of 28.1 years (i.e., half of the original amount de-
     cays in 28.1 years) to evaluate   a.   How much time must
     elapse after an atomic explosion before the resulting
     Strontium 90 is reduced to 10% of its initial amount?

15. An automobile cooling system contains 16 quarts of solu-
     tion, half of which is dirty 3-year-old antifreeze.   In
     hopes of saving time flushing out the mess, the owner
     of the automobile decides to pour fresh water in while
     the old solution is running out.   (Assume that the old
     solution runs out at a rate of 8 quarts per minute and
     the fresh water runs in at the same rate.)   He keeps the
     engine running so that the solution is continually
     stirred to a uniform shade of brown.   How long does it
     take to reduce the solution to only one per cent dirty
     3-year-old anti-freeze?

*16. The solution of Problem 7 depends upon the values of L,
      R, E, and i_0.   Think of L, E and i_0 as fixed while
      R is an adjustable parameter and call the solution
      $i(t; R)$.   Prove that $\lim_{R \to 0} i(t;R) = i(t;0) = Et/L + i_0$
      for each t.

*17. Write a short essay elaborating on the following state-
      ment: The trivial differential equation $x' = h(t)$ can
      be solved even when h is discontinuous, provided h
      is at least nice enough to be the derivative of some
      function.   A useful reference is Boas [1960], pp. 117 ff.
      (or Landau [1934], p. 112, or Fulks [1969], p. 98).   You
      might include a discussion of existence of a solution of
      the equation $x' = h(t)$ where

      (a) $h(t) = \begin{cases} 0 & \text{for} \quad t \le 0 \\ 1 & \text{for} \quad t > 0, \end{cases}$

      (b) $h(t) = \begin{cases} 2t \sin \frac{1}{t} - \cos \frac{1}{t} & \text{for} \quad t \ne 0 \\ 0 & \text{for} \quad t = 0. \end{cases}$

3. SEPARABLE EQUATIONS

An equation of the form

$$M(t) + N(x)x' = 0, \qquad (1)$$

where M and N are given functions, is said to have its variables (x and t) separated. And any equation which can be rewritten in this form is said to be separable.

For example, the equation

$$x' = a(t)h(x), \qquad (1')$$

where the functions a and h are given, is separable provided $h \neq 0$; for it can be rewritten as $a(t) - \frac{1}{h(x)} x' = 0$.

We shall generally assume M and N continuous with $N \neq 0$.

The method of solution called separation of variables is illustrated in the following example.

Example 1. Solve

$$x' = (at+b)x \quad \text{with} \quad x(0) = 1,$$

where a and b are constants. (Actually this equation is linear and hence could be solved by the method of Section 2.) Assume, at first, that a solution x exists on some interval. Since x must be a continuous function with $x(0) = 1$, it follows that $x(t) > 0$ for all t sufficiently close to 0. Thus we can divide the differential equation by x, to "separate" the variables, as long as x remains positive:

$$(at+b) - \frac{1}{x} x' = 0.$$

Now replace t by a dummy variable s and integrate from
t_0 = 0 to t,   obtaining

$$\int_0^t (as+b)ds - \int_0^t \frac{x'(s)}{x(s)} ds = 0.$$

If we make the change of variable $\xi = x(s)$ in the second
integral we get

$$\int_0^t (as+b)ds - \int_1^x \xi^{-1}d\xi = 0,$$

or, as long as x = x(t) remains positive,

$$at^2/2 + bt - \ln x = 0.$$

Since the logarithm function has an inverse -- the exponential
function -- we immediately find

$$x(t) = e^{at^2/2+bt}.$$

We have shown that the only candidate for a solution
near t_0 = 0 is the function x defined above.   It is left
as an easy exercise for the reader to verify that x is a
solution and, in fact, is valid for all t.   Problem 1(a).
Thanks to the positivity of x, it can also be shown that the
solution is unique for all t.   Problem 1(b).

Justification of the Method.   Let us now apply to the general
case, Eq. (1), an argument similar to that used in the example.
Let the given functions M and N be continuous with $N \neq 0$.
And let

$$x(t_0) = x_0. \tag{2}$$

Assume at first that there is a solution x of (1)
and (2) on some open interval J containing t_0.   Then x
is continuous and, since x' = -M(t)/N(x),   x' is continuous

on J. Thus we obtain from (1)

$$\int_{t_0}^{t} M(s)ds + \int_{t_0}^{t} N(x(s))x'(s)ds = 0 \quad \text{for all} \quad t \quad \text{in} \quad J.$$

With the change of variable $\xi = x(s)$, this becomes

$$\int_{t_0}^{t} M(s)ds + \int_{x_0}^{x} N(\xi)d\xi = 0.$$

Now define $g(u) \equiv -\int_{x_0}^{u} N(\xi)d\xi.$ Then

$$g(x) = \int_{t_0}^{t} M(s)ds,$$

where g is a continuously differentiable function with the
property $g'(u) = -N(u) \neq 0$. Hence, from the "inverse function theorem", it follows that g^{-1} exists and is differentiable, so that x is uniquely determined by

$$x(t) = g^{-1}[\int_{t_0}^{t} M(s)ds] \quad \text{for} \quad t \quad \text{in} \quad J. \tag{3}$$

In other words, the function x defined by (3) is the only
possible candidate for a solution of (1) and (2) on J.

    The reader should verify by substitution (Problem 3)
that if a function x is defined by (3) on some interval,
then x is a solution of (1) and (2). To show this one must
recall, from the inverse function theorem, that

$$\frac{d}{dv}[g^{-1}(v)] = \frac{1}{g'(g^{-1}(v))} . \tag{4}$$

    Even though Eq. (3) gives the exact solution of (1) and
(2), we must recognize that only in very simple examples will
$\int M(s)ds$ and $\int N(\xi)d\xi$ be elementary integrals and the inverse function g^{-1} be expressible in simple form.

    Once again it is recommended that, in solving problems,
the reader use the method described in this section and not

the "formula" for the solution given by (3).

Example 2.   Solve
$$x' = 1 + x^2 \quad \text{with} \quad x(0) = 1.$$

This equation is decidedly not linear, but it does satisfy
all the requirements for the method of separation of variables
(everywhere).   From

$$\int_1^x \frac{d\xi}{1+\xi^2} = \int_0^t ds,$$

we obtain

$$\text{Arctan } x - \frac{\pi}{4} = t,$$

or

$$x(t) = \tan\left(\frac{\pi}{4} + t\right).$$

That is, therefore, the only candidate for a solution.   Since
the above steps are all reversible as long as $-3\pi/4 < t < \pi/4$,
it follows that the function x which we have found is a
solution.   Thus we have a unique solution on $J = (-3\pi/4, \pi/4)$,
but no further.   Why?   The fact that the solution cannot be
continued as far back as $-3\pi/4$ or as far forward as $\pi/4$
was completely unpredictable from the form of the differential
equation.   We have seen that such an unpredictable ending of
the solution can not occur for the linear equations of Section 2.

Example 3.   Solve
$$x' + x^2 \sin t = 3(tx)^2 \quad \text{with} \quad x(0) = x_0 > 0.$$

As long as $x(t) \neq 0$, the differential equation is separable
since it can be rewritten as
$$x' = (3t^2 - \sin t)x^2.$$

Using the initial condition we find, for as long as $x(t)$
remains positive,

$$\int_{x_0}^{x} \xi^{-2} d\xi = \int_{0}^{t} (3s^2 - \sin s) ds,$$

or

$$-\frac{1}{x} + \frac{1}{x_0} = t^3 + \cos t - 1,$$

or

$$x(t) = \frac{1}{x_0^{-1} + 1 - t^3 - \cos t}.$$

The reader should verify directly that this is a solution of
the given differential equation and the initial condition
(Problem 5). Our derivation shows that it is unique, i.e.,
no other function can be a solution as long as $x_0^{-1} + 1 - t^3$
- $\cos t > 0$. Moreover the solution is valid at least for all
$t < x_0^{-1/3}$, and generally on a somewhat larger interval. Why?

In case $x_0 < 0$ the solution of Example 2 has the same
form, and is valid as long as $x_0^{-1} + 1 - t^3 - \cos t$ remains
negative, i.e., in some open interval containing 0.

Sometimes it is possible to "formally" apply a method
even when the conditions which would justify that method are
not fulfilled, as in the next examples.

Example 4. Solve

$$x' = x^2 \quad \text{with} \quad x(0) = x_0 = 0.$$

Even though the condition $h \neq 0$ in (1') is not fulfilled,
formal solution with the initial condition $x(0) = x_0$ leads
to

$$x(t) = \frac{x_0}{1 - x_0 t},$$

or $x(t) \equiv 0$ when $x_0 = 0$. It is easily verified by substitution that $x = 0$ is a solution. However we cannot con-

clude, on the basis of what has been done here, that there is
no other solution.   Problem 10.

Example 5.   Solve
$$x' = x^{2/3} \quad \text{with} \quad x(0) = 0.$$

Again the condition   h ≠ 0   is not fulfilled.   However formal
application of the method of separation of variables (Problem
6) yields $x(t) = t^3/27$, which is easily verified to be a
solution.   In this case, not only have we not proved the solution to be unique, indeed the solution is not unique.   For
example, two more solutions, among infinitely many possibilities, are the functions defined by

$$x(t) = 0 \quad \text{for all} \quad t$$

and

$$x(t) \begin{cases} 0 & \text{for } t \le 1, \\ \dfrac{(t-1)^3}{27} & \text{for } t > 1. \end{cases}$$

Problems.

1. (a) Verify by substitution that the candidate for a solu-
         tion, x, obtained in Example 1 really is a solution.

    (b) Noting that   x(t) > 0   for all   t   prove, via the same
         method already used in Example 1, that the solution
         is unique for all   t.   Hint.   Suppose (for contra-
         diction) that there were another solution, x̃.   Then
         there would have to be some point $t_1 \ne 0$ at which
         $\tilde{x}(t_1) = x(t_1) > 0$, and yet   x̃(t) ≠ x(t)   for certain
         values of   t   arbitrarily close to t_1.   Show that
         this is impossible.

2.   Solve the problem in Example 1 by the method given for
     linear equations in Section 2.

3.   Using (4), verify that (3) satisfies (1) and (2).

4.   Sketch a graph of the solution of Example 2 to see what
     happens as t approaches $-3\pi/4$ or $\pi/4$.

5.   Verify the solution found in Example 3 on $J = (-\infty, x_0^{-1/3})$.

6.   Carry out the formal separation of variables calculation
     for Example 5.

7.   Sketch the three solutions given for Example 5 plus a
     fourth solution of your own.

8.   In each of the following use the method of separation of
     variables to find a solution on some interval J.   Deter-
     mine (or conjecture) whether the solution is unique there.
     (a) $x' - x = 1$ with $x(2) = 3$ (Problem 2-4 again).
     (b) $x' = (\sin t)(x+1)$ with $x(t_0) = x_0$ (Problem 2-5
           again).
     (c) $x' = x^2$ with $x(7) = 1$.

     (d) $x^2x' + tx = x$ with $x(-1) = 2$.
     (e) $x' = \sqrt{|x|}$ with $x(0) = 0$.
     (f) $x' = -t/x$.
     (g) $x' = (1 + x^2)/(1 + t^2)$ with $x(1) = -1$.
     (h) $t^2x' = x - tx$.

9.   Solve Problem 8(h) by another method.

*10. A Challenge: The only solution of $x' = x^2$ with $x(0) = 0$
      is $x \equiv 0$.   Try to prove this without reading Chapter II.

Chapter II. <u>UNIQUENESS</u> <u>AND</u> <u>LIPSCHITZ</u> <u>CONDITIONS</u>

<u>FOR</u> <u>ORDINARY</u> <u>DIFFERENTIAL</u> <u>EQUATIONS</u>

We have considered some special cases in which one can explicitly determine a solution of a differential equation with initial condition, <u>and</u> conclude that the solution is unique. Many other special cases can also be handled exactly (cf. Kamke [1944]).

But, in general, differential equations do <u>not</u> have simple solutions. That is, in general, one cannot hope to deduce that some particular simple function <u>is</u> a solution and <u>the</u> <u>only</u> solution of a given differential equation with initial condition. Then we seek instead to obtain information indirectly about the problem.

The two most basic questions to ask about a given problem are: "Does a solution exist?" and "Can there be more than one solution?"

The reader should note that, in most of the examples considered thus far, we have considered the second question first. In the present chapter we shall pursue this uniqueness question further, and shall find rather simple conditions which suffice to assure that a given problem has at most one solution. Later in the book we shall find that the same conditions also assure the existence of a solution.

4. FIRST ORDER SCALAR EQUATIONS

For most differential equations with initial conditions, including the examples which we have considered, it is
not at all obvious in advance that a unique solution is determined. It is certainly not obvious (nor even true) for the
general first order problem of the form

$$x' = f(t,x) \tag{1}$$

with initial condition

$$x(t_0) = x_0. \tag{2}$$

We shall refer to Eq. (1) as a "scalar" equation since it involves only one scalar-valued unknown function, x.

Equation (1) asserts that the rate of change of the unknown function x at the instant t will be determined if
the value of x at t is known. Thus the problem of finding
the unknown function would sound, to the uninitiated, like a
hopeless chicken-and-egg problem: We'd like to find x by
integrating x' as computed from Eq. (1). But in order to
compute x' from Eq. (1) we must first know x.

Nevertheless, we have already encountered examples in
which Eqs. (1) and (2) do determine a unique solution. Thus
we forge on to seek some simple conditions on the equation itself which will assure that a unique solution is determined.

The nature of the uniqueness problem is illustrated by
three simple examples:

Example 1. The linear equation $x' = tx$ with $x(0) = x_0$
has a unique solution, defined by $x(t) = x_0 e^{t^2/2}$, for all t.
(This follows easily from our general method for linear equa-

tions.)

Example 2.   The equation

$$x' = x^{2/3} \quad \text{with} \quad x(0) = 0$$

clearly has a solution $x(t) = 0$ for all   t.   But Example
3-5 showed that this solution is not unique.

Example 3.   The equation

$$x' = x^{4/3} \quad \text{with} \quad x(0) = 0$$

clearly has a solution $x(t) = 0$ for all   t.   And in this
case we will find that the solution is unique.

We should like to have a means for determining uniqueness by examining the differential equation (1) itself.   In
particular, we might ask: Why are Examples 2 and 3 so different?

The analogous general questions regarding existence of
a solution of (1) need not bother us yet.   For in the examples
and problems presented in the first 19 sections of this book
we will be able to explicitly find solutions.   In particular,
we have explicitly found the zero solutions in Examples 2 and
3 above.

In the present section we introduce methods and notation which will be used again later, and prove a simple but
useful "uniqueness theorem" for Eq. (1) with initial condition
(2).   More precisely, we shall show that a certain problem can
have at most one solution.

Our proof will not exhibit a candidate for a solution
nor will it even guarantee that a solution exists.   However,

having proved uniqueness for a particular problem, one then has the right to seek a solution by any means whatsoever -- even a random guess or a dubious calculation. If we find a solution, then we've found the solution.

We consider Eq. (1) with initial condition (2). Let f be a given real-valued function defined on some "open set" $D \subset \mathbb{R}^2$, and assume that $(t_0, x_0) \in D$.

The term open set is defined in Appendix 1. However, the reader who is unfamiliar with this concept can think of D as an open rectangle, say $D = (\alpha, \beta) \times (\gamma, \delta)$. An open rectangle will generally be quite adequate, and the concept is simpler than that of a more general open set. In Examples 1, 2 and 3 one can take $D = \mathbb{R}^2$.

We can now state precisely what we mean by a solution, and by a unique solution.

Definition. A solution of Eqs. (1) and (2) is a continuous, real-valued function x on an interval $J = (\alpha_1, \beta_1)$ such that

(i) $t_0 \in J$ and $x(t_0) = x_0$, and

(ii) for all t in J, $(t, x(t)) \in D$ and $x'(t) = f(t, x(t))$.

Note that this definition implies that x must be differentiable on J and, if f is continuous (as will usually be the case), then it follows that x' must also be continuous.

We shall frequently use the following fact: If f is continuous and x is a continuous function such that $(t, x(t)) \in D$ for all t in J, then x is a solution of

Eqs. (1) and (2) on J   if and only if

$$x(t) = x_0 + \int_{t_0}^{t} f(s,x(s))ds \qquad \text{for all t in J.} \qquad (3)$$

Verify this important assertion using the fundamental theorems of calculus.

A solution of Eqs. (1) and (2) has been defined as a function x on a certain interval J.   Thus if we change J, the domain of x, then strictly speaking, we are no longer talking about the same function.   For instance, in Example 1, one solution is defined by

$$x(t) = x_0 e^{t^2/2} \qquad \text{for}\quad -2 < t < 2,$$

and another is defined by

$$x(t) = x_0 e^{t^2/2} \qquad \text{for}\quad -1 < t < \pi.$$

And yet we want to say the solution is "unique".

Actually there is no problem if we recall the definition mentioned in Section 1:

Definition.   A solution x of Eqs. (1) and (2) is said to be unique if it agrees with every other solution wherever both are defined.

If we do not know that a solution exists, we will say Eqs. (1) and (2) have at most one solution on an interval J if it can be shown that any two solutions (which might exist) would have to agree wherever both are defined on J.

Now, with an example for motivation, we shall lead up to the simple test for uniqueness which is the main goal of this section.   This example looks more complicated, but it will actually turn out to be a little simpler than Example 3.

Example 4. Consider the first order equation

$$x' = 2 \sin t \sin x \quad \text{with} \quad x(t_0) = x_0, \qquad (4)$$

where t_0 and x_0 are given real numbers. For this special case of Eqs. (1) and (2) one can take $D = \mathbb{R}^2$.

The differential equation can be solved by separation of variables so long as $\sin x(t)$ does not become zero. In case $x_0 = 0$, or $\pm\pi$, or $\pm 2\pi, \ldots$, although separation of variables is not applicable, we can see by inspection that $x(t) \equiv$ constant is a solution. But how can we decide in these cases, whether or not there is also some other solution?

The calculations which follow will show that, regardless of the values of t_0 and x_0, the given problem (4), has at most one solution on any open interval J.

As a matter of fact the proof which we shall give works for more general equations. Let D be any open rectangle -- not necessarily $\mathbb{R}^2$. Then we can treat Eq. (1) provided f is continuous and $D_2 f$ exists and is bounded on D. Thus we shall proceed to consider Eq. (1) assuming

$$|D_2 f(t,\xi)| = \left|\frac{\partial f}{\partial \xi}(t,\xi)\right| \leq K \quad \text{on} \quad D. \qquad (5)$$

Note that these conditions are satisfied for our particular example. Indeed (4) satisfies condition (5) with $K = 2$.

In (5) and on future occasions we use a "dummy variable" such as ξ in place of x. This enables us to discuss the properties of a given function, f, without reference or restriction to some unknown function, x.

Suppose Eqs. (1) and (2) have two solutions x and $\tilde{x}$

(pronounced "x-tilda") on some interval $J = (\alpha_1, \beta_1)$. It follows from (3) that for all t in J,

$$x(t) = x_0 + \int_{t_0}^{t} f(s, x(s))ds$$

and

$$\tilde{x}(t) = x_0 + \int_{t_0}^{t} f(s, \tilde{x}(s))ds.$$

Subtracting these two equations we have

$$x(t) - \tilde{x}(t) = \int_{t_0}^{t} [f(s, x(s)) - f(s, \tilde{x}(s))]ds$$

for all t in J. From this it follows that

$$|x(t) - \tilde{x}(t)| \leq \left| \int_{t_0}^{t} |f(s, x(s)) - f(s, \tilde{x}(s))|ds \right|.$$

Why do we keep the absolute value signs outside this last integral?

Now apply the mean value theorem, Theorem A2-A (i.e., Appendix 2, Theorem A), to conclude that, for each fixed s, there is some number, θ, between $x(s)$ and $\tilde{x}(s)$ such that

$$f(s, x(s)) - f(s, \tilde{x}(s)) = D_2 f(s, \theta)[x(s) - \tilde{x}(s)].$$

Then, since $|D_2 f(s, \theta)| < K$ throughout D, it follows that

$$|f(s, x(s)) - f(s, \tilde{x}(s))| \leq K|x(s) - \tilde{x}(s)|.$$

Hence, for all t in J,

$$|x(t) - \tilde{x}(t)| \leq K \left| \int_{t_0}^{t} |x(s) - \tilde{x}(s)|ds \right|. \tag{6}$$

We shall complete the proof by showing, as a consequence of inequality (6), that $|x(t) - \tilde{x}(t)| = 0$ for all t in J. Then it will follow that $x = \tilde{x}$, i.e., there cannot be two different solutions on J. The cases $t_0 \leq t < \beta_1$ and

$\alpha_1 < t \leq t_0$ are treated separately:

For $t_0 \leq t < \beta_1$, define $Q(t) = K\int_{t_0}^{t} |x(s) - \tilde{x}(s)|\,ds$.

Thus (6) is transformed into the linear "differential inequality"

$$Q'(t) = K|x(t) - \tilde{x}(t)| \leq KQ(t)$$

or

$$Q'(t) - KQ(t) \leq 0,$$

with $Q(t_0) = 0$. This inequality will be "solved" by the same method used for linear differential <u>equations</u> in Section 2. Multiply both sides by the (positive) integrating factor e^{-Kt} to obtain

$$\frac{d}{dt}[Q(t)e^{-Kt}] \leq 0.$$

Integration, or application of the mean value theorem, between t_0 and t then gives

$$Q(t)e^{-Kt} - Q(t_0)e^{-Kt_0} \leq 0.$$

Since $Q(t_0) = 0$ and $Q(t) \geq 0$, it follows that $Q(t) = 0$. Thus from (6),

$$|x(t) - \tilde{x}(t)| \leq Q(t) = 0 \quad \text{for} \quad t_0 \leq t < \beta_1.$$

This proves that $x(t) = \tilde{x}(t)$ for $t_0 \leq t < \beta_1$.

In case $\alpha_1 < t \leq t_0$, with $Q(t) \equiv K\int_{t_0}^{t} |x(s) - \tilde{x}(s)|\,ds$ as before, (6) gives

$$Q'(t) \leq -KQ(t),$$

or

$$Q'(t) + KQ(t) \leq 0,$$

with $Q(t_0) = 0$. Now multiply by e^{Kt} to obtain

$$\frac{d}{dt}[Q(t)e^{Kt}] \leq 0,$$

with $Q(t_0) = 0$. Thus, integrating from t to t_0 (the positive direction),

$$Q(t_0)e^{Kt_0} - Q(t)e^{Kt} \leq 0.$$

Then, since $Q(t_0) = 0$ and since in this case $Q(t) \leq 0$, it again follows that $Q(t) \equiv 0$, which proves that $x(t) = \tilde{x}(t)$ for $\alpha_1 < t \leq t_0$.

If we try to apply the above argument to prove uniqueness for Example 3 (or even Example 1) we run into difficulty. The trouble is that condition (5), the boundedness of D_2f, fails. In Examples 1 and 3 we have, respectively,

$$D_2f(t,\xi) = t \quad \text{and} \quad D_2f(t,\xi) = \frac{4}{3}\xi^{1/3}.$$

In one D_2f is unbounded as $|t|$ becomes large, and in the other D_2f is unbounded as $|\xi|$ becomes large.

Actually this difficulty is not serious. We can overcome it by considering appropriate smaller open rectangles instead of the entire set $D = \mathbb{R}^2$. The following theorem uses this idea to remove condition (5) from the stated requirements.

Theorem A.   If f and D_2f are continuous on D and if $(t_0,x_0) \in D$, then Eqs. (1) and (2) have at most one solution on any open interval J containing t_0.

Remarks.   This is probably the best known and most easily applied of all uniqueness theorems. It certainly establishes uniqueness of the solutions in Examples 1 and 3 and in Problem 3-10.

But why does Theorem A not also assert uniqueness for
Example 2, which would of course be false? In Example 2,
$f(t,\xi) = \xi^{2/3}$ which is continuous on $\mathbb{R}^2$. But D_2f is undefined at points $(t,0)$, so it is certainly not continuous
there.

Proof of Theorem A. Suppose (for contradiction) that there
are two solutions x and $\tilde{x}$ on (α_1,β_1) with $x \neq \tilde{x}$. We
must, of course, have $x(t_0) = x_0 = \tilde{x}(t_0)$. Let us assume
$x(t) \neq \tilde{x}(t)$ for some t in (t_0,β_1). (The case of $x(t) \neq$
$\tilde{x}(t)$ for some t in (α_1,t_0) is handled similarly. Problem
2.)

Let

$$t_1 = \inf \{t \in (t_0,\beta_1): x(t) \neq \tilde{x}(t)\},$$

where "inf" stands for infimum or greatest lower bound. Then,
since x and $\tilde{x}$ are both continuous, it follows that $x(t_1) =$
$\tilde{x}(t_1) \equiv x_1$. Also $t_1 < \beta_1$, since we have assumed $x(t) \neq \tilde{x}(t)$
for some t in (t_0,β_1).

Now choose positive numbers a and b such that

$$A = [t_1-a, t_1+a] \times [x_1-b, x_1+b] \subset D.$$

Since D_2f is continuous on the closed bounded rectangle A,
it follows from Theorem A2-I that

$$|D_2f(t,\xi)| \leq K \quad \text{for all} \quad (t,\xi) \quad \text{in} \quad A,$$

for some K. Thus condition (5) is fulfilled on the new open
set $D_* = (t_1-a,t_1+a) \times (x_1-b,x_1+b)$.

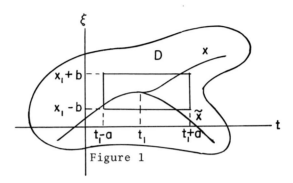

Figure 1

But, by continuity, the points $(t,x(t))$ and $(t,\tilde{x}(t))$ must remain in D_* on some open interval J_* about t_1. Thus we can apply the computation of Example 4 with (t_1,x_1) playing the role of (t_0,x_0). Specifically, we obtain condition (6) with t_1 in place of t_0. It follows that $x(t) = \tilde{x}(t)$ on J_*. But this contradicts the definition of t_1. □

In the proof of uniqueness we did not actually use the partial derivative D_2f itself. What we used was inequality (6) which is a consequence of the boundedness of D_2f. It will sometimes be useful to have isolated out the actual condition on f which is used. This is called a "Lipschitz condition".

Definition. Let $A \subset \mathbb{R}^2$. Then a function f is said to satisfy a Lipschitz condition on A if there exists a constant K (called a Lipschitz constant) such that

$$|f(t,\xi) - f(t,\tilde{\xi})| \le K|\xi - \tilde{\xi}| \tag{7}$$

whenever (t,ξ) and $(t,\tilde{\xi}) \in A$. We shall also say f is

Lipschitzian on A.

If condition (5) holds on A then we certainly get the
Lipschitz condition (7). On the other hand a Lipschitz con-
dition is slightly weaker than condition (5), Problem 11.
Hence the following theorem is slightly stronger than Theorem
A.

Theorem B. Let f be continuous on D and Lipschitzian on
each closed bounded rectangle A ⊂ D, and let $(t_0, x_0) \in D$.
Then Eqs. (1) and (2) have at most one solution on any open
interval J containing t_0.

Remarks. The Lipschitz constant K for f will, in
general, be different for different closed bounded rectangles
A.

The proof of Theorem B is virtually the same as that
of Theorem A. Actually it is slightly simpler since in the
proof of Theorem A we first had to essentially produce the
Lipschitz condition from the properties of $D_2 f$ in order to
get inequality (6).

Problems.

1. Show that the equation $x' = x^{2/3}$, of Example 2, with
    initial condition $x(0) = 1$ has a solution for all t in
    $(-\infty, \infty)$, but the solution is not unique on this interval.
    On what interval (if any) is it unique?

2. Complete the proof of Theorem A by treating the case
    $x(t) \neq \tilde{x}(t)$ for some t in (α_1, t_0). Hint: Define
    $t_1 = \sup\{t \in (\alpha_1, t_0): x(t) = \tilde{x}(t)\}$, where "sup" stands

for supremum or least upper bound.

3. For each of the following, find an appropriate set   D,
    and prove, using the methods or theorems of this section,
    that the problem has at most one solution in   D.

    (a) $x' + \frac{1}{t}x = 3t$ with $x(1) = 2$.

    (b) $x' = (x + 1)\sin t$ with $x(t_0) = x_0$.

    (c) $(\cos t)x' + (\sin t)x = 1$ with $x(0) = 1$.

    (d) $x' = 1 + x^2$ with $x(0) = 1$.

    (e) $x^2x' + tx = x$ with $x(-1) = 2$.

    (f) $2t \ln|x| - 1 + \frac{t^2 + 1}{x} x' = 0$ with $x(2) = -e$.

    (g) $N' = kN - kN^2/P$,   where   k   and   P   are positive con-
          stants, with $N(0) = N_0$ arbitrary.

    (h) $x' = (1 + t^3)x^{4/3}$ with $x(0) = 0$.

4. The equation $x + (t + t^2x)x' = 0$ with $x(1) = 0$ has
    the solution $x = 0$ for all   t.   On what interval can you
    prove that this is unique?

5. Show that if $x_0 \neq 0$ in Example 4 then $x(t)$ can never
    become zero.

6. Show in detail why Theorem B is not contradicted by the
    nonuniqueness encountered in Example 2.

7. By considering the equation $x' = 1 + x^{2/3}$ with $x(0) = 0$,
    show that a Lipschitz condition is not a necessary condi-
    tion for uniqueness.   Hint: Apply the method of solution
    of Section 3.

8. Consider "Bernoulli's equation", $x' + a(t)x = h(t)x^n$,

where   n   is a constant and   a   and   h   are continuous.
Let $x(t_0) = 0$.

(a)   Show that if $n \geq 1$, then the only solution is
$x = 0$.

(b)   Show that if $0 < n < 1$, the solution may not be
unique.

*9.   Consider the differential equation $x' = x^{p/q}$ where   p
and   q   are positive integers with   q   odd.   (This includes the equations of Examples 2 and 3 as special
cases.)

(a)   If $x(0) = 0$, for what p/q is the solution, $x = 0$,
unique?

(b)   If $x(0) = x_0 > 0$, solve and determine where the
solution is valid for various values of p/q.

*10.   Complete the following alternate proof for Example 4.   If
x and $\tilde{x}$ are two solutions of Eqs. (1) and (2), let
$v(t) \equiv [x(t) - \tilde{x}(t)]^2$.   Then

$$v'(t) = 2[x(t) - \tilde{x}(t)][f(t,x(t)) - f(t,\tilde{x}(t))] \leq 2Kv(t).$$

11.   Show that the condition   "D_2f exists and is continuous
on   A" is not a necessary condition for a Lipschitz condition.   This will show that Theorem B is stronger than
Theorem A.   Hint.   Consider $f(t,\xi) = |\xi|$ on R^2.

*12.   Let   f   be defined on some open rectangle $D \subset R^2$.   As-
sume D_2f is unbounded somewhere on   D.   Then prove that
f   cannot be Lipschitzian on   D.

5.   SYSTEMS OF EQUATIONS

Frequently in applications one encounters problems involving more than one unknown function and more than one differential equation.

Example 1.   The two electric currents, $x_1(t)$ and $x_2(t)$, in the circuit of Figure 1 satisfy a pair of simultaneous differential equations

$$2x_1' + 4(x_1 - x_2) = 10 \sin 2t$$
$$2x_2' + 6x_2 - 4(x_1 - x_2) = 0.$$

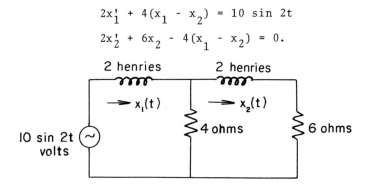

Figure 1

These differential equations can be rewritten in the form

$$x_1' = -2x_1 + 2x_2 + 5 \sin 2t$$
$$x_2' = 2x_1 - 5x_2.$$

More generally, let us consider a system of n first order equations involving n unknown functions, $x_1,\ldots,x_n$,

$$x_1' = f_1(t,x_1,\ldots,x_n)$$
$$\vdots$$
$$x_n' = f_n(t,x_1,\ldots,x_n) \, ,$$

where each f_i is a given real-valued function defined on

some open set $D \subset \mathbb{R}^{1+n}$. Again you may assume for simplicity that D is an "open rectangle", say $D = (\alpha,\beta) \times (\gamma_1,\delta_1) \times \cdots \times (\gamma_n,\delta_n)$.

We can write the above system of n equations more briefly as

$$x_i' = f_i(t,x_1,\ldots,x_n) \quad \text{for} \quad i = 1,\ldots,n.$$

We shall consider this system together with n initial conditions

$$x_i(t_0) = x_{0i} \quad \text{for} \quad i = i,\ldots,n.$$

where $(t_0,x_{01},\ldots,x_{0n})$ is a given point in D.

It is convenient to introduce vector notation and represent the n equations even more compactly as

$$x' = f(t,x) \tag{1}$$

with initial conditions

$$x(t_0) = x_0. \tag{2}$$

The symbols x_0, $x(t)$, $x'(t)$, and $f(t,\xi)$ now stand for vectors in $\mathbb{R}^n$ which we shall represent as columns, e.g.,

$$x(t) = \begin{pmatrix} x_1(t) \\ \vdots \\ x_n(t) \end{pmatrix}.$$

To save space we will often write $x(t) = \text{col}\,(x_1(t),\ldots,x_n(t))$. Similarly, $f(t,\xi)$ may be written $\text{col}\,(f_1(t,\xi),\ldots,f_n(t,\xi))$ where $\xi = \text{col}\,(\xi_1,\ldots,\xi_n)$.

Instead of saying "f is an n-vector-valued function on D" we shall say more briefly f maps $D \to \mathbb{R}^n$ or $f: D \to \mathbb{R}^n$.

Addition of vectors and multiplication by scalars satisfy the usual rules of associativity, commutativity, and dis-
tributivity for a linear vector space. These are as follows.

Linear Vector Space Properties. Whenever ξ, η, and $\zeta \in \mathbb{R}^n$
and c, c_1, and $c_2 \in \mathbb{R}$, then

    (i) $\xi + \eta = \eta + \xi$ (commutative law of addition),

    (ii) $(\xi + \eta) + \zeta = \xi + (\eta + \zeta)$ (associative law of
        addition),

    (iii) There exists a unique vector 0, called the zero
        vector, such that $\xi + 0 = \xi$ for all ξ in $\mathbb{R}^n$,

    (iv) For each ξ in $\mathbb{R}^n$ there exists a unique vector
        $-\xi$ in $\mathbb{R}^n$ such that $\xi + (-\xi) = 0$,

    (v) $c_1(c_2\xi) = (c_1c_2)\xi$ (associative law of scalar
        multiplication),

    (vi) $(c_1 + c_2)\xi = c_1\xi + c_2\xi$ $\Big\}$ (distributive laws),

    (vii) $c(\xi + \eta) = c\xi + c\eta$

    (viii) $1\xi = \xi$.

In conditions (iii) and (iv) the symbols 0 and $-\xi$
stand for col $(0,\ldots,0)$ and col $(-\xi_1,\ldots,-\xi_n)$ respectively.
Note that we will be using the same symbol, 0, to stand for
both the number zero and the zero vector.

Subtraction of vectors is defined by

$$\xi - \eta \equiv \xi + (-1)\eta,$$

so that $(\xi - \eta)_i = \xi_i - \eta_i$ for i = 1,\ldots,n.

An n-vector-valued function y is said to be continuous (or differentiable or integrable) if each of its components, y_i, is continuous (or differentiable or integrable

respectively). If y is differentiable or integrable on
[a,b] we write, respectively, $y' = \text{col } (y_1', \ldots, y_n')$ or

$$\int_a^b y(s)ds = \text{col } (\int_a^b y_1(s)ds, \ldots, \int_a^b y_n(s)ds).$$

The reader should verify that differentiation and integration
of vector-valued functions are linear operations. That is,
if n-vector-valued functions y and z are differentiable
or Riemann integrable, then so is $c_1 y + c_2 z$ and

$$\frac{d}{dt} (c_1 y + c_2 z) = c_1 y' + c_2 z' \tag{3}$$

or

$$\int_a^b (c_1 y + c_2 z)(s)ds = c_1 \int_a^b y(s)ds + c_2 \int_a^b z(s)ds \tag{4}$$

for arbitrary scalar constants c_1 and c_2. Problem 1.

As Problem 2, verify two other basic properties of
vector integration and differentiation -- the following analogs of the fundamental theorems of calculus:

If y is a continuous function mapping an interval
$J \to \mathbb{R}^n$ and if $t_0, t \in J$, then

$$\frac{d}{dt} \int_{t_0}^t y(s)ds = y(t) \tag{5}$$

(where a one-sided derivative is understood if t happens to
be an endpoint of the interval J).

If y mapping $[a,b] \to \mathbb{R}^n$ has a continuous (or merely
integrable) derivative, y', on [a,b] (implying one-sided
derivatives at a and b), then

$$\int_a^b y'(s)ds = y(b) - y(a). \tag{6}$$

With this notation available we are now ready to define
a "solution" of (1) and (2) and then to present an easily ap-

plied condition for uniqueness.

<u>Definition</u>.   A <u>solution</u> of (1) and (2) is a continuous func-
tion x mapping an open interval $J \to \mathbb{R}^n$ such that

$\quad$ (i) $t_0 \in J$ and $x(t_0) = x_0$, and

$\quad$ (ii) for all t in J, $(t,x(t)) \in D$ and $x'(t) =$
$\qquad$ $f(t,x(t))$.

$\quad$ This is a natural generalization of the definition
given for scalar equations.   In fact it reads almost word-for-
word the same as the definition in Section 4.   The difference
lies in the meaning of the symbols f, x, and x_0.
$\quad$ Uniqueness is defined exactly as in Section 4.
$\quad$ If f is continuous and $x: J \to \mathbb{R}^n$ is a continuous
function such that $(t,x(t)) \in D$ for all t in J, then x
is a solution of Eqs. (1) and (2) on J if and only if

$$x(t) = x_0 + \int_{t_0}^{t} f(s,x(s))ds \quad \text{for all} \quad t \quad \text{in} \quad J. \quad (7)$$

This analog of 4-(3) is an easy consequence of Eqs. (5) and
(6).

<u>Theorem A</u>.   If f and its partial derivatives $D_{1+j}f_i$ for
i, j = 1,...,n are continuous on D and if $(t_0,x_0) \in D$,
then Eqs. (1) and (2) have at most one solution on any open
interval J.

$\quad$ We postpone temporarily the proof of this analog of
Theorem 4-A.
$\quad$ Note that in Example 1

$$f(t,\xi) = col \ (-2\xi_1 + 2\xi_2 + 5 \sin 2t, \ 2\xi_1 - 5\xi_2),$$

and f can be considered to map $\mathbb{R}^3 \to \mathbb{R}^2$. The function f is continuous on $D = \mathbb{R}^3$; and the partial derivatives required in Theorem A are

$$D_2 f_1 = -2, \quad D_3 f_1 = 2, \quad D_2 f_2 = 2, \quad \text{and} \quad D_3 f_2 = -5,$$

which are also clearly continuous. Hence the system in Example 1 has at most one solution if $x(t_0) = x_0 = \text{col }(x_{01}, x_{02})$ is given.

As a tool for proving uniqueness theorems, and for other purposes later, we introduce a "norm" for vectors. If $\xi \in \mathbb{R}^n$ we shall define the <u>norm</u> of ξ to be

$$||\xi|| = \sum_{i=1}^{n} |\xi_i| = |\xi_1| + \cdots + |\xi_n|. \tag{8}$$

The norm of a vector as defined in (8) can be considered as an extension of the concept of absolute value because of the following properties.

<u>Properties of the Norm</u>. Let the norm of a vector in $\mathbb{R}^n$ be defined by Eq. (8). Then

  (i) $||\xi|| \geq 0$ for all ξ in $\mathbb{R}^n$,

  (ii) $||\xi|| = 0$ if and only if $\xi = 0$ (the zero vector),

  (iii) $||c\xi|| = |c| \cdot ||\xi||$ for all ξ in $\mathbb{R}^n$ and c in $\mathbb{R}$, and

  (iv) $||\xi + \eta|| \leq ||\xi|| + ||\eta||$ for all ξ, η in $\mathbb{R}^n$ (the "triangle inequality").

The verification of these properties is left as Problem 3, for those readers who have not already encountered them.

Remark. Equation (8) is by no means the only suitable definition for extending the concept of absolute value to vectors in $\mathbb{R}^n$. Other possible definitions include

$$||\xi||_\infty = \max_{i=1,\ldots,n} |\xi_i| \qquad (8a)$$

and

$$||\xi||_2 = [\sum_{i=1}^{n} \xi_i^2]^{1/2}. \qquad (8b)$$

Both of these "norms" can also be shown to have the properties (i) through (iv) listed above. Problem 4. We shall not use these norms.

Regarding integration of vector-valued functions, the important property of the norm is this. If y is a continuous (or merely integrable) n-vector-valued function on $[a,b]$ then

$$\left|\left| \int_a^b y(s)ds \right|\right| \leq \int_a^b ||y(s)||ds. \qquad (9)$$

This is proved as follows:

$$\left|\left| \int_a^b y(s)ds \right|\right| = \sum_{i=1}^{n} \left| \int_a^b y_i(s)ds \right| \leq \sum_{i=1}^{n} \int_a^b |y_i(s)|ds$$

$$= \int_a^b \sum_{i=1}^{n} |y_i(s)|ds = \int_a^b ||y(s)||ds.$$

Using the norm notation, let us now extend the meaning of "Lipschitz condition" to vector-valued functions f mapping a subset of $\mathbb{R}^{1+n}$ into $\mathbb{R}^n$, as occurs in Eq. (1). Anticipating Section 6 where we shall want a Lipschitz condition for a function from $\mathbb{R}^{1+n}$ into $\mathbb{R}$, let f map into $\mathbb{R}^m$. Then we will be able to take $m = 1$, or $m = n$, or other values as needed.

Definition. Let $A \subset \mathbb{R}^{1+n}$. Then a function $f: A \to \mathbb{R}^m$ is

said to satisfy a <u>Lipschitz</u> <u>condition</u> on A if there exists
a constant K (called a <u>Lipschitz</u> <u>constant</u>) such that

$$||f(t,\xi) - f(t,\tilde{\xi})|| \leq K||\xi - \tilde{\xi}|| \tag{10}$$

for all (t,ξ) and $(t,\tilde{\xi})$ in A. We also say f is <u>Lip-</u>
<u>schitzian</u> on A.

Note that the two norms involved in (10) will be different if $m \neq n$. Specifically, (10) means

$$\sum_{i=1}^{m} |f_i(t,\xi) - f_i(t,\tilde{\xi})| \leq K \sum_{j=1}^{n} |\xi_j - \tilde{\xi}_j|.$$

Under the hypotheses of Theorem A, that each $D_{1+j}f_i$
is continuous, it follows that f is Lipschitzian on each
closed bounded $(n+1)$-dimensional "rectangle"

$$A = [\bar{\alpha},\bar{\beta}] \times [\bar{\gamma}_1,\bar{\delta}_1] \times \cdots \times [\bar{\gamma}_n,\bar{\delta}_n] \subset D.$$

We show this as follows.

Let A be a closed bounded $(n+1)$-dimensional rectangle
in D and let any (t,ξ) and $(t,\tilde{\xi})$ in A be given. Then
it follows from the mean value theorem, Theorem A2-D, that,
for each i,

$$f_i(t,\xi) - f_i(t,\tilde{\xi}) = \sum_{j=1}^{n} (D_{1+j}f_i)(t,\theta)(\xi_j - \tilde{\xi}_j)$$

where $(t,\theta) \in A$. This means that (t,θ) is also in D, as
it must be in order that $(D_{1+j}f_i)(t,\theta)$ be defined.

Since A is closed and bounded, it follows that each
of the continuous functions $D_{1+j}f_i$ is bounded on A
(Theorem A2-I). Let B be the largest of the bounds for
$|D_{1+j}f_i|$ $(i, j = 1,...,n)$. Then we have for each i

$$|f_i(t,\xi) - f_i(t,\tilde{\xi})| \leq \sum_{j=1}^{n} B|\xi_j - \tilde{\xi}_j| = B||\xi - \tilde{\xi}||.$$

Consequently

$$\sum_{i=1}^{n} |f_i(t,\xi) - f_i(t,\tilde{\xi})| \leq nB||\xi - \tilde{\xi}||,$$

which is Lipschitz condition (10) with $K = nB$.

Now, instead of proving Theorem A, we shall prove the following slightly more general theorem -- a vector analog of Theorem 4-B.

Theorem B. Let f be continuous on D and Lipschitzian on each closed bounded (n+1)-dimensional rectangle A in D, and let $(t_0, x_0) \in D$. Then Eqs. (1) and (2) have at most one solution on any open interval J containing t_0.

　　Proof. (Compare with the proof of Theorem 4-A.) Suppose (for contradiction) that there are two solutions x and $\tilde{x}$ on (α_1, β_1) with $x \neq \tilde{x}$. Let us assume $x(t) \neq \tilde{x}(t)$ for some t in (t_0, β_1). (The case of $x(t) \neq \tilde{x}(t)$ for some t in (α_1, t_0) is handled similarly.) Let

$$t_1 = \inf \{t \in (t_0, \beta_1): x(t) \neq \tilde{x}(t)\}.$$

Then, since x and $\tilde{x}$ are both continuous, it follows that $t_1 < \beta_1$ and $x(t_1) = \tilde{x}(t_1) \equiv x_1$.

　　Now choose positive numbers a and b such that

$$A = [t_1-a, t_1+a] \times \underset{i=1}{\overset{n}{\times}} [x_{1i}-b, x_{1i}+b] \subset D.$$

Compare Figure 4-1. Then it follows that f satisfies a Lipschitz condition (10) on A.

　　Now regard x and $\tilde{x}$ as solutions of Eqs. (1) and (2) with the new open set

$$D_* = (t_1-a, \; t_1+a) \times \bigtimes_{i=1}^{n} (x_{1i}-b, \; x_{1i}+b)$$

playing the role of D and with the point (t_1,x_1) playing
the role of (t_0,x_0).

By continuity, the points $(t,x(t))$ and $(t,\tilde{x}(t))$
must remain in D_* on some open interval J_* about t_1.
Moreover it follows from (7), with t_1 in place of t_0, that
for all t in I_*

$$x(t) = x_1 + \int_{t_1}^{t} f(s,x(s))ds$$

and

$$\tilde{x}(t) = x_1 + \int_{t_1}^{t} f(s,\tilde{x}(s))ds.$$

Subtraction of these two equations gives

$$x(t) - \tilde{x}(t) = \int_{t_1}^{t} [f(s,x(s)) - f(s,\tilde{x}(s))]ds.$$

Then, applying inequality (9) and the Lipschitz condition,
we obtain for all t in J_*,

$$||x(t) - \tilde{x}(t)|| \leq \left| \int_{t_1}^{t} ||f(s,x(s)) - f(s,\tilde{x}(s))||ds \right| \tag{11}$$

$$\leq K \left| \int_{t_1}^{t} ||x(s) - \tilde{x}(s)||ds \right|.$$

But (11) is the analog of inequality 4-(6). Proceed-
ing as in that case, one finds $||x(t) - \tilde{x}(t)|| = 0$ or
$x(t) = \tilde{x}(t)$ for all t in J_*. This contradicts the definition of t_1. □

The system of differential equations in Example 1 is
called a "linear" system. In general, system (1) is said to
be linear if it has the form

$$x'_i = \sum_{j=1}^{n} a_{ij}(t)x_j + h_i(t) \quad \text{for} \quad i = 1,\ldots,n, \quad (12)$$

where each a_{ij} and each h_i $(i, j = 1,\ldots,n)$ is a given function. For a linear system we will usually have $D = J \times \mathbb{R}^n$, for some open interval $J = (\alpha,\beta)$.

For such a linear system,

$$f_i(t,\xi) = \sum_{j=1}^{n} a_{ij}(t)\xi_j + h_i(t) \quad \text{for} \quad i = 1,\ldots,n.$$

So

$$D_{1+j}f_i(t,\xi) = a_{ij}(t).$$

Thus the conditions for uniqueness in Theorem A will be satisfied if each a_{ij} and each h_i is continuous on I, and we have the following.

Theorem C. Let each a_{ij} and each h_i be a continuous function mapping $J \to \mathbb{R}$. Then Eqs. (12) and (2) have at most one solution on any open subinterval of J.

Problems

1. Prove that differentiation and integration of vector-valued functions are linear operations, as asserted by Eqs. (3) and (4).

2. Prove the validity of Eqs. (5) and (6) by using the corresponding properties of scalar functions.

3. Prove that the "norm" defined by Eq. (8) has the properties (i) through (iv) listed.

*4. Prove that $||\cdot||_\infty$ and $||\cdot||_2$ as defined by (8a) and (8b) also have the properties (i) through (iv). The proof of property (iv) for $||\cdot||_2$ is nontrivial, but can be found in books on linear algebra or advanced calculus.

5. For any norm having the properties (i) through (iv)
    listed, prove that

$$||\xi - \eta|| \geq \left| ||\xi|| - ||\eta|| \right|.$$

6. If $f(t, \xi_1, \xi_2) \equiv \text{col } (t\xi_1+\xi_2^2, 1+2\xi_2-5e^{\xi_1})$ on
    $D = (-1,2) \times (-2,-1) \times (0,1)$, what can you say about uni-
    queness of solutions of the system $x' = f(t,x)$?

7. (a) Prove that f in Example 1 is Lipschitzian on $\mathbb{R}^3$,
         with $K = 7$.

    (b) Prove that f in Problem 6 is Lipschitzian on its
         given domain D.

6. HIGHER ORDER EQUATIONS

        The theorems of the previous section are easily applied
to scalar equations of order $n \geq 2$. We illustrate the method
by analyzing a certain second order equation. We will not
only prove uniqueness in this example, but will actually find
the unique solution.

Example 1. Consider an object of mass $m > 0$ bouncing up and
down on the end of a spring hanging from the ceiling. Assume
the restoring force of the spring is proportional to the
amount it is stretched from its normal length ℓ with propor-
tionality constant k (the "spring constant"). Let $y(t)$
be the position of the mass below the fixed support at time
t. Then the net downward force on m due to gravity pulling
down and the spring pulling up is $mg - k[y(t) - \ell]$, where g

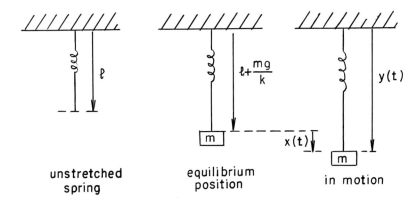

Figure 1

is the acceleration of gravity. In addition, let us assume a
frictional force -by'(t) proportional to the speed of the
mass and in the direction opposing the motion.   Then the equation of motion is

$$my''(t) = mg - k[y(t) - \ell] - by'(t).$$

This equation will be simplified if we define $x = y -$
$(\ell + mg/k)$.

    The resulting simpler equation is

$$mx'' + bx' + kx = 0 \qquad\qquad (1)$$

where m, b, and k are positive constants. Equation (1) is
called a "second order, linear, homogeneous equation with constant coefficients." We shall show first that, regardless of
the values of m, b, and k, if t_0, x_0, and v_0 are arbitrary real numbers, then Eq. (1) has at most one solution such
that

$$x(t_0) = x_0 \quad \text{and} \quad x'(t_0) = v_0. \tag{2}$$

In other words, if the position and velocity of the object are specified at some instant, there can be at most one solution.

Defining $v = x'$ we convert the given second order equation (1) into an equivalent system of two first order equations:

$$\begin{aligned} x' &= v, \\ v' &= -\frac{k}{m}x - \frac{b}{m}v. \end{aligned} \tag{3}$$

The reader should convince himself that this system is equivalent to Eq. (1) in the sense that any solution of (3) provides a solution of (1) and conversely. Problem 1.

Thus it suffices to show that system (3) with the initial conditions

$$x(t_0) = x_0, \quad v(t_0) = v_0$$

has at most one solution. We can take $D = \mathbb{R}^3$ so that $(t_0, x_0, v_0) \in D$ is clear. Then we find that f, defined by

$$f(t, \xi_1, \xi_2) = \text{col} \left(\xi_2, -\frac{k}{m}\xi_1 - \frac{b}{m}\xi_2 \right)$$

satisfies all the conditions of Theorem 5-A. It follows that system (3) with the given initial conditions has at most one solution. Thus Eqs. (1) and (2) also have at most one solution.

The standard approach for actually finding solutions of homogeneous, linear differential equations with constant co-efficients is to seek exponential solutions, i.e., solutions of the form $e^{\lambda t}$ where λ is a constant. Substituting

$x(t) = e^{\lambda t}$ into Eq. (1), we find

$$(m\lambda^2 + b\lambda + k)e^{\lambda t} = 0.$$

Since $e^{\lambda t}$ cannot vanish we require

$$m\lambda^2 + b\lambda + k = 0.$$

This quadratic equation, called the <u>characteristic</u> <u>equation</u>
for (1), has solutions

$$\lambda_1 = \frac{-b + (b^2 - 4mk)^{1/2}}{2m} \quad \text{and} \quad \lambda_2 = \frac{-b - (b^2 - 4mk)^{1/2}}{2m}.$$

Let us assume that $b^2 > 4mk$ so that λ_1 and λ_2 are real
and distinct. (We will treat the other cases later.) Then
both $e^{\lambda_1 t}$ and $e^{\lambda_2 t}$ represent solutions of Eq. (1). In
fact, because Eq. (1) is linear and homogeneous, one can
readily verify that a linear combination of $e^{\lambda_1 t}$ and $e^{\lambda_2 t}$,

$$x(t) = c_1 e^{\lambda_1 t} + c_2 e^{\lambda_2 t}, \tag{4}$$

defines a solution for any constants c_1 and c_2. Problem 2.
    Now let arbitrary initial conditions of the form

$$x(t_0) = x_0, \quad x'(t_0) = v_0$$

be specified. Assuming x is given by (4), these conditions
become

$$c_1 e^{\lambda_1 t_0} + c_2 e^{\lambda_2 t_0} = x_0,$$

$$c_1 \lambda_1 e^{\lambda_1 t_0} + c_2 \lambda_2 e^{\lambda_2 t_0} = v_0.$$

We want to solve for c_1 and c_2. Since the determinant
of the coefficients of the unknowns, c_1 and c_2, is

$$\begin{vmatrix} e^{\lambda_1 t_0} & e^{\lambda_2 t_0} \\ \lambda_1 e^{\lambda_1 t_0} & \lambda_2 e^{\lambda_2 t_0} \end{vmatrix} = (\lambda_2 - \lambda_1) e^{(\lambda_1 + \lambda_2) t_0} \neq 0,$$

it follows that one <u>can</u> always solve for c_1 and c_2. In other words, there does exist <u>a</u> solution of Eq. (1) satisfying the given initial conditions (2); and, from the uniqueness argument, we know that this is <u>the</u> solution.

The procedure used in Example 1 suggests a simple corollary of Theorem 5-A for the n'th order equation

$$x^{(n)} = f(t, x, x', \dots, x^{(n-1)}) \tag{5}$$

with initial conditions

$$x(t_0) = b_0, \quad x'(t_0) = b_1, \quad \dots, \quad x^{(n-1)}(t_0) = b_{n-1}. \tag{6}$$

Assuming that f is a given function mapping $D \to \mathbb{R}$ for some open set $D \subset \mathbb{R}^{1+n}$ and that $(t_0, b_0, b_1, \dots, b_{n-1}) \in D$, we define a solution of Eqs. (5) and (6) as follows.

<u>Definition</u>. A <u>solution</u> of (5) and (6) is a function x mapping an open interval $J \to \mathbb{R}$ such that

(i) $t_0 \in J$ and (6) is satisfied, and

(ii) for all t in J, $(t, x(t), x'(t), \dots, x^{(n-1)}(t))$
$\in D$ and (5) is satisfied.

It is implicit in this definition that $x^{(n)}$ exists on J, and this in turn implies that $x, x', \dots,$ and $x^{(n-1)}$ are all continuous. Thus if the given function f is continuous, it follows from (5) that $x^{(n)}$ is also continuous on J.

<u>Theorem A</u>.   If f, D_2f, D_3f, ...,   and $D_{1+n}f$ are continuous on D and $(t_0, b_0, b_1, ..., b_{n-1}) \in D$, then Eqs. (5) and (6) have at most one solution on any open interval J.

<u>Proof</u>.   Define   n   new unknown functions as follows

$$x_1 = x, \quad \text{and} \quad x_i = x^{(i-1)} \quad \text{for} \quad i = 2, ..., n, \quad (7)$$

and consider the system

$$x_1' = x_2$$
$$\vdots$$
$$x_{n-1}' = x_n$$
$$x_n' = f(t, x_1, x_2, ..., x_n). \tag{8}$$

System (8) with initial conditions

$$x_i(t_0) = b_{i-1} \quad \text{for} \quad i = 1, ..., n \tag{9}$$

is equivalent to Eq. (5) with initial conditions (6) in the sense that any solution of one yields, through (7), a solution of the other.   (Verify this.)   Since Eqs. (8) and (9) have at most one solution, by Theorem 5-A it follows that Eqs. (5) and (6) have at most one solution on J.   ☐

As in Secs. 4 and 5 the conditions on f can be relaxed to continuity and a Lipschitz condition.

<u>Theorem B</u>.   Let f be continuous on D and Lipschitzian on each closed, bounded, (n+1)-dimensional rectangle in D, and let $(t_0, b_0, b_1, ..., b_{n-1}) \in D$.   Then Eqs. (5) and (6) have at most one solution on any open interval J.

The proof is similar to that of Theorem A except that

we invoke Theorem 5-B instead of 5-A. The key point is contained in Problem 3.

We shall say Eq. (5) is a _linear_ n'th order equation
if it takes the form

$$a_n(t)x^{(n)} + a_{n-1}(t)x^{(n-1)} + \cdots + a_1(t)x' + a_0(t)x = h(t),$$

where $a_n, \ldots, a_1, a_0$, and h are given functions, with
$a_n(t) \neq 0$. In this case we can, without loss of generality,
assume $a_n(t) \equiv 1$. For otherwise we could divide through by
$a_n(t)$ and then relabel the coefficients. Thus we consider
the equation

$$x^{(n)} + a_{n-1}(t)x^{(n-1)} + \cdots + a_1(t)x' + a_0(t)x = h(t). \quad (10)$$

For a linear equation we take $D = (\alpha,\beta) \times \mathbb{R}^n$.

The following uniqueness theorem for Eqs. (10) and (6)
is a corollary of Theorem A. It would have nicely covered
uniqueness for Eq. (1).

<u>Theorem C</u>. Let $a_0, a_1, \ldots, a_{n-1}$, and h be continuous
functions on $(\alpha,\beta) \to \mathbb{R}$, and let $(t_0,b_0,b_1,\ldots,b_{n-1})$ be an
arbitrary point in $(\alpha,\beta) \times \mathbb{R}^n$. Then there is at most one
solution of Eqs. (10) and (6) on any subinterval of (α,β).

The proof is left as Problem 4.

The trick of transforming a single higher-order equation into a system of first-order equations is the foundation
of the proofs of Theorems A, B, and C. The next example illustrates how this idea can also be applied to a _system_ of
higher-order equations.

Example 2. Consider the pair of differential equations

$$x_1'' = - \frac{kM}{(x_1^2+x_2^2)^{3/2}} x_1 \quad \text{and} \quad x_2'' = - \frac{kM}{(x_1^2+x_2^2)^{3/2}} x_2 \quad (11)$$

as representing the motion of the earth at (x_1, x_2) about the sun, assumed stationary at $(0,0)$ and having mass M. Here k is the "constant of gravitation". (We assume there are no other celestial bodies.) We shall not undertake to solve these equations. We will not even prove that a solution exists. But we shall prove that there can be at most one solution having a specified position and velocity at t_0. More precisely, we shall show that if

$$x_1(t_0) = x_{10}, \quad x_2(t_0) = x_{20},$$

$$v_1(t_0) = v_{10}, \quad \text{and} \quad v_2(t_0) = v_{20} \quad (12)$$

are given with $x_{10}^2 + x_{20}^2 \neq 0$, then, as long as the earth does not collide with the sun, it can have at most one trajectory.

Once again, we replace the given differential equations (11) by an equivalent system of first order equations. This time we get the four first order equations,

$$x_1' = v_1, \quad x_2' = v_2,$$

$$v_1' = - \frac{kM}{(x_1^2+x_2^2)^{3/2}} x_1, \quad \text{and} \quad v_2' = - \frac{kM}{(x_1^2+x_2^2)^{3/2}} x_2. \quad (13)$$

To relate to the notation of Section 5, we define the vector-valued function f by

$$f(t, \xi_1, \xi_2, \xi_3, \xi_4) = \begin{pmatrix} \xi_3 \\ \xi_4 \\ -kM\xi_1/(\xi_1^2 + \xi_2^2)^{3/2} \\ -kM\xi_2/(\xi_1^2 + \xi_2^2)^{3/2} \end{pmatrix}. \qquad (14)$$

Then f is continuous and has continuous derivatives as long as $\xi_1^2 + \xi_2^2 > 0$. Now, with an appropriate choice of the open set D, one can apply Theorem 5-A. Problem 6. Thus there can be at most one solution, as long as the earth and sun do not collide.

   Remarks. In connection with Example 2 it might be appropriate to insert some comments on the relationship between "real-world" phenomena and the associated mathematical models. In the above discussion is the statement: "We will not even prove that a solution exists." A natural reaction of the non-mathematician to this statement is: "Why bother? Everybody knows a solution exists for the motion of the earth about the sun. And, for that matter, everyone knows it is unique."

   The flaw in such reasoning is simply this: While we may "know" that the earth travels around the sun, we also "know" just as surely that the mathematical model presented in Example 2 is wrong. Indeed, we can be sure that, no matter how much we improve it, it will still be wrong. For example: we have ignored the influence of all other celestial bodies; we have assumed that the earth and the sun are homogeneous spherical solids[*]; we have ignored "relativity"; and we have assumed that the sun is not influenced by the motion of the earth. This last objection wouldn't count (see Problem 7) if the gravitational interaction between the earth and the sun really propagated instantaneously as we've assumed. But it is no longer believed that this is the case. In other words, our equations of motion should also incorporate time delays.

Every mathematical model should be considered (at best) as an approximation to some real-world phenomenon. If we have confidence that the real phenomenon which we are trying to study is well behaved, then we should test our model to make certain it is also well behaved. The study of existence and uniqueness of solutions of differential equations can be regarded as one way of testing mathematical models. If no solution exists for a differential equation we are using as a model, or if uniqueness, or some other property which we demand, fails to hold -- then we better change our model.

As a matter of fact there is an even more basic need for existence and uniqueness theorems. We need to discover what kind of questions one should ask for the given differential equations. For example, given the linear, second order Eq. (1), how do we know whether to specify as initial conditions

$$x(t_0) = x_0, \tag{15}$$

---

[*] Actually we assumed them to be point particles, but this defect can be remedied by defining a collision as

$$x_1^2(t) + x_2^2(t) \le R \quad \text{for some} \quad R > 0 \quad \text{instead of} \quad R = 0.$$

or the conditions $x(t_0) = x_0$ and $x'(t_0) = v_0$ of (2), or

$$x(t_0) = x_0, \quad x'(t_0) = v_0, \quad \text{and} \quad x''(t_0) = w_0, \qquad (16)$$

or something else. Why are (15) and (16) inappropriate?
Problem 8.

It is sometimes assumed that "physical intuition" suggests the appropriateness of the initial conditions (2).
But, quite the contrary, it is actually the theory of differential equations which shows that these are appropriate
"physical" initial conditions -- assuming the physical problem can indeed be represented by a second order, ordinary
differential equation such as Eq. (1).

Section 9 will discuss still another type of question
one might pose for an equation like (1).

Problems.

1.   Prove that the differential equation (1) with $m \neq 0$ is
     equivalent to system (3).

2.   Verify that (4) defines a solution of Eq. (1) regardless
     of the values of the constants c_1 and c_2.

3.   Recalling the meaning of the Lipschitz conditions from
     5-(10), show that if K is a Lipschitz constant for f
     on A, then $K_1 = K + 1$ is a Lipschitz constant for the
     right hand side of system (8) on A.

4.   Prove Theorem C.

5. State a uniqueness result for

   (a) $x'' + (\cot t)x' + (\cos t)x = 0$.

   (b) $x'' = g - \frac{b}{m}(x')^2$ where g, b, and m are positive
       constants. (This equation is sometimes used to re-
       present the motion of a falling body of mass m at
       height x, where g is the acceleration of gravity
       and b is a coefficient of friction.)

   (c) $x'' = - \frac{kM}{x^2}$ where k and M are positive constants.
       (This equation represents the straight-line free fall
       of an object at a great distance x from the center
       of the earth, where M is the mass of the earth and
       k is the constant of gravitation. It is a special
       case of system (11). Why?)

   (d) $x'' + x(x' - 1)^{2/3} = 0$.

   (e) $1 + (x')^2 + xx'' = 0$ with $x(2) = 2$ and $x'(2) = 1/2$.

*6. Verify the applicability of Theorem 5-A to system (13)
    when $D \equiv \{(t, \xi_1, \xi_2, \xi_3, \xi_4) \in \mathbb{R}^5 : \xi_1^2 + \xi_2^2 > 0\}$. The
    ambitious student will want to verify that this set D
    is open as defined in Appendix 1.

*7. Actually the sun is not stationary, as assumed in Example
    2. In the (idealized) two-body problem the sun moves be-
    cause of the influence of the earth. Write the equations
    of motion without assuming the sun stationary. Let
    $(y_1(t), y_2(t))$ be the position of the earth and
    $(z_1(t), z_2(t))$ be the position of the sun. Show from
    your equations that $my_i' + Mz_i' = c_i$, a constant for
    $i = 1, 2$. The constant vector $\mathrm{col}\ (c_1, c_2)$ is the velo-
    city of the "center of mass" of the system. Now define

$x_i = y_i - z_i$ for $i = 1, 2$ and show that your equations
reduce to those in Example 2 except that M is replaced
with M + m.

8. Discuss the solution of Eq. (1) subject to

   (a) condition (15),

   (b) conditions (16).

7. COMPLEX SOLUTIONS

In the study of linear differential equations with constant coefficients, such as $mx'' + bx' + kx = 0$, it will be
helpful to be able to consider complex-valued functions as
possible solutions. This is the case even if the coefficients are all real and we are ultimately interested only in
real-valued solutions.

If the set of all complex numbers is denoted by $\mathbb{C}$,
then a complex-valued function of a real variable is any function f mapping $A \to \mathbb{C}$ for some set $A \subset \mathbb{R}$.

Definitions. Let A be a subset of $\mathbb{R}$.

   (i) If f_1 and f_2 are two functions mapping $A \to \mathbb{C}$
       and if c is a complex constant, we define new
       functions $f_1 + f_2$, cf_1, and $f_1 f_2$ mapping $A \to \mathbb{C}$
       in the natural manner:

$$(f_1 + f_2)(t) \equiv f_1(t) + f_2(t),$$
$$(cf_1)(t) \equiv cf_1(t), \qquad \text{and}$$
$$(f_1 f_2)(t) \equiv f_1(t)f_2(t)$$

       for t in A.

(ii)   If $f: A \to \mathbb{C}$, we define two functions mapping $A \to \mathbb{R}$ denoted by   Re f   and   Im f, called the real and imaginary parts of   f, by

$$(\text{Re } f)(t) \equiv \text{Re } [f(t)], \quad (\text{Im } f)(t) \equiv \text{Im } [f(t)]$$

$$\text{for } t \text{ in } A.$$

(iii)   We say $f: A \to \mathbb{C}$ is continuous (or differentiable or integrable) if   Re f   and   Im f   are both continuous (or differentiable or integrable).   If   f is differentiable, we define $f': A \to \mathbb{C}$ (also denoted by   Df   or   df/dt) by

$$f'(t) = (\text{Re } f)' + i(\text{Im } f)'(t) \quad \text{for } t \text{ in } A.$$

If   f   is integrable on   [a,b], we define

$$\int_a^b f(s)ds = \int_a^b (\text{Re } f)(s)ds + i\int_a^b (\text{Im } f)(s)ds.$$

Under definition (iii), the familiar rules of differential and integral calculus carry over to complex-valued functions of a real variable.   In particular, let f_1 and f_2 be differentiable (or integrable) functions mapping $A \to \mathbb{C}$ for some $A \subset \mathbb{R}$.   Then, for any complex constants c_1 and c_2, the new function $c_1 f_1 + c_2 f_2$ is differentiable (or integrable) and $f_1 f_2$ is differentiable on   A, and

$$(c_1 f_1 + c_2 f_2)' = c_1 f_1' + c_2 f_2' \tag{1}$$

and

$$(f_1 f_2)' = f_1' f_2 + f_1 f_2' \quad \text{on } A \tag{2}$$

or, if   A = [a,b],

$$\int_a^b (c_1 f_1 + c_2 f_2)(s)ds = c_1 \int_a^b f_1(s)ds + c_2 \int_a^b f_2(s)ds. \tag{3}$$

Equations (1) and (3) were encountered earlier, as Eqs. 5-(3)
and 5-(4), for vector-valued functions. To prove (1), (2),
and (3), one considers the real and imaginary parts of each
equation. Problem 1.

If f_1 and f_2 are two differentiable functions mapping $J \to \mathbb{C}$ for some interval $J \subset \mathbb{R}$ and if

$$f_1'(t) = f_2'(t) \qquad \text{for all} \quad t \quad \text{in} \quad J, \tag{4}$$

then there exists some $c \in \mathbb{C}$ such that

$$f_1(t) = f_2(t) + c \qquad \text{for all} \quad t \quad \text{in} \quad J. \tag{5}$$

The proof again depends on consideration of the real and imaginary parts. Problem 2.

The fundamental theorems of calculus are easily obtained using the corresponding theorems for the real and imaginary parts. These results are as follows, just as for
vector-valued functions in Section 5.

If f is a continuous function mapping an interval
$J \to \mathbb{C}$ and if $t_0, t \in J$, then

$$\frac{d}{dt} \int_{t_0}^{t} f(s)ds = f(t) \tag{6}$$

(understanding, as usual, a one-sided derivative if t is
an endpoint of J).

If $f: [a,b] \to \mathbb{C}$ has a continuous (or merely integrable) derivative on $[a,b]$, then

$$\int_{a}^{b} f'(s)ds = f(b) - f(a). \tag{7}$$

With the aid of these results we can now extend the
uniqueness theorems for linear differential equations and

linear differential systems to the case of complex-valued functions.

Theorem $\underline{A}$.   The uniqueness assertion of Theorem 5-C (or Theorem 6-C) remains valid if the given functions a_{ij} and h_i for i, j = 1,...,n and the initial values, $x_{01},\ldots,x_{0n}$ (or a_k and h for k = 0,...,n-1 and $b_0,\ldots,b_{n-1}$), are complex valued. Now a solution x maps an interval into $\mathbb{C}^n$ (or into $\mathbb{C}$).

Proof.   Admitting complex values as indicated, con-sider the linear system

$$x_k' = \sum_{j=1}^{n} a_{kj}(t)x_j + h_k(t) \quad \text{for} \quad k = 1,\ldots,n \quad (8)$$

on an interval (α,β), with initial conditions

$$x_k(t_0) = x_{0k} \quad \text{for} \quad k = 1,\ldots,n. \quad (9)$$

For any solution x, mapping a subinterval of (α,β) into $\mathbb{C}^n$ let us introduce $u_k(t) = \mathrm{Re}\, x_k(t)$ and $v_k(t) = \mathrm{Im}\, x_k(t)$ so that

$$x_k(t) = u_k(t) + iv_k(t) \quad \text{for} \quad k = 1,\ldots,n.$$

Now system (8) will be satisfied if and only if the real and imaginary parts of both sides of (8) agree, i.e.,

$$\left.\begin{array}{l} u_k' = \sum\limits_{j=1}^{n} \mathrm{Re}\, a_{kj}(t)u_j - \sum\limits_{j=1}^{n} \mathrm{Im}\, a_{kj}(t)v_j + \mathrm{Re}\, h_k(t) \\[3mm] v_k' = \sum\limits_{j=1}^{n} \mathrm{Im}\, a_{kj}(t)u_j + \sum\limits_{j=1}^{n} \mathrm{Re}\, a_{kj}(t)v_j + \mathrm{Im}\, h_k(t) \end{array}\right\} \quad (10)$$

for k = 1,...,n. System (10) consists of 2n real equations for the 2n real unknown functions $u_1,\ldots,u_n$, $v_1,\ldots,v_n$.

And the associated real initial conditions, obtain from (9), are

$$u_k(t_0) = \text{Re } x_{0k}, \quad v_k(t_0) = \text{Im } x_{0k} \quad \text{for } k = 1,\ldots,n. \quad (11)$$

The fact that Eqs. (10) and (11) have at most one solution follows from Theorem 5-A or 5-C since the coefficients in (10) are all real and continuous. This then assures that (8) and (9) have at most one complex vector-valued solution.

The required uniqueness for a complex linear n'th order scalar equation

$$x^{(n)} + a_{n-1}(t)x^{(n-1)} + \cdots + a_1(t)x' + a_0(t)x = h(t)$$

with

$$x(t_0) = b_0, \quad x'(t_0) = b_1, \quad \ldots, \quad x^{(n-1)}(t_0) = b_{n-1}$$

can now be shown. One simply transforms this (complex) scalar equation into a system of n (complex) linear first order equations as was done in the proof of Theorem 6-A or 6-C. $\square$

The complex-valued functions which will be of greatest interest to us are complex exponentials of the form $f(t) = e^{\lambda t}$ for λ complex, say $\lambda = \mu + i\omega$ where μ and ω are real. We have

$$f(t) = e^{\lambda t} = e^{\mu t}(\cos \omega t + i \sin \omega t), \quad (12)$$

so that

$$(\text{Re } f)(t) = e^{\mu t} \cos \omega t \quad \text{and} \quad (\text{Im } f)(t) = e^{\mu t} \sin \omega t.$$

Lemma B. If λ is a complex constant, then the function f defined by

$$f(t) = e^{\lambda t} \quad \text{for } t \in \mathbb{R}$$

is continuous and differentiable. Moreover

$$f'(t) = \lambda e^{\lambda t};$$
(13)

and, if $\lambda \neq 0$ and $a, b \in \mathbb{R}$,

$$\int_a^b e^{\lambda t} \, dt = \frac{1}{\lambda} (e^{\lambda b} - e^{\lambda a}).$$
(14)

The proof is left to the reader as Problem 4. For (13) one must differentiate (12). Equation (14) is then proved with the aid of (13).

Example 1. Consider again the second order differential equation

$$mx'' + bx' + kx = 0,$$
(15)

where m, b, and k are positive constants. Example 6-1 developed the solution of Eq. (15) with initial conditions

$$x(t_0) = x_0, \qquad x'(t_0) = v_0$$
(16)

under the assumption that $b^2 > 4mk$.

Let us now assume $b^2 < 4mk$. If, as before, we seek exponential solutions $e^{\lambda t}$ we are led again to the characteristic equation

$$m\lambda^2 + b\lambda + k = 0.$$
(17)

But now the solutions of (17) are a pair of conjugate complex numbers, $\lambda_1 = \mu + i\omega$ and $\lambda_2 = \mu - i\omega$, where

$$\mu = -b/2m, \qquad \omega = (4mk - b^2)^{1/2}/2m.$$

This implies complex-valued solutions of Eq. (15), $e^{\lambda_1 t}$ and $e^{\lambda_2 t}$. (Note that the calculation of the derivatives of $e^{\lambda t}$

which led to Eq. (17) are justified for complex λ by Lemma
B.)   Further solutions of (15) are obtained as linear combinations of $e^{\lambda_1 t}$ and $e^{\lambda_2 t}$,

$$x(t) = c_1 e^{\lambda_1 t} + c_2 e^{\lambda_2 t}$$

$$= c_1 e^{\mu t}(\cos \omega t + i \sin \omega t) + c_2 e^{\mu t}(\cos \omega t - i \sin \omega t),$$

where the constants c_1 and c_2 may be real or complex.   If
we introduce new constants

$$C_1 \equiv c_1 + c_2 \quad \text{and} \quad C_2 \equiv ic_1 - ic_2,$$

this family of solutions takes the form

$$x(t) = C_1 e^{\mu t}\cos \omega t + C_2 e^{\mu t}\sin \omega t. \tag{18}$$

Again it appears that C_1 and C_2 may be real or complex
numbers.

Now let us consider Eq. (15) in conjunction with the
initial conditions (16) where x_0 and v_0 are real.   Is it
possible to choose C_1 and C_2 so that $x(t)$ as defined by
(18) satisfies (16)?

Substitution of (18) into Eqs. (16) leads to two conditions on the constants C_1 and C_2:

$$C_1 e^{\mu t_0} \cos \omega t_0 + C_2 e^{\mu t_0} \sin \omega t_0 = x_0,$$

$$C_1 (\mu e^{\mu t_0} \cos \omega t_0 - \omega e^{\mu t_0} \sin \omega t_0) \tag{19}$$

$$+ C_2 (\mu e^{\mu t_0} \sin \omega t_0 + \omega e^{\mu t_0} \cos \omega t_0) = v_0.$$

To determine whether these equations are solvable for C_1
and C_2, we calculate the determinant of the coefficients,

$$
\left|
\begin{array}{cc}
e^{\mu t_0}\cos \omega t_0 & e^{\mu t_0}\sin \omega t_0 \\
\mu e^{\mu t_0}\cos \omega t_0 - \omega e^{\mu t_0}\sin \omega t_0) & (\mu e^{\mu t_0}\sin \omega t_0 + \omega e^{\mu t_0}\cos \omega t_0)
\end{array}
\right|
$$

$$= \omega e^{2\mu t_0}.$$

Since $\omega e^{2\mu t_0} \neq 0$, regardless of the value of t_0, it follows that Eqs. (19) can always be solved for C_1 and C_2. Putting the resulting values of C_1 and C_2 into (18) we obtain a function, x, on $\mathbb{R}$ which satisfies (15) and (16). And Theorem A assures us that this is the unique solution of (15) and (16).

Note that C_1 and C_2 obtained from Eqs. (19) will be real numbers because x_0, v_0, μ, ω, and t_0 are all real. Hence the unique solution (18) of Eqs. (15) and (16) will be a real-valued function, as of course it must be if it is to represent the position of an object bouncing on the end of a spring. This solution (and the remaining case $b^2 = 4mk$) will be discussed further in Section 10.

Finally it should be observed that (18) with unspecified values the constants C_1 and C_2 represents the general solution of Eq. (15). This follows from the fact that, given any solution, $\tilde{x}$ on an interval J, (no matter how obtained) one can always choose some t_0 in J and define $x_0 \equiv \tilde{x}(t_0)$, $v_0 \equiv \tilde{x}'(t_0)$. Now solve Eqs. (19) using these values of t_0, x_0, and v_0, and put the resulting values of C_1 and C_2 into (18). This gives another solution, x, of (15) and (16). From uniqueness, it then follows that $x = \tilde{x}$.

Another useful result from calculus which carries over to complex-valued functions is the following.

<u>Lemma C</u>. Let f: [a,b] → ¢ be continuous (or merely inte-
grable). Then

$$\left| \int_a^b f(t)dt \right| \leq \int_a^b |f(t)|dt.$$

Proof. Since f is integrable it follows that Re f
and Im f are integrable, on [a,b]. Let

$$\int_a^b f(t)dt = re^{i\theta} \quad \text{where} \quad r \geq 0.$$

Then, using Eq. (3),

$$\left| \int_a^b f(t)dt \right| = r = e^{-i\theta} \int_a^b f(t)dt = \int_a^b e^{-i\theta}f(t)dt.$$

But since this last integral must be real we have

$$\left| \int_a^b f(t)dt \right| = \text{Re} \int_a^b e^{-i\theta}f(t)dt = \int_a^b \text{Re}[e^{-i\theta}f(t)]dt$$

$$\leq \int_a^b |\text{Re}[e^{-i\theta}f(t)]|dt \leq \int_a^b |e^{-i\theta}f(t)|dt = \int_a^b |f(t)|dt. \quad \square$$

<u>Problems</u>.

1. Verify Eqs. (1), (2), and (3).

2. Prove that (4) implies (5).

3. Prove (6) and (7).

4. Prove Lemma B.

5. (Important for future reference.) Derive the identities

$$\cos \theta = \frac{e^{i\theta} + e^{-i\theta}}{2} \quad \text{and} \quad \sin \theta = \frac{e^{i\theta} - e^{-i\theta}}{2i}$$

for θ real.

8.   A VALUABLE LEMMA

The type of trick used for handling inequality 4-(6) in the basic uniqueness proof dates back at least to Peano [1885]. It was formalized into a lemma by Gronwall [1918] and then generalized by Reid [1930]. This lemma, which we shall now present essentially in Reid's form, will be used repeatedly in later chapters.

Lemma A (Reid's Lemma). Let C be a given constant and k a given non-negative continuous function on an interval J. Let $t_0 \in J$. Then if $v: J \to [0,\infty)$ is continuous and

$$v(t) \leq C + \left| \int_{t_0}^{t} k(s)v(s)ds \right| \qquad \text{for all}\ \ t\ \ \text{in}\ \ J, \qquad (1)$$

it follows that

$$v(t) \leq C e^{\left| \int_{t_0}^{t} k(s)ds \right|} \qquad \text{for all}\ \ t\ \ \text{in}\ \ J. \qquad (2)$$

Remarks. The hypothesis, (1), involves the unknown function, v, on both the left and right hand sides. However, the conclusion, (2), gives an unambiguous upper bound for v(t).

Note that if C = 0 and $k(s) \equiv K$, then (1) is equivalent to inequality 4-(6) with $v(t) = |x(t) - \tilde{x}(t)|$. Then Lemma A gives v(t) = 0 for all t in J as desired.

Proof of Lemma A. The procedure is similar to that used in Example 4-4.

If $t \geq t_0$ (with $t \in J$), inequality (1) can be rewritten as

$$k(t)v(t) - k(t)[C + \int_{t_0}^{t} k(s)v(s)ds] \leq 0.$$

Introducing $Q(t) \equiv C + \int_{t_0}^{t} k(s)v(s)ds$, this becomes

$$Q'(t) - k(t)Q(t) \leq 0.$$

Multiply through by the integrating factor
$\exp\{-\int_{t_0}^{t} k(s)ds\} > 0$ to find

$$\frac{d}{dt}[Q(t)e^{-\int_{t_0}^{t} k(s)ds}] \leq 0.$$

Integrating this last inequality from t_0 to t, and noting
that $Q(t_0) = C$, we find

$$Q(t)e^{-\int_{t_0}^{t} k(s)ds} - C \leq 0$$

or

$$Q(t) \leq Ce^{\int_{t_0}^{t} k(s)ds}.$$

Now substitute this estimate for $Q(t)$ into (1) to obtain

$$v(t) \leq Q(t) \leq Ce^{\int_{t_0}^{t} k(s)ds},$$

which is inequality (2).

If $t \leq t_0$, the proof that (1) yields (2) is left to
the reader as Problem 1.   □

To illustrate the utility of this lemma we shall prove
an elementary but valuable theorem regarding the growth of
errors in the solution of a system of differential equations.
If we think of a system of n equations

$$x' = f(t,x), \tag{3}$$

as representing some physical process, then the n initial
conditions

$$x(t_0) = x_0 \tag{4}$$

become input data.   If the conditions $x(t_0) = x_0$ have been
measured somehow then of course they have only been evaluated
to the accuracy of the measuring equipment.   Assuming we actually measured $\tilde{x}_0$ instead of x_0 the question is this:

If $\tilde{x}_0$ is "close" to x_0 will the corresponding solution $\tilde{x}$ be close to the solution x?

Theorem B asserts that the answer is "yes" if f satisfies a Lipschitz condition and if $t-t_0$ is not too large.

<u>Theorem B</u>.   Let f map $D \to \mathbb{R}^n$ for some open set $D \subset \mathbb{R}^{1+n}$
and let f be continuous and satisfy the Lipschitz condition

$$||f(t,\xi) - f(t,\tilde{\xi})|| \leq K||\xi - \tilde{\xi}||$$

on D.   Let $(t_0,x_0) \in D$ and $(t_0,\tilde{x}_0) \in D$ and let x and
$\tilde{x}$ be solutions of Eq. (3) on an interval J such that
$x(t_0) = x_0$ and $\tilde{x}(t_0) = \tilde{x}_0$.   Then·

$$||x(t) - \tilde{x}(t)|| \leq ||x_0 - \tilde{x}_0||e^{K|t-t_0|} \qquad \text{for all t in J. (5)}$$

<u>Proof</u>.   For t in J

$$x(t) = x_0 + \int_{t_0}^{t} f(s,x(s))ds$$

and

$$\tilde{x}(t) = \tilde{x}_0 + \int_{t_0}^{t} f(s,\tilde{x}(s))ds.$$

Subtracting these two equations we obtain the estimate

$$||x(t) - \tilde{x}(t)|| = ||x_0 - \tilde{x}_0 + \int_{t_0}^{t} [f(s,x(s)) - f(s,\tilde{x}(s))]ds||$$

$$\leq ||x_0 - \tilde{x}_0|| + \left|\int_{t_0}^{t} K||x(s) - \tilde{x}(s)||ds\right|$$

for all t in J.   But from this Lemma A gives inequality (5)
at once.   ☐

The next theorem applies this result to a system of n linear equations with variable coefficients

$$x_i' = \sum_{j=1}^{n} a_{ij}(t)x_j + h_i(t) \qquad \text{for} \quad i = 1,\ldots,n. \qquad (6)$$

__Theorem C__. Let each a_{ij} and each h_i be a continuous function mapping some interval $(\alpha,\beta) \to \mathbb{R}$ and let x and $\tilde{x}$ be two solutions of (6) on (α_1,β_1) where $\alpha < \alpha_1 < \beta_1 < \beta$. Then, if $x(t_0) = x_0$ and $\tilde{x}(t_0) = \tilde{x}_0$,

$$||x(t) - \tilde{x}(t)|| \leq ||x_0 - \tilde{x}_0||e^{K|t-t_0|} \qquad \text{for} \quad \alpha_1 < t < \beta_1$$

for some constant $K > 0$.

__Proof.__ Since each a_{ij} is continuous on the closed bounded interval $[\alpha_1,\beta_1]$, there must exist some $B > 0$ such that

$$|a_{ij}(t)| \leq B \qquad \text{for} \quad \alpha_1 \leq t \leq \beta_1, \qquad i, j = 1,\ldots,n.$$

But then system (6) is a special case of (3) with

$$|f_i(t,\xi) - f_i(t,\tilde{\xi})| = |\sum_{j=1}^{n} a_{ij}(t)[\xi_j - \tilde{\xi}_j]| \leq B||\xi - \tilde{\xi}||$$

for all (t,ξ) and $(t,\tilde{\xi})$ in $[\alpha_1,\beta_1] \times \mathbb{R}^n$ and $i = 1,\ldots,n$. Thus

$$||f(t,\xi) - f(t,\tilde{\xi})|| \leq nB||\xi - \tilde{\xi}||,$$

and the assertion of the theorem follows from Theorem B. $\square$

Lemma A is one of the best known examples of a theorem on "integral inequalities". This result and generalizations of it have been exploited widely in the study of ordinary differential equations, functional differential equations, partial differential equations, and integral equations.

A minor generalization of Lemma A which will be of use in Chapters IV and VI is presented as a corollary:

Corollary D.   Let M and k be given non-negative functions with k continuous on an interval J. Let $t_0 \in J$ and assume $M(t)$ increases as $|t-t_0|$ increases. Then if v is any non-negative continuous function such that

$$v(t) \leq M(t) + \left| \int_{t_0}^{t} k(s)v(s)ds \right| \qquad \text{for all } t \text{ in } J, \qquad (7)$$

it follows that

$$v(t) \leq M(t)e^{\left| \int_{t_0}^{t} k(s)ds \right|} \qquad \text{for all } t \text{ in } J. \qquad (8)$$

Proof.   Let $t_0 \leq t \leq t_1$ with $t_1 \in J$. Then from (7) we have

$$v(t) \leq M(t_1) + \int_{t_0}^{t} k(s)v(s)ds,$$

where t_1 is now regarded as a constant. Thus Lemma A gives

$$v(t) \leq M(t_1)e^{\int_{t_0}^{t} k(s)ds},$$

and (8) follows by putting $t = t_1$. The proof is similar when $t_1 \leq t \leq t_0$. □

Problems

1.   Complete the proof of Lemma A by treating the case $t \leq t_0$. (Be careful of the algebraic signs at each step of your proof.)

2.   If we add the hypothesis "v is continuously differentiable", what is wrong with the following simpler "proof" for Lemma A? For $t \geq t_0$ inequality (1) becomes

$$v(t) \leq C + \int_{t_0}^{t} k(s)v(s)ds.$$

Differentiate this with respect to t and rearrange to find

$$v'(t) - k(t)v(t) \leq 0 \quad \text{with} \quad v(t_0) = v_0 \leq C.$$

Now multiply by $\exp\{-\int_{t_0}^{t} k(s)ds\}$ and integrate from t_0 to t, getting

$$v(t)e^{-\int_{t_0}^{t} k(s)ds} - v(t_0) \leq 0.$$

This immediately gives (2).

3. Consider a scalar linear equation,

$$x' + a(t)x = h(t),$$

where a and h are continuous on $\mathbb{R}$ with $|a(t)| \leq B$ for all t. Compute the difference between two solutions, x and $\tilde{x}$, in terms of $x(t_0) = x_0$ and $\tilde{x}(t_0) = \tilde{x}_0$ proceeding directly from the general solution (Problem 2-2). How does your result compare with the estimate obtained using Theorem B?

4. Consider the system of equations for the electric currents in Example 5-1,

$$x_1' = -2x_1 + 2x_2 + 5 \sin 2t$$
$$x_2' = 2x_1 - 5x_2.$$

Assume the system has two solutions on $J = \mathbb{R}$, $x(t) = \text{col } (x_1(t), x_2(t))$ satisfying initial conditions $x(t_0) = \text{col } (x_{01}, x_{02})$, and $\tilde{x}(t) = \text{col } (\tilde{x}_1(t), \tilde{x}_2(t))$ satisfying

$\tilde{x}(t_0) = col\ (\tilde{x}_{01}, \tilde{x}_{02})$. Then show, using Theorem B, that for all t in $\mathbb{R}$

$$|x_1(t) - \tilde{x}_1(t)| + |x_2(t) - \tilde{x}_2(t)|$$

$$\leq (|x_{01} - \tilde{x}_{01}| + |x_{02} - \tilde{x}_{02}|)e^{7|t-t_0|}.$$

5. Obtain an estimate for the rate of growth of the error in the solution of x''' - x'' = 0 due to incorrect initial data.

*6. For the equation (1-t)x' = x - 1 with x(0) = 1, it is easily seen that a solution is defined by x(t) = 1 for all t < 1.

(a) Prove that this solution is unique.

(b) Nevertheless, if these equations represent any real-world physical system, then that system will surely explode as t → 1. Show clearly that this is true. Or is it?

(c) Why does the assertion in (b) not contradict Theorem B?

*7. Discuss the assertions and questions of Problem 6 for the equation $(1-t)^{1/3}x' = x - 1$ with x(0) = 1.

9. A BOUNDARY VALUE PROBLEM

Sections 4 through 8 dealt only with "initial value problems". From these one might jump to the conclusion that the solution to an n'th order equation, or system of n first order equations, will be unique whenever n auxiliary conditions are specified.

To dispel this notion, and to re-emphasize that existence and uniqueness should not be considered obvious, we will
briefly discuss another type of problem for differential equations. This important topic -- "boundary value problems" --
will not be studied in this text beyond the brief introduction
it gets in the present section.

Consider the second order linear homogeneous differential equation

$$x'' + \omega^2 x = 0, \tag{1}$$

where ω is a given positive constant. The general solution
of (1) was shown in Example 7-1 to be given by

$$x(t) = c_1 \cos \omega t + c_2 \sin \omega t, \tag{2}$$

where c_1 and c_2 are constants (called C_1 and C_2 in
Example 7-1).

Let us seek a function $x: [\alpha, \beta] \rightarrow \mathbb{R}$, where $\alpha < \beta$,
which satisfies Eq. (1) together with certain "boundary condi-
tions". The latter will be restrictions on x or x' at α
and β, the end points or boundaries of the interval. Thus,
instead of imposing two conditions at one point, t_0, we will
now impose one condition at each of two points, α and β.

When we talk about a solution of a differential equa-
tion on a closed interval, such as $[\alpha, \beta]$ it should be understood that any derivatives are interpreted as one-sided derivatives at the end points of the interval.

For simplicity we shall take $\alpha = 0$. This, it can be
shown, does not restrict the generality of our conclusions.

Keep in mind that every solution must be a special case
of (2).

Example 1.   Find a solution of Eq. (1) on $[0,\beta]$ which satisfies the boundary conditions

$$x(0) = 0 \quad \text{and} \quad x(\beta) = 0. \tag{3}$$

Putting $t = 0$ in Eq. (2), the first boundary condition, $x(0) = 0$, reduces to the requirement $c_1 = 0$.   Thus any
solution of (1) and (3) must be of the form

$$x(t) = c_2 \sin \omega t.$$

Now impose the second boundary condition $x(\beta) = 0$, to find

$$c_2 \sin \omega \beta = 0.$$

If $\omega \beta$ is not an integral multiple of π then $\sin \omega \beta \neq 0$
and we must have $c_2 = 0$.   But if $\omega \beta = n\pi$ for some integer
$n = 1, 2, \ldots$, then $\sin \omega \beta = 0$, and any value of c_2 will be
acceptable.   Thus

(i)   If $\omega \beta \neq n\pi$ for $n = 1, 2, \ldots$ the solution of
Eqs. (1) and (3) is unique, $x(t) \equiv 0$.

(ii)   If $\omega \beta = n\pi$ for some $n = 1, 2, \ldots$, then Eqs. (1)
and (3) have infinitely many solutions of the form
$x(t) = c_2 \sin \omega t$.

Physically, one can think of Eq. (1) as describing the
motion of an object bouncing on the end of a spring with no
damping -- an idealized special case of Example 7-1.   Then the
boundary conditions (3) specify that the object must be at its
equilibrium position at the instants $t = 0$ and $t = \beta$.

Example 2.   Let the boundary conditions for Eq. (1) be

$$x(0) = 0 \quad \text{and} \quad x(\beta) = 1. \tag{4}$$

Then again the first of these conditions alone reduces the possible solutions to

$$x(t) = c_2 \sin \omega t.$$

The second boundary condition in (4) now becomes

$$c_2 \sin \omega \beta = 1. \tag{5}$$

If $\omega \beta \neq n\pi$ for $n = 1, 2, \ldots$, then this yields

$$c_2 = \frac{1}{\sin \omega \beta} .$$

But if $\omega \beta = n\pi$ for some $n = 1, 2, \ldots$, there is no way to satisfy Eq. (5). Thus

(i) If $\omega \beta \neq n\pi$ for $n = 1, 2, \ldots$, then Eqs. (1) and (4) have a <u>unique</u> solution $x(t) = (\sin \omega t)/(\sin \omega \beta)$.

(ii) If $\omega \beta = n\pi$ for some $n = 1, 2, \ldots$, then Eqs. (1) and (4) have <u>no solution</u>.

Admittedly, in these two examples it is only in the exceptional cases

$$\beta = \frac{n\pi}{\omega} \quad \text{for} \quad n = 1, 2, \ldots$$

that one fails to get a unique solution. And, indeed, it is easy to find examples which always have unique solutions (Problem 1).

But, using a nonlinear equation, we can also exhibit examples in which the solution of a boundary value problem is never unique, regardless of the value of β.

Example 3.   The equation

$$x'' + x^2 = 0 \tag{6}$$

with boundary conditions

$$x(0) = 0, \quad x(\beta) = 0 \tag{7}$$

clearly has the trivial solution $x(t) \equiv 0$ on $[0,\beta]$. Moreover, if we write (6) as $x'' = f(t,x) = -x^2$, f and $D_2 f$ are
continuous everywhere.

Nevertheless, we will now show that, regardless of the
values of $\beta > 0$, there is also a nontrivial solution of (6)
and (7).

In an attempt to find a nontrivial solution let us
multiply both sides of (6) by x' to obtain $x'x'' + x^2 x' = 0$.
This implies

$$(x')^2/2 + x^3/3 = \text{constant.}$$

For convenience let us write the constant as $c^3/3$. Then we
have

$$x' = \pm(2/3)^{1/2}(c^3 - x^3)^{1/2}.$$

To get a nontrivial solution of (6) and (7) we must have
$x'(0) > 0$. Why? Thus replace the $\pm$ sign by $+$ (for sufficiently small $t > 0$). Then, by separation of variables,

$$\int_0^{x(t)} (c^3 - \xi^3)^{-1/2} d\xi = (2/3)^{1/2} t. \tag{8}$$

Let us try to construct a solution having the general form
indicated in Figure 1. It should increase as $t \to \beta/2$ to the
value $x(\beta/2) = c$ with $x'(\beta/2) = 0$. Then it should decrease
symmetrically as t goes from $\beta/2$ to β. All this can be
achieved if we can choose c in (8) such that $x(\beta/2) = c$,

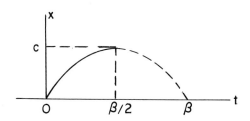

Figure 1

i.e.,

$$\int_0^c (c^3 - \xi^3)^{-1/2}d\xi = \beta/6^{1/2}.$$

Introducing $u = \xi/c$ this condition becomes

$$\int_0^1 (1 - u^3)^{-1/2}du = \beta(c/6)^{1/2}. \tag{9}$$

It is left as an exericse for the reader to show that the improper integral converges, and hence the appropriate value of
c can be determined from (9) for each given $\beta > 0$. Problem
2. Then Eq. (8) determines the desired solution $x(t) \neq 0$
implicitly on $[0, \beta/2]$.

Problems

1. The equation of motion for a free falling body at a
   distance $x(t)$ above the earth's surface is

$$x'' = -g,$$

   where g is a positive constant. Show that a unique solution is determined by any given boundary conditions of
   the form

$$x(0) = x_0 \quad \text{and} \quad x(\beta) = x_\beta \quad \text{where} \quad \beta > 0.$$

2. Prove the convergence of the improper integral in (9).

Chapter III.   <u>THE</u> <u>LINEAR</u> <u>ORDINARY</u> <u>EQUATION</u> <u>OF</u> <u>ORDER</u> <u>n</u>

The examples in Chapters I and II have introduced special methods for finding exact solutions of some special equations. Generally speaking, the discovery of exact solutions of a differential equation (when this is possible) does depend upon the use of ad hoc methods or tricks, often requiring considerable ingenuity.

However in the case of linear equations and linear systems a well organized general theory exists. And in the special case of linear equations with constant coefficients,

$$a_n x^{(n)} + a_{n-1} x^{(n-1)} + \cdots + a_1 x' + a_0 x = h(t),$$

where $h(t)$ is reasonably well behaved, the general theory provides exact solutions.

Linear systems and linear equations of order n are important in physics, engineering, and other applied sciences. In this chapter we shall concentrate on the scalar linear equation of order n. Systems of linear first order equations will be discussed in Chapter IV.

10. CONSTANT COEFFICIENTS (THE HOMOGENEOUS CASE)

This chapter is primarily devoted to the n'th order linear equation with constant coefficients

$$a_n x^{(n)} + a_{n-1} x^{(n-1)} + \cdots + a_1 x' + a_0 x = h(t), \qquad (1)$$

where $a_n \neq 0$. We shall often seek a solution of (1) which also satisfies the initial conditions

$$x(t_0) = b_0, \quad x'(t_0) = b_1, \quad \ldots, \quad x^{(n-1)}(t_0) = b_{n-1}. \qquad (2)$$

In the present section we shall concentrate on the homogeneous case of Eq. (1) -- the case when $h(t) = 0$ -- which we write as

$$a_n y^{(n)} + a_{n-1} y^{(n-1)} + \cdots + a_1 y' + a_0 y = 0. \qquad (3)$$

A useful tool for the systematic discovery of solutions of (3) is the set of "polynomials" in the differential operator D. Given an ordinary polynomial,

$$p(\lambda) = a_n \lambda^n + a_{n-1} \lambda^{n-1} + \cdots + a_1 \lambda + a_0,$$

we define the symbolic polynomial operator in D,

$$p(D) = a_n D^n + a_{n-1} D^{n-1} + \cdots + a_1 D + a_0 I,$$

by the requirement that

$$p(D)z = a_n D^n z + a_{n-1} D^{n-1} z + \cdots + a_1 Dz + a_0 z \qquad (4)$$

for every function $z = z(t)$ having n derivatives. (The symbol "I" used above stands for the identity operator since $Iz = z$ always.) Actually life will be simpler, and it will be quite adequate for our purposes, if we assume henceforth

that the functions z used in the definition of p(D) are
infinitely often differentiable.

Note that with the above notation we can express Eqs.
(1) and (3) respectively as

$$p(D)x = h(t) \quad \text{and} \quad p(D)y = 0.$$

It is important to observe that the definition of p(D)
makes sense if the coefficients, $a_n, \ldots, a_1$, and a_0, and the
functions z are assumed to be complex valued. Henceforth,
let us assume them to be complex valued unless stated otherwise. The reason for this will soon become clear.

A basic property of p(D) is its _linearity_. That is,
if z_1 and z_2 are functions (possibly complex valued) and
c_1 and c_2 are constants (possibly complex), then

$$p(D)(c_1 z_1 + c_2 z_2) = c_1 p(D) z_1 + c_2 p(D) z_2. \tag{5}$$

This follows easily from (4) and the linearity of each D^j,

$$D^j(c_1 z_1 + c_2 z_2) = c_1 D^j z_1 + c_2 D^j z_2.$$

Now assume we have another polynomial, $q(\lambda) = b_m \lambda^m +$
$b_{m-1} \lambda^{m-1} + \cdots + b_1 \lambda + b_0$, with complex coefficients, so that
the effect of

$$q(D) = b_m D^m + b_{m-1} D^{m-1} + \cdots + b_1 D + b_0 I$$

is also well defined on every complex-valued z. We shall
then define _addition_ and _multiplication_ of the operators p(D)
and q(D) by the equations

$$[p(D) + q(D)]z = p(D)z + q(D)z$$

and   (6)

$$[p(D)q(D)]z = p(D)[q(D)z]$$

for _every_ infinitely differentiable function z.

The principal usefulness of the polynomial operator notation lies in the following results. Let $r(\lambda)$ and $s(\lambda)$ be the ordinary polynomials defined by

$$r(\lambda) \equiv p(\lambda) + q(\lambda) \quad \text{and} \quad s(\lambda) \equiv p(\lambda)q(\lambda).$$

Then it can be shown that

$$p(D) + q(D) = r(D)$$

and   (7)

$$p(D)q(D) = s(D).$$

Problem 1. These equations should not be considered obvious. What they assert is that, for every infinitely differentiable function z,

$$p(D)z + q(D)z = r(D)z$$

and, more subtly,   (7')

$$p(D)[q(D)z] = s(D)z.$$

From Eqs. (7) it follows that addition and multiplication of polynomial operators in D satisfy the usual associative, commutative and distributive laws. This implies that _a polynomial in_ D _can be factored just like an ordinary polynomial._ In particular, let

$$p(D) = a_n D^n + a_{n-1} D^{n-1} + \cdots + a_1 D + a_0 I$$

and let $\lambda_1, \lambda_2, \ldots, \lambda_n$ be the n roots (not necessarily distinct) of $p(\lambda) = 0$.   Then, as we know,

$$p(\lambda) = a_n(\lambda-\lambda_1)(\lambda-\lambda_2) \cdots (\lambda-\lambda_n).$$

Hence

$$p(D) = a_n(D-\lambda_1 I)(D-\lambda_2 I) \cdots (D-\lambda_n I). \qquad (8)$$

Here is the reason for emphasizing that the various results about polynomial operators are valid in the case of complex coefficients: Even though we will usually start with an equation (1) or (3) having real coefficients, it is quite likely that some roots of $p(\lambda) = 0$ will be complex, and hence some factors of $p(D)$ will have complex coefficients.

Example 1.   Let us return once more to the second order, linear differential equation

$$mx'' + bx' + kx = 0 \qquad (9)$$

with real coefficients and with $m \neq 0$.   The initial conditions, when needed, will be

$$x(t_0) = x_0 \quad \text{and} \quad x'(t_0) = v_0. \qquad (10)$$

Equation (9) is equivalent to

$$p(D)x = (mD^2 + bD + kI)x = 0.$$

If λ_1 and λ_2 are the roots of $m\lambda^2 + b\lambda + k = 0$, then, by (8), Eq. (9) is equivalent to

$$m(D-\lambda_1 I)(D-\lambda_2 I)x = 0.$$

Introducing $x_1 \equiv (D-\lambda_2 I)x$, we have $(D-\lambda_1 I)x_1 = 0$, or

$$x_1' - \lambda_1 x_1 = 0,$$

which implies $x_1(t) = c_1 e^{\lambda_1 t}$, where c_1 is a constant. To find x, it remains to solve

$$x' + \lambda_2 x = c_1 e^{\lambda_1 t}. \tag{11}$$

Note that to this point, we have not said whether λ_1 and λ_2 are real and distinct, or complex conjugates, or identically equal real numbers. But now, in solving Eq. (11), we must distinguish between the cases of (i) distinct roots and (ii) identical roots, i.e., a "double root".

(i) If $\lambda_2 \neq \lambda_1$ (i.e., if $b^2 \neq 4mk$), use of the integrating factor $e^{-\lambda_2 t}$ leads to the solution of (11),

$$x(t) = \frac{c_1}{\lambda_1 - \lambda_2} e^{\lambda_1 t} + c_2 e^{\lambda_2 t}.$$

(Problem 2). By renaming the constant c_1, we can rewrite this in the form

$$x(t) = c_1 e^{\lambda_1 t} + c_2 e^{\lambda_2 t} \tag{12}$$

found in Examples 6-1 and 7-1. And, in these cases, we have seen that (12) is the general solution of Eq. (9).

(ii) In case $\lambda_2 = \lambda_1 = \lambda$ (i.e., if $b^2 = 4mk$), the same procedure gives, as the solution of (11),

$$x(t) = c_1 t e^{\lambda t} + c_2 e^{\lambda t}. \tag{13}$$

(Problem 2 continued). The reader is asked to prove that the constants c_1 and c_2 in (13) will always be (uniquely) determined by the initial conditions (10) for any t_0. Problem 3.

Let us now assume that m and k are positive numbers

and that $b \geq 0$, this being the natural case in physical applications.   Then the behavior of solutions can be classified
in terms of the magnitude of the "damping coefficient" b.

If $b^2 < 4mk$ the system is said to be underdamped.   We
then have $\lambda_1 = \mu + i\omega$ and $\lambda_2 = \mu - i\omega$, where

$$\mu = -b/2m < 0 \quad \text{and} \quad \omega = (4mk-b^2)^{1/2}/2m > 0, \tag{14}$$

and the solution (12) can be rewritten (recall Example 7-1) as

$$x(t) = C_1 e^{\mu t} \cos \omega t + C_2 e^{\mu t} \sin \omega t, \tag{15}$$

where C_1 and C_2 are real if x_0 and v_0 in (10) are
real.   This is damped oscillatory motion and is illustrated in
Figure 1.   The solution is contained in an "envelope" between

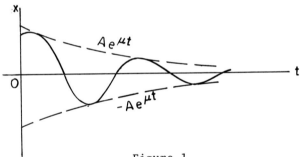

Figure 1

$Ae^{\mu t}$ and $-Ae^{\mu t}$ for $A = (C_1^2 + C_2^2)^{1/2}$; and the solution has
infinitely many zeros spaced π/ω units apart.   Problem 4(a).

If $b^2 > 4mk$ Eq. (9) is said to be overdamped, and the
solution is given by (12) where λ_1 and λ_2 are distinct
negative numbers.   The shape of the solution curve now depends on the initial conditions (10) as illustrated in Figure
2.   It can be shown that, unless it is identically zero, the
solution can become zero for at most one value of $t > t_0$.
Problem 4(b).

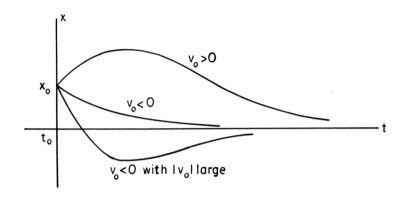

Figure 2

If $b^2 = 4mk$ Eq. (9) is said to be <u>critically</u> <u>damped</u>, and the solution is given by (13) with $\lambda = -b/2m < 0$.   In this case the possible types of solution resemble those in Figure 2 for the overdamped case.   Problem 4(c).

Note that in the case of overdamping, the solution (12) contains one exponential, say $e^{\lambda_1 t}$, which decays more slowly than in either the underdamped or critically damped cases, i.e., $\lambda_1 > -b/2m$.   Thus it is possible that the solution in this case may "die out" more slowly than in the case of a smaller damping coefficient, such as in critical damping.

In an automobile suspension system, for example, the shock absorbers provide damping.   It is desirable that the system have critical damping or slight overdamping so that the vehicle does not keep on bouncing after hitting a bump.   In fact, if after hitting an isolated bump or rut your car bounces through the "equilibrium level" more than once then your shock absorbers are bad.

Turning now to the n'th order homogeneous equation (3) we shall begin, as in Examples 6-1 and 7-1, by seeking exponential solutions of the form $e^{\lambda t}$. Substituting $y(t) = e^{\lambda t}$ into Eq. (3) one finds

$$(a_n \lambda^n + a_{n-1} \lambda^{n-1} + \cdots + a_1 \lambda + a_0)e^{\lambda t} = 0.$$

Since $e^{\lambda t}$ is never zero, we must require

$$a_n \lambda^n + a_{n-1} \lambda^{n-1} + \cdots + a_1 \lambda + a_0 = 0. \tag{16}$$

This n'th degree polynomial equation is called the <u>characteristic equation</u> for (3) or (1). For each root, $\lambda_1, \lambda_2, \ldots,$ or λ_n of (16),

$$y_j(t) = e^{\lambda_j t} \qquad (j = 1, \ldots, \text{ or } n)$$

defines a solution of (3). And it follows from (5), the linearity of $p(D)$, that an arbitrary linear combination of these,

$$y(t) = c_1 e^{\lambda_1 t} + c_2 e^{\lambda_2 t} + \cdots + c_n e^{\lambda_n t}, \tag{17}$$

also defines a solution. We thus obtain a solution with n arbitrary constants, and we can <u>hope</u> that it will always be possible to choose $c_1, \ldots, c_n$ to satisfy any given initial conditions of the form

$$y(t_0) = b_0, \quad y'(t_0) = b_1, \quad \ldots, \quad y^{(n-1)}(t_0) = b_{n-1}. \tag{18}$$

But how shall we proceed if Eq. (16) has multiple roots, so that there are only, say, $q < n$ distinct roots, $\lambda_1, \lambda_2, \ldots, \lambda_q$? In this case Eq. (17) would effectively involve no more than q arbitrary constants.

Such a situation was encountered in Example 1, case

(ii).   There the characteristic equation $m\lambda^2 + b\lambda + k = 0$
with $b^2 = 4mk$ had only one root, $\lambda = -b/2m$.   This was, of
course, a double root (i.e., a root of multiplicity 2).   In
that case we found that the differential equation, which was
equivalent to

$$(D - \lambda I)^2 y = 0,$$

had the solution

$$y(t) = c_1 t e^{\lambda t} + c_2 e^{\lambda t},$$

for all values of the constants c_1 and c_2.   That is, in
addition to $e^{\lambda t}$, another solution was defined by $te^{\lambda t}$.

        We shall find that this is a prototype for the general
case.   Lemmas A and B will establish that if Eq. (16) has a
root λ of multiplicity   m,   then $e^{\lambda t}$, $te^{\lambda t}$, $t^2 e^{\lambda t}$,..., and
$t^{m-1} e^{\lambda t}$ all give solutions of (3).

Lemma A.   Let $\lambda \in \mathbb{C}$, and let   m   be a positive integer.   Then
$(D - \lambda I)^m (t^k e^{\lambda t}) = 0$ for all   t   in $\mathbb{R}$ for $k = 0, 1,\ldots,m-1$.

        Proof.   Whether λ is real or complex,

$$(D - \lambda I)(t^k e^{\lambda t}) = kt^{k-1} e^{\lambda t} + \lambda t^k e^{\lambda t} - \lambda t^k e^{\lambda t} = kt^{k-1} e^{\lambda t}.$$

Proceeding by induction, we find

$$(D - \lambda I)^k (t^k e^{\lambda t}) = k! e^{\lambda t},$$

and hence, since $m > k$,

$$(D - \lambda I)^m (t^k e^{\lambda t}) = 0. \quad \square$$

Lemma B.   Let λ_j with multiplicity m_j for $j = 1,\ldots,q$
be the roots of (16).   (Thus $m_1 + m_2 + \cdots + m_q = n$.)   Then
the function defined by

$$y(t) = P_1(t)e^{\lambda_1 t} + P_2(t)e^{\lambda_2 t} + \cdots + P_q(t)e^{\lambda_q t}, \qquad (19)$$

where P_j is an arbitrary polynomial of degree $m_j - 1$ or lower, is a solution of (3) for all t in $\mathbb{R}$.

Proof.   Since we can express the "characteristic poly-nomial", $p(\lambda) = a_n\lambda^n + a_{n-1}\lambda^{n-1} + \cdots + a_1\lambda + a_0$, in factored form as

$$p(\lambda) = a_n(\lambda-\lambda_1)^{m_1}(\lambda-\lambda_2)^{m_2}\ldots(\lambda-\lambda_q)^{m_q},$$

it follows that $p(D)$ can be factored similarly.   So the dif-ferential equation (3) can be re-written as

$$a_n(D-\lambda_1 I)^{m_1}(D-\lambda_2 I)^{m_2}\ldots(D-\lambda_q I)^{m_q}y = 0. \qquad (3')$$

Moreover the factors in this product of operators can be written in any order.   Thus it follows from Lemma A that each function $t^k e^{\lambda_j t}$, for $k = 0, 1, \ldots, m_j - 1$ and $j = 1, \ldots, q$, defines a solution of Eq. (3).   But y as defined in (19) is just a linear combination of these functions, and hence is also a solution of (3). $\square$

The function defined by (19) involves $m_1 + m_2 + \cdots + m_q = n$ arbitrary constants, just as it did in the special case (17) with n distinct roots of the char-acteristic equation.   However, this does not _prove_ that (19) is the general solution of Eq. (3).   The proof that indeed it is will come in the next section.   What remains to be shown is that the n arbitrary constants in (19) can _always_ be chosen so that y satisfies n arbitrary initial conditions (18).

Example 2.   Let us apply Lemma B to find solutions of

$$y^{(4)} - 6y^{(3)} + 12y'' - 8y' = 0.$$

The characteristic equation,

$$\lambda^4 - 6\lambda^3 + 12\lambda^2 - 8\lambda = 0,$$

has solutions $\lambda = 0, 2, 2, 2$. Thus solutions of the homogeneous equation are defined by

$$y(t) = c_1 + c_2 e^{2t} + c_3 te^{2t} + c_4 t^2 e^{2t},$$

for arbitrary constants c_1, c_2, c_3, and c_4.

Remark. Usually, the coefficients in Eq. (3) will be real numbers. Then the coefficients in the characteristic equation (16) will be real, and it follows that any complex roots must occur in conjugate pairs with the same multiplicities. Thus if $\lambda_j = \mu + i\omega$ is a complex root with multiplicity $m_j = m$, for some $j = 1, \ldots,$ or q, then for some other index k we must have $\lambda_k = \overline{\lambda}_j = \mu - i\omega$ and $m_k = m$. In this case it will be convenient to replace the corresponding terms in (19),

$$P_j(t)e^{\lambda_j t} + P_k(t)e^{\lambda_k t}$$

where P_j and P_k are polynomials of degree $m-1$ or lower, with

$$Q(t)e^{\mu t} \cos \omega t + R(t)e^{\mu t} \sin \omega t, \tag{20}$$

where Q and R are polynomials of degree $m-1$ or lower. The reader should verify (Problem 5) that this is always possible and that

$$Q(t) = P_j(t) + P_k(t) \quad \text{and} \quad R(t) = iP_j(t) - iP_k(t). \tag{21}$$

Problems

1. Prove Eqs. (7), i.e., (7'),

   (a) for the special case $p(D) = a_1 D + a_0 I$ and
       $q(D) = b_1 D + b_0 I$,

   (b) in general.

2. Carry out the solution of Eq. (11) in cases (i) $\lambda_2 \neq \lambda_1$
   and (ii) $\lambda_2 = \lambda_1 = \lambda$. The answers should be given by (12)
   and (13) respectively.

3. Prove that the constants c_1 and c_2 in (13) are always
   (uniquely) determined if initial conditions (10) are
   given.

4. Assume $m > 0$, $b \geq 0$, $k > 0$ in Eq. (9) and x_0 and v_0
   real in Eq. (10).

   (a) For the underdamped case, $b^2 < 4mk$, show that (15) can
       be rewritten as
       $$x(t) = Ae^{\mu t} \sin (\omega t + \phi),$$
       where $A = (C_1^2 + C_2^2)^{1/2}$ and $\phi = \arctan C_1/C_2$, and
       thereby verify the assertions about Figure 1.

   (b) For the overdamped case, $b^2 > 4mk$, verify that solu-
       tions can take the three shapes indicated in Figure 2.
       Specifically, show that if $x(t_0) = x_0$ and
       $x'(t_0) = v_0$ then
       $$x(t) = \frac{\lambda_2 x_0 - v_0}{\lambda_2 - \lambda_1} e^{\lambda_1 (t - t_0)} + \frac{v_0 - \lambda_1 x_0}{\lambda_2 - \lambda_1} e^{\lambda_2 (t - t_0)}.$$
       From this prove that (unless x is identically zero)
       $x(t)$ has at most one zero for $t > t_0$; and in order
       that $x(t)$ have a zero for some $t > t_0$ one must

have v_0 and x_0 of opposite sign with $v_0/x_0 < \lambda_2$, i.e., $|v_0|$ sufficiently large.

(c)   For the critically-damped case, $b^2 = 4mk$,

$\lambda_1 = \lambda_2 = \lambda = -b/2m < 0$. Show that the solution is now

$$x(t) = (v_0 - \lambda x_0)(t-t_0)e^{\lambda(t-t_0)} + x_0 e^{\lambda(t-t_0)}.$$

Again show that (if x is not identically zero) $x(t)$ can be zero at most once for $t > t_0$ and this occurs if $v_0/x_0 < \lambda$. Why can you be sure that $x(t) \rightarrow 0$ as $t \rightarrow \infty$?

5.   If λ_k and λ_j are complex conjugate roots of Eq. (16) each having multiplicity m, verify that the corresponding terms in (19) can be rewritten as in (20).

6.   Find a solution involving n arbitrary constants for each of the following n'th order equations. Your solution should be in the form (19) if all λ_j are real, or modified as in (20) if some λ_j are complex.

(a) $y'' - 9y = 0$.

(b) $y'' + 9y = 0$.

(c) $y'' - 6y' + 9y = 0$.

(d) $y'' + 4y' + 9y = 0$.

(e) $y''' + 3y'' + 3y' + y = 0$.

(f) $y^{(4)} - 2y''' + 2y'' - 2y' + y = 0$.

(g) $y^{(4)} + 2y'' + y = 0$.

7.   Solve the following equations of Problem 6 with the indicated initial conditions.

(a)   with $y(0) = 0$, $y'(0) = 6$.

(b) with $y(\pi/2) = 2$, $y'(\pi/2) = -1$.

(c) with $y(1) = -1$, $y'(1) = 2$.

*8. The method of solution illustrated in Example 1 can some-
     times be applied to equations with variable coefficients.

     (a) Show that the equation

          $$t^2x'' + (t^3+t)x' + (t^2-1)x = 2t^3$$

          is equivalent to

          $$(D+tI)(D+t^{-1}I)x = 2t \quad \text{as long as } t \neq 0,$$

          where the product of operators is defined as in Eq.
          (4).

     (b) Solve the equation in part (a).

     (c) Show that $(D+tI)(D+t^{-1}I) \neq (D+t^{-1}I)(D+tI)$, so that
          multiplication of operators with variable coeffici-
          ents is <u>not</u> <u>commutative</u>, in general.

11. <u>LINEAR</u> <u>INDEPENDENCE</u> <u>AND</u> <u>WRONSKIANS</u>

     The previous section developed a procedure for finding
a solution involving n arbitrary constants, $c_1,\ldots,c_n$, of
a linear, homogeneous n'th order differential equation with
constant coefficients. It remains to show that this procedure
yields the general solution.

     The concepts, methods, and theorems in this section
actually apply to a more general linear differential equation
with variable coefficients,

   $$x^{(n)} + a_{n-1}(t)x^{(n-1)} + \cdots + a_1(t)x' + a_0(t)x = h(t). \quad (1)$$

The assumption made here that the coefficients of $x^{(n)}$ is
"one" is merely a matter of convenience. In case the leading

term is $a_n(t)x^{(n)}$ where $a_n(t) \neq 0$ one could divide the equation through by $a_n(t)$ to put it into the form of Eq. (1).

Our ultimate objective is to find a (unique) solution of Eq. (1) which satisfies given initial conditions

$$x(t_0) = b_0, \quad x'(t_0) = b_1, \quad \ldots, \quad x^{(n-1)}(t_0) = b_{n-1}, \qquad (2)$$

where $t_0, b_0, \ldots,$ and b_{n-1} are given.

We shall soon find that the analysis of Eq. (1) is related to the study of the associated homogeneous equation

$$y^{(n)} + a_{n-1}(t)y^{(n-1)} + \cdots + a_1(t)y' + a_0(t)y = 0. \qquad (3)$$

The basic feature of the linear homogeneous equation (3) is that any linear combination of solutions is again a solution. Furthermore any solution of (3) can be added to any solution of (1) to produce another solution of (1). These properties, and more, are described in detail in the following "superposition" theorem. It is really nothing more than an assertion of the linearity of a "polynomial operator" in D with variable coefficients. (Cf. Eq. 10-(5) for the constant coefficient case.)

Theorem A (Superposition). Let $h(t) = k_1 h_1(t) + \cdots + k_m h_m(t)$ where $k_1, \ldots,$ and k_m are constants, and for $j = 1, \ldots, m$ let x_j be a (particular) solution of

$$x^{(n)} + a_{n-1}(t)x^{(n-1)} + \cdots + a_0(t)x = h_j(t)$$

on (α, β) . Let $y_1, \ldots,$ and y_ℓ be any solutions of (3) on (α, β) . Then

$$x \equiv k_1 x_1 + \cdots + k_m x_m + c_1 y_1 + \cdots + c_\ell y_\ell$$

is a solution of (1) on (α, β) for every choice of the (complex) constants $c_1, \ldots, c_\ell$.

     Proof.   Substituting x into the left hand side of
(1), we find

$$x^{(n)} + \cdots + a_0(t)x$$

$$= \sum_{j=1}^{m} k_j [x_j^{(n)} + \cdots + a_0(t)x_j] + \sum_{i=1}^{\ell} c_i [y_i^{(n)} + \cdots + a_0(t)y_i]$$

$$= \sum_{j=1}^{m} k_j h_j(t) + 0 = h(t). \qquad \square$$

     We can expect, from the uniqueness theorems and from
the examples considered thus far, that the general solution
of Eq. (1) or (3) will be a function containing n arbitrary
constants.   But one must be careful not to jump to conclusions
too quickly from this.

Example 1.   It is easily verified that the linear, homogeneous
differential equation

$$y''' + 4y' = 0$$

has solutions $y_1(t) = 1$, $y_2(t) = \cos 2t$, and $y_3(t) = \sin^2 t$.
Thus

$$y(t) = c_1 + c_2 \cos 2t + c_3 \sin^2 t$$

is a solution with three arbitrary constants.   But is it the
general solution?   In other words, can we always choose c_1,
c_2, and c_3 so as to satisfy three given initial conditions
of the form $y(t_0) = b_0$, $y'(t_0) = b_1$, $y''(t_0) = b_2$?
     The answer is "no" because we don't have three inde-
pendent constants at our disposal.   Since $\cos 2t \equiv 1 - 2\sin^2 t$,

our solution can be rewritten as

$$y(t) = c_1 + c_2 + (c_3 - 2c_2)\sin^2 t.$$

Then there are really only two arbitrary constants, $(c_1 + c_2)$
and $(c_3 - 2c_2)$.

The trouble is that our three original solutions, y_1,
y_2, and y_3, were not "independent".

To make matters precise we now digress to introduce
some concepts from linear algebra, which in themselves are independent of differential equations.

<u>Definitions</u>. A set of functions, $v_1, \ldots, v_k$, each mapping
$A \to \mathbb{C}$ for some $A \subset \mathbb{R}$ is said to be <u>linearly</u> <u>dependent</u> if
there exist complex constants, $c_1, \ldots, c_k$, at least one of
which is not zero, such that the linear combination
$c_1 v_1 + \cdots + c_k v_k$ is the zero function on A, i.e.,

$$c_1 v_1(t) + \cdots + c_k v_k(t) = 0 \quad \text{for all}\quad t \quad \text{in}\quad A.$$

Otherwise the functions are said to be <u>linearly</u> <u>independent</u>.
In other words, $v_1, \ldots, v_k$ is a linearly independent set if
$c_1 v_1 + \cdots + c_k v_k$ is the zero function only when
$c_1 = c_2 = \cdots = c_k = 0$.

The three functions in Example 1 are linearly dependent
on any set A since

$$1y_1(t) + (-1)y_2(t) + (-2)y_3(t) = 0 \quad \text{for all}\quad t \quad \text{in}\quad \mathbb{R}.$$

<u>Example 2</u>. Define v_1 and v_2: $(-1,1) \to \mathbb{C}$ by

$$v_1(t) = \begin{cases} t^3 & \text{for} \quad -1 < t \le 0 \\ 0 & \text{for} \quad 0 < t < 1 \end{cases} \quad \text{and} \quad v_2(t) = \begin{cases} 0 & \text{for} \quad -1 < t \le 0 \\ t^3 & \text{for} \quad 0 < t < 1. \end{cases}$$

Then v_1 and v_2 are linearly independent because if $c_1 v_1(t) + c_2 v_2(t) = 0$ for $-1 < t < 1$ we have only to set $t = -1/2$ to conclude that we must have $c_1 = 0$ and set $t = 1/2$ to see that we must also have $c_2 = 0$.

A particular determinant function which is closely related to the concept of linear independence is given a special name.

Definition. If $v_1, \ldots, v_n$ are functions mapping $(\alpha, \beta) \to \mathbb{C}$ and each has $n-1$ derivatives on (α, β), then the Wronskian is the function mapping $(\alpha, \beta) \to \mathbb{C}$ defined by

$$W(t) = W(v_1, \ldots, v_n)(t) = \begin{vmatrix} v_1(t) & \cdots & v_n(t) \\ v_1'(t) & \cdots & v_n'(t) \\ & \vdots & \\ v_1^{(n-1)}(t) & \cdots & v_n^{(n-1)}(t) \end{vmatrix}.$$

We have already encountered the Wronskian in special cases in Examples 6-1 and 7-1. The following theorems extend the method used in those examples.

Theorem B.

    (i) Let $y_1, \ldots, y_n$ be any (complex-valued) functions on (α, β) each having $n-1$ derivatives. If $W(t) = W(y_1, \ldots, y_n)(t_1) \ne 0$ for some t_1 in (α, β), then $y_1, \ldots, y_n$ are linearly independent functions. (The converse is false.)

(ii) More specifically, let $y_1,\ldots,y_n$ be solutions
of Eq. (3) on (α,β), where $a_{n-1},\ldots,a_1$, and a_0
are continuous functions (possibly complex-valued)
on (α,β). Then $y_1,\ldots,y_n$ are linearly independent if and only if $W(t) \neq 0$ for every t in
(α,β).

Proof. (i) Let $W(t_1) \neq 0$ for some t_1 in (α,β),
and assume that, for some choice of constants $c_1,\ldots,c_n$ one
has

$$c_1 y_1(t) + \cdots + c_n y_n(t) = 0 \quad \text{for} \quad \alpha < t < \beta.$$

Then it follows also that

$$c_1 y_1'(t) + \cdots + c_n y_n'(t) = 0$$
$$\vdots$$
$$c_1 y_1^{(n-1)}(t) + \cdots + c_n y_n^{(n-1)}(t) = 0$$

Replacing t by t_1, the above becomes a system of n
linear algebraic equations for the numbers $c_1,\ldots,c_n$; and the
determinant of the coefficients is $W(t_1) \neq 0$. Thus
$c_1 = c_2 = \cdots = c_n = 0$ is the only solution.

To see that the converse of (i) is false, it suffices
to consider $n = 2$ with $y_1 = v_1$ and $y_2 = v_2$, the functions
of Example 2. Then it is easy to verify that $W(t) =$
$W(v_1,v_2)(t) = 0$ for $-1 < t < 1$. However, as shown in
Example 2, v_1 and v_2 are linearly independent.

(ii) Now consider the special case when $y_1,\ldots,y_n$
are solutions of some n'th order, linear, homogeneous differential equation (3) with continuous coefficients. The "if"
part of assertion (ii) is covered by (i).

To prove the "only if" assertion, let us assume that $W(t_1) = 0$ for some t_1 in (α, β). Then a theorem from algebra asserts that the system of equations for $c_1, \ldots, c_n$,

$$c_1 y_1(t_1) + \cdots + c_n y_n(t_1) = 0,$$
$$c_1 y_1'(t_1) + \cdots + c_n y_n'(t_1) = 0,$$
$$\vdots$$
$$c_1 y_1^{(n-1)}(t_1) + \cdots + c_n y_n^{(n-1)}(t_1) = 0,$$

has a nontrivial solution, i.e., a solution for $c_1, \ldots, c_n$ not all zero. Using these values of $c_1, \ldots, c_n$, consider the solution of Eq. (3) defined by

$$y \equiv c_1 y_1 + \cdots + c_n y_n \quad \text{on} \quad (\alpha, \beta).$$

Then, since $y(t_1) = 0$, $y'(t_1) = 0, \ldots,$ and $y^{(n-1)}(t_1) = 0$, the uniqueness theorem asserts that $y(t) \equiv 0$. In other words the functions $y_1, \ldots, y_n$ are linearly dependent. Thus $y_1, \ldots, y_n$ can be linearly independent only if their Wronskian is nonzero for all t in (α, β). $\square$

Corollary C. Let $y_1, \ldots, y_n$ be solutions of Eq. (3) on (α, β), where the coefficient functions, $a_{n-1}, \ldots, a_1,$ and a_0 are continuous. Then $W(t) = W(y_1, \ldots, y_n)(t)$ is either zero for all t in (α, β) or different from zero for all t in (α, β).

Proof. If $W(t) \neq 0$ for some t in (α, β) then $y_1, \ldots, y_n$ are linearly independent. But this implies that $W(t) \neq 0$ for all t in (α, β). $\square$

We are now ready to describe the form of the general solution of Eq. (1).

Theorem $\underline{D}$.   If Eq. (3) with continuous coefficients has   n
linearly independent solutions $y_1, \ldots, y_n$ on an interval
(α, β) and if $\tilde{x}$ is a (particular) solution of Eq. (1) on
(α, β) then every solution of Eq. (1) on (α, β) is the sum
of $\tilde{x}$ and some linear combination of $y_1, \ldots, y_n$.   That is,
the general solution of Eq. (1) on (α, β) is

$$x = \tilde{x} + c_1 y_1 + \cdots + c_n y_n, \tag{4}$$

where $c_1, \ldots, c_n$ are unspecified complex constants.   Moreover
the constants $c_1, \ldots, c_n$ are determined (uniquely) if initial
conditions (2) are specified using any point
$(t_0, b_0, \ldots, b_{n-1}) \in (\alpha, \beta) \times \text{\textbf{R}}^n$.

Remark.   Nothing we have said thus far proves that   n
linearly independent solutions of Eq. (3) $\underline{\text{exist}}$.   This will
eventually be proved under the same assumptions (continuous
coefficients) in Section 20.   For the present we will be able
to apply Theorem D only to certain special cases in which we
can exhibit the desired set of   n   linearly independent solutions.   The most important such special case will be the case
of constant coefficients.

$\underline{\text{Proof}} \ \underline{\text{of}} \ \underline{\text{Theorem}} \ \underline{D}$.   Theorem A asserts that (4) is a
solution of (1) for every choice of the constants $c_1, \ldots, c_n$.
It remains to show that if the initial values (2) are
specified for any t_0 in (α, β) then a set of constants
$c_1, \ldots, c_n$ can be determined (uniquely) so that (4) represents
$\underline{a}$ solution, and hence, by uniqueness, $\underline{\text{the}}$ solution.   We require

$$
\left.
\begin{aligned}
c_1 y_1(t_0) + \cdots + c_n y_n(t_0) &= b_0 - \tilde{x}(t_0), \\[2mm]
c_1 y_1'(t_0) + \cdots + c_n y_n'(t_0) &= b_1 - \tilde{x}'(t_0), \\[1mm]
&\;\;\vdots \\[2mm]
c_1 y_1^{(n-1)}(t_0) + \cdots + c_n y_n^{(n-1)}(t_0) &= b_{n-1} - \tilde{x}^{(n-1)}(t_0).
\end{aligned}
\right\} \quad (5)
$$

This is a system of n linear algebraic equations in n unknowns, $c_1, \ldots, c_n$. Since the determinant $W(y_1, \ldots, y_n)(t_0) \neq 0$ by Theorem B (ii), it follows that one can solve for $c_1, \ldots, c_n$ (uniquely). And the resulting function (4) satisfies Eqs. (1) and (2).

Finally, to show that (4) gives the general solution of Eq. (1), let $\hat{x}$ be any solution on some interval (α, β). Choose some t_0 in (α, β) and reconsider Eqs. (5) with $b_0, b_1, \ldots,$ and b_{n-1} replaced by the numbers $\hat{x}(t_0)$, $\hat{x}'(t_0), \ldots,$ and $\hat{x}^{(n-1)}(t_0)$ respectively. Solve for $c_1, \ldots, c_n$ and put these numbers into Eq. (4). Then, the resulting solution x must satisfy the same initial conditions (2) as does $\hat{x}$. Hence, by uniqueness, x must be identical with $\hat{x}$. $\square$

Corollary E. Eq. (3) cannot have more than n linearly independent solutions on (α, β).

Proof. Suppose (for contradiction) that $y_1, \ldots, y_{n+1}$ are linearly independent solutions of (3) on (α, β). Then clearly, $y_1, \ldots, y_n$ are linearly independent. Applying Theorem D, we conclude that

$$
y_{n+1} = c_1 y_1 + \cdots + c_n y_n
$$

for some appropriate choice of constants $c_1, \ldots, c_n$. This contradicts the linear independence of $y_1, \ldots, y_{n+1}$. Why? □

Problems

1. Determine whether the sets of functions defined as follows are linearly dependent or linearly independent.

(a) $v_1(t) = \sin t$, $v_2(t) = \cos t$ for $0 < t < \pi$.

(b) $v_1(t) = \sin t$, $v_2(t) = \cos t$, $v_3(t) = 0$ for $0 < t < \pi$.

(c) $v_1(t) = t^2$, $v_2(t) = 1 + t^2$ for $-1 < t < 1$.

(d) $v_1(t) = t$, $v_2(t) = e^t$, $v_3(t) = \sin 2t$ for $0 \le t \le 1$.

(e) $v_1(t) = e^{-2t}$, $v_2(t) = e^{2t}$, $v_3(t) = \cosh 2t$ for $1 \le t \le 2$.

2. If the functions v_1 and v_2 defined in Example 2 were both solutions of some second order linear homogeneous equation (3) with continuous coefficients, then Theorem B would be contradicted. Show directly that <u>neither</u> v_1 nor v_2 can be a solution of such an equation.

3. The linear homogeneous equation $y'' + 2y' + y = 0$ has (by the method of Section 10) solutions $y_1(t) = e^{-t}$ and $y_2(t) = te^{-t}$.

(a) Prove that y_1 and y_2 are linearly independent on every interval (α, β).

(b) Prove that another pair of linearly independent solutions of this equation is defined by $y_1(t) = (1-t)e^{-t}$ and $y_2(t) = (\pi+17t)e^{-t}$. Thus the latter pair is equally acceptable for use in Theorem D.

(c) Find still another pair of linearly independent solutions of the equation.

4. Use Theorem D and the results of Problem 3 to find the
general solution of   x" + 2x' + x = t.   Hint.   A particular solution can be found in the form $\tilde{x}(t) = A + Bt$ by
an appropriate choice of   A   and   B.

5. Suppose that $e^{(\mu+i\omega)t}$ and $e^{(\mu-i\omega)t}$ (where μ and ω
are real constants with $\omega \neq 0$) both define solutions of
some linear homogeneous ordinary differential equation.
(We do not specify the order of the equation   nor whether
its coefficients are constants.)   Prove that $y_1(t) =$
$e^{\mu t}$ cos ωt and $y_2(t) = e^{\mu t}$ sin ωt are necessarily
linearly independent solutions of that same equation.

6. Suppose that $e^{(\mu+i\omega)t}$ (where μ and ω are real constants and $\omega \neq 0$) defines a solution of some linear homogeneous ordinary differential equation with real-valued
coefficients.   Then prove that y_1 and y_2, as defined
in Problem 5, are linearly independent solutions of that
same equation.

7. Find the general solution, valid for   t > 0, of

$$t^2 y" + b_1 ty' + b_0 y = 0,$$

where b_1 and b_0 are real constants such that
$(b_1-1)^2 > 4b_0$.   This is an example of "Euler's equation".
Hint: Try to find a solution of the form $y(t) = t^\lambda$ for
constant λ.

8.   Define $y_1(t) = t^2$ and $y_2(t) = t + t^4$ for $-1 < t < 1$.
     Prove that y_1 and y_2 are linearly independent and yet
     $W(y_1, y_2)(0) = 0$. Verify that y_1 and y_2 both satisfy
     the linear homogeneous equation

$$y''' - 8t^2 y'' + 40ty' - 64y = 0 \quad \text{for} \quad -1 < t < 1.$$

     Why does this not contradict Theorem B (and Corollary C)?

9.   Prove that there cannot exist any third order, linear,
     homogeneous equation of the form (3) with continuous co-
     efficients, $a_2(t)$, $a_1(t)$, $a_0(t)$, which is satisfied by
     $y(t) = t^3$ on $(-1,1)$.

10. Consider Eq. (3) with $n = 2$. Show that if y_1 and y_2
     are any two solutions on $J = (\alpha, \beta)$ then

$$\frac{d}{dt} W(y_1, y_2)(t) = -a_1(t) W(y_1, y_2)(t), \text{ so that}$$

$$W(y_1, y_2)(t) = W(y_1, y_2)(t_0) e^{-\int_{t_0}^{t} a_1(s)ds} \quad \text{for all} \quad t, t_0 \text{ in } J.$$

     This result (known as Abel's formula) provides another
     proof of Corollary C for $n = 2$. How? (An analogous re-
     result holds for arbitrary n, and it can be obtained
     using a little knowledge of the theory of determinants.)

12.   CONSTANT COEFFICIENTS   (GENERAL SOLUTION FOR SIMPLE h)

     We now return to the equation with constant coeffi-
cients

$$x^{(n)} + a_{n-1} x^{(n-1)} + \cdots + a_1 x' + a_0 x = h(t), \tag{1}$$

the initial conditions

$$x(t_0) = b_0, \quad x'(t_0) = b_1, \ldots, \quad x^{(n-1)}(t_0) = b_{n-1}, \quad (2)$$

and the associated homogeneous equation

$$y^{(n)} + a_{n-1}y^{(n-1)} + \cdots + a_1 y' + a_0 y = 0. \quad (3)$$

Again we remark that there is no loss of generality in assuming $a_n = 1$.

In Section 10 we obtained a solution of (3) involving n arbitrary constants. Now we will apply the theory of Section 11 to show that this is indeed the general solution of (3), which is needed to find the general solution of (1).

Theorem A.   Let the roots of the characteristic equation for (1) and (3),

$$\lambda^n + a_{n-1}\lambda^{n-1} + \cdots + a_1 \lambda + a_0 = 0, \quad (4)$$

be λ_j with multiplicity m_j for $j = 1, \ldots, q$. (Thus $m_1 + \cdots + m_q = n$.) Then Eq. (3) has n linearly independent solutions of the form

$$y_{jk}(t) = t^k e^{\lambda_j t} \quad \text{for} \quad k = 0, 1, \ldots, m_{j-1} \quad \text{and} \quad j = 1, \ldots, q;$$

and the general solution of Eq. (1) is given, for all t, by

$$x(t) = \tilde{x}(t) + P_1(t)e^{\lambda_1 t} + P_2(t)e^{\lambda_2 t} + \cdots + P_q(t)e^{\lambda_q t}, \quad (5)$$

where $\tilde{x}$ is any particular solution of Eq. (1) and P_j is an arbitrary polynomial of degree $m_j - 1$ for each j. (Since P_j involves m_j arbitrary constants, the expression (5) for x involves n arbitrary constants.)

Proof.   According to Theorem 11-D, it will suffice to show that the functions y_{jk} are linearly independent, on

some interval (α, β).   We could proceed by trying to show
that the Wronskian of these   n   functions is not zero.   But
that is not the easiest method, as Problem 1 will illustrate.

For simplicity we shall treat the case $q = 3$ so that
the <u>distinct</u> roots of Eq. (4) are λ_1, λ_2, and λ_3.   The
reader will see how to generalize the proof to arbitrary   q.

Suppose (for contradiction) that the   n   functions
y_{jk} are linearly dependent.   Then

$$P_1(t)e^{\lambda_1 t} + P_2(t)e^{\lambda_2 t} + P_3(t)e^{\lambda_3 t} = 0, \tag{6}$$

where P_1, P_2, and P_3 are some polynomials, not all zero,
with the degree of each P_j no greater than $m_j - 1$.   Without
loss of generality we can assume $P_3(t) \not\equiv 0$, for otherwise we
merely relabel the λ_j's.   If P_3 has degree ℓ, then

$$P_3(t) = ct^{\ell} + \text{lower order terms},$$

with $c \neq 0$.   Multiply Eq. (6) by $e^{-\lambda_1 t}$ to obtain

$$P_1(t) + P_2(t)e^{(\lambda_2 - \lambda_1)t} + P_3(t)e^{(\lambda_3 - \lambda_1)t} = 0.$$

Then, since P_1 is a polynomial of degree at most $m_1 - 1$,
differentiating this identity m_1 times annihilates P_1 and
leaves a new identity

$$Q_2(t)e^{(\lambda_2 - \lambda_1)t} + Q_3(t)e^{(\lambda_3 - \lambda_1)t} = 0. \tag{7}$$

Moreover, each Q_j is a polynomial of the same degree as P_j
(at most $m_j - 1$), and in particular

$$Q_3(t) = (\lambda_3 - \lambda_1)^{m_1} ct^{\ell} + \text{lower order terms}.$$

Now multiply Eq. (7) by $e^{(\lambda_1 - \lambda_2)t}$ and differentiate the resulting identity m_2 times to annihilate $Q_2(t)$. This leaves

$$R_3(t)e^{(\lambda_3-\lambda_2)t} = 0, \tag{8}$$

where R_3 is the polynomial

$$R_3(t) = (\lambda_3-\lambda_1)^{m_1}(\lambda_3-\lambda_2)^{m_2}ct^{\ell} + \text{lower order terms.}$$

Since the exponential in Eq. (8) is not zero, we must have $R_3(t) \equiv 0$ and in particular $c = 0$ -- a contradiction. ☐

Generally speaking, in practical problems the coefficients $a_{n-1}, \ldots, a_0$ in Eq. (1) or (3) will be real constants. However, as we have noted, it is possible to have some or all λ_j complex, with the complex roots always occurring in conjugate pairs. In this case the following form of Theorem A is more useful.

Theorem B. Let $a_{n-1}, \ldots, a_0$ be real constants and let the roots of the characteristic equation (4) be the complex conjugate pairs

$\mu_j \pm i\omega_j$ with multiplicity m_j for $j = 1, \ldots, r$, and the real roots

λ_j with multiplicity m_j for $j = 2r+1, \ldots, q$.

(Thus $\sum_{j=1}^{r} 2m_j + \sum_{j=2r+1}^{q} m_j = n$.) Then the general solution of Eq. (1) is given, for all t, by

$$x(t) = \tilde{x}(t) + \sum_{j=1}^{r} [Q_j(t) \cos \omega_j t + R_j(t) \sin \omega_j t]e^{\mu_j t}$$

$$\tag{9}$$

$$+ \sum_{j=2r+1}^{q} P_j(t)e^{\lambda_j t},$$

where $\tilde{x}$ is any particular solution of Eq. (1) and each P_j,
Q_j, and R_j is an arbitrary polynomial of degree m_j-1.

   Proof.   Let $\mu \pm i\omega$ be a pair of complex conjugate
solutions of (4) with multiplicity m.   Then we can write
(according to Problem 7-5)

$$e^{\mu t} \cos \omega t = \frac{e^{(\mu+i\omega)t} + e^{(\mu-i\omega)t}}{2}$$

and

$$e^{\mu t} \sin \omega t = \frac{e^{(\mu+i\omega)t} - e^{(\mu-i\omega)t}}{2i} .$$

Thus it follows that, for each $k = 0,1,\ldots,m-1$,

$$t^k e^{\mu t} \cos \omega t \quad \text{and} \quad t^k e^{\mu t} \sin \omega t$$

define solutions of Eq. (3).

   There remains the question of linear independence before we can invoke Theorem 11-D to complete the proof.

   Suppose that some linear combination of

$$t^k e^{\mu_j t} \cos \omega_j t, \quad t^k e^{\mu_j t} \sin \omega_j t$$

(for $k = 0,\ldots,m_j-1$; $j = 1,\ldots,r$)   and

$$t^k e^{\lambda_j t}$$

(for $k = 0,\ldots,m_j-1$; $j = 2r+1,\ldots,q$)   vanishes identically.
For each $j = 1,\ldots,r$ and $k = 0,\ldots,m_j-1$, pair up the terms
in this sum which involve $t^k e^{\mu_j t} \cos \omega_j t$ and $t^k e^{\mu_j t} \sin \omega_j t$,
say

$$u_{kj}(t) = b t^k e^{\mu_j t} \cos \omega_j t + d t^k e^{\mu_j t} \sin \omega_j t. \qquad (10)$$

These terms can be rewritten as

$$u_{kj}(t) = (\frac{b}{2} + \frac{d}{2i}) t^k e^{(\mu_j + i\omega_j)t} + (\frac{b}{2} - \frac{d}{2i}) t^k e^{(\mu_j - i\omega_j)t}.$$

Thus our vanishing linear combination takes the form

$$\sum_{j=1}^{q} P_j(t) e^{\lambda_j t}.$$

Hence by Theorem A, in addition to having zero coefficients
for the terms $t^k e^{\lambda_j t}$ for $j = 2r+1,\ldots,q$, we must have from
each u_{kj} for $j = 1,\ldots,r$

$$\frac{b}{2} + \frac{d}{2i} = 0 \quad \text{and} \quad \frac{b}{2} - \frac{d}{2i} = 0.$$

This implies $b = 0$ and $d = 0$ in (10).   □

Remark. The polynomials in the general solution (9)
involve a total of n arbitrary constants; and, if initial
conditions (2) are specified, these constants can be found
just as in the proof of Theorem 11-D.

Example 1. Find the general solution of

$$(D+3I)(D-2I)^4 (D^2+4D+5I)(D^2+4I)^2 y = 0.$$

The roots of the characteristic equation are

$$\lambda_1 = -3, \text{ with multiplicity } m_1 = 1,$$

$$\lambda_2 = 2, \text{ with multiplicity } m_2 = 4,$$

$$\lambda_3 = -2 \pm i, \text{ each with multiplicity } m_3 = 1,$$

$$\lambda_4 = \pm 2i, \text{ each with multiplicity } m_4 = 2.$$

Thus the general solution is given by

$$y(t) = c_1 e^{-3t} + (c_2 + c_3 t + c_4 t^2 + c_5 t^3) e^{2t} + c_6 e^{-2t} \cos t$$

$$+ c_7 e^{-2t} \sin t + (c_8 + c_9 t) \cos 2t + (c_{10} + c_{11} t) \sin 2t.$$

Theorems A and B provide the complete general solution of Eq. (3) in all cases, and they provide the general solution of Eq. (1) <u>if</u> <u>we</u> <u>can</u> <u>somehow</u> <u>find</u> <u>some</u> <u>particular</u> <u>solution</u> of Eq. (1). The next example illustrates a systematic procedure for finding a particular solution of Eq. (1) <u>provided</u> h is of a certain simple form.

<u>Example</u> <u>2</u>. Find the general solution of

$$x'' + x' - 2x = 3e^{-2t} + te^{-2t} + 5 \cos 3t. \tag{11}$$

As usual, we begin by considering the characteristic equation $\lambda^2 + \lambda - 2 = 0$. This has roots $\lambda_1 = 1$ and $\lambda_2 = -2$ each of multiplicity one. Thus the general solution of Eq. (11) is given by

$$x(t) = \tilde{x}(t) + c_1 e^t + c_2 e^{-2t}$$

where $\tilde{x}$ is "any particular solution".

But now observe that the terms on the right side of Eq. (11),

$$h(t) = 3e^{-2t} + te^{-2t} + 5 \cos 3t,$$

are themselves terms of the type which occur in Eq. (9). Thus there ought to be some linear, <u>homogeneous</u> differential equation with constant coefficients for which h itself is a <u>solution</u>. Indeed, we see from Theorem B that

$$(D+2I)^2 (D+3iI)(D-3iI)h = 0.$$

We might say that the operator $(D+2I)^2(D^2+9I)$ annihilates h.

If we now rewrite Eq. (11) as

$$(D-I)(D+2I)x = h(t),$$

then "multiplication" from the left by $(D+2I)^2(D^2+9I)$ gives

$$(D+2I)^2(D^2+9I)(D-I)(D+2I)x = 0.$$

In other words, the general solution of the nonhomogeneous, second order equation (11) must also be a solution of the new homogeneous sixth order equation

$$(D-I)(D+2I)^3(D^2+9I)x = 0.$$

Thus it must be that

$$x(t) = c_1 e^t + c_2 e^{-2t} + c_3 t e^{-2t} + c_4 t^2 e^{-2t} + c_5 \cos 3t$$
$$+ c_6 \sin 3t$$

for some choice of the constants c_1, c_2, c_3, c_4, c_5, and c_6.

But we know already that, regardless of the values of c_1 and c_2, the first two terms will always satisfy the homogeneous equation $y'' + y' - 2y = 0$. So, to find a particular solution of Eq. (11), we should seek constants A, B, C, and D such that

$$\tilde{x}(t) = Ate^{-2t} + Bt^2 e^{-2t} + C \cos 3t + D \sin 3t$$

satisfies Eq. (11). For this we compute

$$\tilde{x}'(t) = A(e^{-2t} - 2te^{-2t}) + B(2te^{-2t} - 2t^2 e^{-2t})$$
$$- 3C \sin 3t + 3D \cos 3t,$$

and

$$\tilde{x}''(t) = A(-4e^{-2t} + 4te^{-2t}) + B(2e^{-2t} - 8te^{-2t} + 4t^2e^{-2t})$$

$$- 9C \cos 3t - 9D \sin 3t.$$

Substitution into Eq. (11) thus leads to the condition

$$3e^{-2t} + te^{-2t} + 5 \cos 3t = -3Ae^{-2t} + 2Be^{-2t} - 6Bte^{-2t}$$

$$+ (-2C + 3D - 9C)\cos 3t + (-2D - 3C - 9D)\sin 3t,$$

or

$$-3A + 2B = 3, \quad -6B = 1, \quad -11C + 3D = 5, \quad -3C - 11D = 0.$$

Solving these simultaneous algebraic equations, we find

$$A = -\frac{10}{9}, \quad B = -\frac{1}{6}, \quad C = -\frac{11}{26}, \quad D = \frac{3}{26}.$$

Finally then, the general solution of Eq. (11) is

$$x(t) = -\frac{10}{9}te^{-2t} - \frac{1}{6}t^2e^{-2t} - \frac{11}{26}\cos 3t + \frac{3}{26}\sin 3t$$

$$+ c_1e^t + c_2e^{-2t}.$$

The above method for finding $\tilde{x}$ is sometimes called the method of "undetermined coefficients". The coefficients which were, at first, undetermined were A, B, C, and D. We now give a precise description of the procedure as a theorem.

Theorem C. Rewrite Eq. (1) in the form $p(D)x = h(t)$. Assume there exists another polynomial operator, $q(D)$, with constant coefficients such that $q(D)h = 0$. (This will be the case if and only if h is equivalent to a sum of terms of the types in Eq. (9).) Then the general solution of the new homogeneous equation

$$q(D)p(D)x = 0$$

will include terms which represent the general solution of
Eq. (3) plus terms which, with appropriate coefficients, give
a particular solution of Eq. (1).

The proof would look very similar to Example 2.

Example 3.   Find the general solution of

$$x'' + 4x = \sin 2t. \tag{12}$$

The characteristic equation, $\lambda^2 + 4 = 0$, has roots $\lambda = \pm 2i$.
Thus the general solution of the homogeneous equation is

$$y(t) = c_1 \cos 2t + c_2 \sin 2t.$$

But since (D^2+4I) annihilates $h(t) = \sin 2t$, every solution
of Eq. (12), $(D^2+4I)x = h(t)$, must also be a solution of the
new homogeneous equation

$$(D^2+4I)^2 x = 0.$$

Thus

$$x(t) = c_1 \cos 2t + c_2 \sin 2t + At \cos 2t + Bt \sin 2t,$$

for appropriate choices of the undetermined coefficients, A
and B.   Hence a particular solution of the nonhomogeneous
equation can be found in the form

$$\tilde{x}(t) = At \cos 2t + Bt \sin 2t.$$

Substitution leads us to set $A = -1/4$, $B = 0$.   Thus the general solution is

$$x(t) = -\frac{1}{4}t \cos 2t + c_1 \cos 2t + c_2 \sin 2t. \qquad (13)$$

This example illustrates the phenomenon of "resonance". The homogeneous equation $y'' + 4y = 0$ has solutions $\cos 2t$ and $\sin 2t$ which oscillate at the "frequency" $1/\pi$ cycles per unit time, say $1/\pi$ cycles per second.   (Note that $\cos \omega t$ and $\sin \omega t$ describe periodic functions with period $2\pi/\omega$, and hence with frequency $\omega/2\pi$ cycles per unit of time.)

The function $h(t) = \sin 2t$ on the right hand side of the equation can be thought of as a "forcing" or "driving" function.   Example 3 illustrates the fact that when the driving function is periodic with the same frequency as the "natural frequency" of the equation, then we get a solution which oscillates with ever growing and unbounded amplitude.

In a more practical example of resonance the oscillations do not grow without bound, but are limited by "friction" or "resistance" in the physical system.

Example 4.   Let positive constants   R, L, and   C   be respectively the resistence in ohms, the inductance in henries, and the capacitance in farads in the tuning circuit of a radio receiver.   Figure 1.   If the voltage difference between the antenna and the ground due to a certain radio broadcasting station is $v(t) = a \sin \omega_1 t$, then the current, $x(t)$, in amps through the circuit satisfies the differential equation

$$Lx'' + Rx' + \frac{1}{C}x = v'(t) = a\omega_1 \cos \omega_1 t. \qquad (14)$$

The solutions of Eq. (14) are found and discussed below, with details left to the reader as Problem 5.

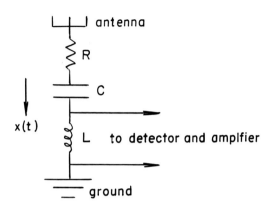

Figure 1

Applying the methods of this section, the reader should
find the general solution of Eq. (14) to be

$$x(t) = \frac{a}{(\omega_1 L - 1/\omega_1 C)^2 + R^2} [-(\omega_1 L - \frac{1}{\omega_1 C})\cos \omega_1 t + R \sin \omega_1 t]$$

$$+ c_1 e^{\mu t} \cos \omega t + c_2 e^{\mu t} \sin \omega t,$$

(15)

where

$$\mu = -\frac{R}{2L} \quad \text{and} \quad \omega = (\frac{1}{LC} - \frac{R^2}{4L^2})^{1/2}.$$

The constants c_1 and c_2 could be determined if the initial
conditions were known. However, since $\mu < 0$, the terms with
coefficients c_1 and c_2 die out as $t \to \infty$ regardless of
the values of c_1 and c_2. These terms, called "transients",
are not of much interest in the analysis of a radio receiver
circuit.

The remaining terms describe the "steady state" current,
$x_{ss}(t)$, which can be rewritten as

$$x_{ss}(t) = \frac{a}{[(\omega_1 L - 1/\omega_1 C)^2 + R^2]^{1/2}} \sin(\omega_1 t - \phi) \qquad (16)$$

where ϕ = Arctan $[(\omega_1 L - 1/\omega_1 C)/R]$. Note that $x_{ss}(t)$ oscillates at the same frequency as the applied voltage, $v(t)$, but with a "phase shift", ϕ.

The amplitude of the oscillations of x_{ss} depends on the values of ω_1, L, C, and R (as well as a); and this is what enables a radio receiver to distinguish between the signels from various transmitters operating at different frequencies. Suppose that the signals from two transmitters at frequencies of 636 kilohertz (kilocycles per second) and 796 kilohertz are reaching the antenna with equal strength. These frequencies imply values of

$$\omega_1 = 636,000 \times 2\pi = 4 \times 10^6 \quad \text{and} \quad \omega_2 = 796,000 \times 2\pi = 5 \times 10^6.$$

Thus let us consider the applied voltage (between antenna and ground) to be

$$v(t) = a \sin \omega_1 t + a \sin \omega_2 t. \qquad (17)$$

Putting $v'(t)$ from (17) on the right hand side of Eq. (14) and solving, we find a steady state current of

$$x_{ss}(t) = \frac{a \sin(\omega_1 t - \phi_1)}{[(\omega_1 L - 1/\omega_1 C)^2 + R^2]^{1/2}} + \frac{a \sin(\omega_2 t - \phi_2)}{[(\omega_2 L - 1/\omega_2 C)^2 + R^2]^{1/2}}. \qquad (18)$$

In an AM radio we might have $L = 250 \times 10^{-6}$ henries and $R = 2$ ohms. If we want to "receive" the broadcast at 636 kilohertz, we would "tune the radio" by adjusting the value of C so that $\omega_1 L - 1/\omega_1 C = 0$ -- in this case to

$C = 250 \times 10^{-12}$ farads -- in order to maximize the desired component of x_{ss} in (18). Then we would find that x_{ss} has two terms -- the one from the 636-kilohertz transmitter having amplitude $a/R = a/2$ and that from the 796-kilohertz transmitter having an amplitude of only $a/[(\omega_2 L - 1/\omega_2 C)^2 + R^2]^{1/2} = a/450$. Thus we would "receive" the desired broadcast and "reject" the other.

Problems

1. If λ_1, λ_2, and λ_3 are distinct (real or complex) numbers, prove that $e^{\lambda_1 t}$, $e^{\lambda_2 t}$, and $e^{\lambda_3 t}$ are linearly independent on any interval by computing their Wronskian. How does this proof compare to the method used in the proof of Theorem A?

2. Find the general solutions for each of the following:

   (a) $x''' + 6x'' + 12x' + 8x = 0$.

   (b) $x'' + x = \cos t$.

   (c) $x' + x = \sin t$ (Problem 2-3, but now much easier).

   (d) $x'' - x' - 2x = 1 + \sin 2t + e^t$.

   (e) $x''' - 4x'' + 4x' = e^t$.

   (f) $x''' - 4x'' + 4x' = e^{2t}$.

   (g) $x''' + x' - 10x = t^2$.

   (h) $x''' + x'' - 10x' = t^2$.

   (i) $x'' - x' - 2x = te^{-t} + 2e^{-t} - e^{2t}$.

   (j) $x'' + 4x' + 5x = e^{-2t} \cos t$. Describe the behavior of the (general) solution as $t \to \infty$.

   (k) $x'' + 3x' + 2x = \sin t \cos t$.

   (ℓ) $x'' + 4x = \sin^2 t$.

3. Find the (unique) solutions of the first three equations
    of Problem 2 which satisfy the following initial condi-
    tions:

    (a)   with $x(0) = 2$, $x'(0) = 0$, $x''(0) = -10$.

    (b)   with $x(\pi/2) = 0$, $x'(\pi/2) = 1$.

    (c)   with $x(\pi) = 2$.

4. Find the simplest form of a particular solution which you
    know will work for each of the following.   Do not evaluate
    the coefficients.

    (a) $x'' + x = t \sin t$.

    (b) $x''' + 9x'' + 27x' + 27x = t + t^2 e^{-3t}$.

5. In Example 4,

    (a)   Find the general solution (15) of Eq. (14).

    (b)   Show that the steady state terms in (15) can be com-
          bined into the form (16).

    (c)   Verify that the voltage $v(t)$ in (17) leads to the
          steady state current $x_{ss}(t)$ in (18).

    (d)   If $L = 250 \times 10^{-6}$ henries and $R = 2$ ohms, verify
          the appropriate choice of C to best receive the
          636 kilohertz signal, and verify the amplitudes of
          the two terms in $x_{ss}(t)$ in this case.

    (e)   Repeat part (d) using the value of C which would
          best receive the 796 kilohertz signal.

13.   VARIATION OF PARAMETERS

Once again consider the linear equation with variable coefficients

$$x^{(n)} + a_{n-1}(t)x^{(n-1)} + \cdots + a_1(t)x' + a_0(t)x = h(t), \quad (1)$$

perhaps with initial conditions

$$x(t_0) = b_0, \quad x'(t_0) = b_1, \quad \ldots, \quad x^{(n-1)}(t_0) = b_{n-1}, \quad (2)$$

and the associated homogeneous equation

$$y^{(n)} + a_{n-1}(t)y^{(n-1)} + \cdots + a_1(t)y' + a_0(t)y = 0. \quad (3)$$

The functions $a_{n-1}, \ldots, a_1,\ a_0,$ and h are assumed to be continuous on an interval $J = (\alpha, \beta)$.

In the previous section we studied Eq. (1) in the special case when the coefficients $a_{n-1}, \ldots, a_0$ were constants and the function h was itself a solution of some linear, homogeneous differential equation with constant coefficients. For that important special case, Section 12 provided a systematic method for the complete solution of Eq. (1).

Unfortunately, no such simple general method is available for solving linear equations with variable coefficients (when $n \geq 2$). The present section introduces a technique for solving Eq. (1), or at least reducing it to a simpler problem, provided one knows some solution(s) of the homogeneous Eq. (3) to begin with. This method is central in the study of differential equations, and it will be encountered again later in more general situations.

The simplest illustration of the method uses a

familiar equation from Chapter I.

Example 1. Consider the nonhomogeneous, linear first order
equation

$$x' + a(t)x = h(t), \tag{4}$$

where a and h are continuous functions on $J = (\alpha, \beta)$, together with the initial condition

$$x(t_0) = x_0 \tag{5}$$

where $t_0 \in J$. Forgetting the integrating factor method of
Section 2, we shall present an alternative method of solution
which rests upon our first finding the general solution of
the associated homogeneous equation

$$y' + a(t)y = 0. \tag{6}$$

If $y(t_1) = 0$ for some t_1 in J, then, by uniqueness,
$y(t) = 0$ for all t in J. Otherwise $y(t) \neq 0$ for all t,
and we can solve Eq. (6) by the method of separation of vari-
ables. This gives $\ln |y(t)| + \int_{t_0}^{t} a(s)ds = c_1$, or

$$y(t) = ce^{-\int_{t_0}^{t} a(s)ds} \equiv cy_1(t).$$

Considering c as an arbitrary constant, we have the general solution of Eq. (6).

Now the idea in the method called "variation of parameters" (or "variation of constants") is this. Let us seek
a solution of the nonhomogeneous Eq. (4) in the form

$$x(t) = y_1(t)p(t), \tag{7}$$

where the arbitrary constant (or parameter) c in the

solution of Eq. (6) has been replaced by an unknown function
p -- a "varying parameter". This expression for x must be
substituted into Eq. (4) in order to discover the condition(s)
on p. The calculations for this substitution are conveniently
displayed as follows

$$
\begin{array}{c|l}
a(t) & x(t) \ = \ y_1(t)p(t) \\
1 & x'(t) \ = \ y_1'(t)p(t) \ + \ y_1(t)p'(t)
\end{array}
$$

$$0 \quad + \ y_1(t)p'(t) \ = \ h(t). \qquad (8)$$

The quantities a(t) and 1 at the far left are the appropriate multipliers for x and x' respectively before the
columns are totaled to complete the substitution into Eq. (4),
yielding (8). Note that with the columns aligned as above,
the first column to the right of the equal sign corresponds
to substitution of $y_1(t)p$ into the left hand side of Eq.
(4) or (6) with p regarded as a constant. Since y_1 satisfies Eq. (6), this column must total zero.

   From (8) we find $p'(t) = h(t)/y_1(t)$. Hence $p(t) =$
$c + \int_{t_0}^{t} [h(s)/y_1(s)]ds$, or

$$x(t) \ = \ p(t)y_1(t) \ = \ cy_1(t) \ + \ \int_{t_0}^{t} h(s)[y_1(t)/y_1(s)]ds.$$

Clearly $x(t_0) = c$, so $c = x_0$. Thus

$$x(t) \ = \ x_0 e^{-\int_{t_0}^{t} a(s)ds} \ + \ \int_{t_0}^{t} h(s)e^{-\int_{s}^{t} a(u)du} ds \qquad (9)$$

-- precisely the solution obtained in Section 2 (Problem 2-2).

Remarks. In the calculation leading to Eq. (8), the decision

to write $x(t) = y_1(t)p(t)$ instead of the more explicit form
$x(t) = e^{\int_{t_0}^{t} a(s)ds} p(t)$ was an arbitrary choice of convenience.
If this is not clear, rewrite the calculation using the more
explicit form and its derivative to see that the outcome is
precisely the same.

The next example shows how the same trick used in
Example 1 may also work for a second order linear equation.

Example 2. Find the general solution of the "Euler equation"

$$t^2x'' + 3tx' + x = 2t. \tag{10}$$

Observe that this equation can be put into the form of Eq. (1)
on the interval $(0,\infty)$ or on the interval $(-\infty,0)$, but not
on any interval which contains the point 0. Thus we restrict
ourselves to either $(0,\infty)$ or $(-\infty,0)$.

First consider the associated homogeneous equation

$$t^2y'' + 3ty' + y = 0, \tag{11}$$

and, as in Problem 11-7, seek a solution of the form
$y(t) = t^\lambda$ for $t > 0$, where λ is a constant. (For $t < 0$
one could use $y(t) = (-t)^\lambda$.) Substitution into Eq. (10)
gives $\lambda(\lambda-1)t^\lambda + 3\lambda t^\lambda + t^\lambda = 0$, and, since $t \neq 0$, this implies

$$\lambda(\lambda-1) + 3\lambda + 1 = 0,$$

or $\lambda^2 + 2\lambda + 1 = 0$. This quadratic equation has the double
root $\lambda_1 = \lambda_2 = -1$. Thus $y_1(t) = t^{-1}$ is a solution of the
homogeneous Eq. (11).

Now let us try to find a solution of the nonhomogeneous

Eq. (10) in the form $x = y_1 p$. The calculations involved in substituting $x = y_1 p$ into (10) are displayed below in a format analogous to that used in Example 1.   This time it is more economical to use the specific form, t^{-1}, of our solution of the homogeneous equation rather than $y_1(t)$.

$$
\begin{array}{c|l}
1 & x \;=\; t^{-1}p \\
3t & x' = -t^{-2}p + t^{-1}p' \\
t^2 & x'' = 2t^{-3}p - 2t^{-2}p' + t^{-1}p'' \\
\hline
& 0 \;+\; p' + \; tp'' \;=\; 2t. \hspace{2cm} (12)
\end{array}
$$

The result is a first order linear equation for the unknown $q = p'$, namely

$$q' + t^{-1}q = 2.$$

This is solved (with the aid of an integrating factor) to give $tq = t^2 + c$, or

$$p'(t) = q(t) = t + c/t.$$

Hence $p(t) = t^2/2 + c \ln |t| + c_1$ and

$$x(t) = t^{-1}p(t) = \frac{t}{2} + c\frac{1}{t} \ln |t| + c_1 \frac{1}{t}.$$

For any constants, c and c_1, this x is a solution of the original differential equation (10) on $(0,\infty)$.   It happens also to be a solution on $(-\infty,0)$.

To prove that it is the general solution, we would have to show that

$$y_1(t) = \frac{1}{t} \quad \text{and} \quad y_2(t) = \frac{1}{t} \ln |t|$$

represent two linearly independent solutions of the homogeneous Eq. (11).   This is left to the reader as Problem 1.

In problems at the end of this section the reader will encounter further examples of Euler's equation

$$t^n x^{(n)} + a_{n-1} t^{n-1} x^{(n-1)} + \cdots + a_1 t x' + a_0 x = h(t), \quad (13)$$

where $a_{n-1}, \ldots, a_1, a_0$ are constants.

In its most general form, the method of "variation of parameters" aims to study a nonhomogeneous equation (1) of order n with the aid of k linearly independent solutions of the corresponding homogeneous equation (3) where $1 \leq k \leq n$. The most important case is that in which $k = n$.

The Method of Variation of Parameters. If $y_1, \ldots, y_k$ are linearly independent solutions of Eq. (3) with $1 \leq k \leq n$, we shall seek a solution of Eq. (1) in the form

$$x = y_1 p_1 + \cdots + y_k p_k, \quad (14)$$

where $p_1, \ldots, p_k$ are functions yet to be defined. (We know that if $p_1, \ldots, p_k$ were all constants, then (14) would represent just another solution of Eq. (3) and, of course, it would not even be linearly independent of $y_1, \ldots, y_k$.)

Now, as will be seen later, the single requirement that x satisfy Eq. (1) will not uniquely determine $p_1, \ldots, p_k$ in (14). In other words, if there is one choice of functions $p_1, \ldots, p_k$ which provide a solution of Eq. (1), then there are other choices which would also work. This (unproved) assertion implies a freedom to impose certain other conditions on $p_1, \ldots, p_k$. The further conditions will be chosen in a way that simplifies the calculations.

The substitution of (14) into Eq. (1) requires a

computation of successive derivatives of x as follows.

$$x' = y_1'p_1 + \cdots + y_k'p_k + y_1p_1' + \cdots + y_kp_k',$$

$$x'' = y_1''p_1 + \cdots + y_k''p_k + 2y_1'p_1' + \cdots + 2y_k'p_k'$$

$$+ y_1p_1'' + \cdots + y_kp_k'',$$

etc. Clearly these get ever more cumbersome. But notice how much simpler this would become if one imposed the following k-1 conditions on the first derivatives, $p_1',\ldots,p_k'$, of the k unknown functions:

$$y_1p_1' + \cdots + y_kp_k' = 0$$

$$y_1'p_1' + \cdots + y_k'p_k' = 0$$

$$\vdots \tag{15}$$

$$y_1^{(k-2)}p_1' + \cdots + y_k^{(k-2)}p_k' = 0.$$

These conditions and the corresponding calculation of the derivatives of (14) for substitution into Eq. (1) are conveniently remembered and recorded in the following format.

$$a_0(t) \quad\Big|\quad x \ = y_1p_1 + \cdots + y_kp_k,$$

$$a_1(t) \quad\Big|\quad x' = y_1'p_1 + \cdots + y_k'p_k + \boxed{y_1p_1' + \cdots + y_kp_k' = 0,}$$

$$a_2(t) \quad\Big|\quad x'' = y_1''p_1 + \cdots + y_k''p_k + \boxed{y_1'p_1' + \cdots + y_k'p_k' = 0,}$$

etc.

Note that if $k = 1$, as it was in Examples 1 and 2, then (15) is vacuous, i.e., no extra conditions are imposed. If $k = 2$ we impose only the first condition in (15),..., and if $k = n$ we impose $n-1$ conditions.

It is not at all obvious that one can find functions $p_1,\ldots,p_k$ which satisfy the $k-1$ conditions (15) and such that (14) satisfies Eq. (1). Further calculations in specific examples will determine whether this is possible.

Example 3   (with $k = n = 2$). Find the general solution of

$$tx'' - (t+2)x' + 2x = t^3 \quad \text{for} \quad t > 0, \tag{16}$$

using the fact (which is easily verified) that $y_1(t) = e^t$ and $y_2(t) = t^2+2t+2$ are two linearly independent solutions of the associated homogeneous equation. We seek a solution in the form $x = y_1 p_1 + y_2 p_2$. And, since $k = 2$, we impose the single condition (15) $y_1 p_1' + y_2 p_2' = 0$. The calculations are written out as follows:

$$
\begin{array}{c|l}
2 & x = y_1 p_1 + y_2 p_2 \\[2mm]
-t-2 & x' = y_1' p_1 + y_2' p_2 + \boxed{y_1 p_1' + y_2 p_2' = 0} \\[2mm]
t & x'' = y_1'' p_1 + y_2'' p_2 + y_1' p_1' + y_2' p_2' \\
\hline
& \quad 0 \;+\; 0 \;+\; t y_1' p_1' + t y_2' p_2' = t^3,
\end{array}
$$

where the last line is the result of substitution into Eq. (16). So there are two algebraic equations for p_1' and p_2' :

$$y_1 p_1' + y_2 p_2' = 0$$

$$y_1' p_1' + y_2' p_2' = t^2.$$

From these,

$$p_1'(t) = \frac{\begin{vmatrix} 0 & y_2(t) \\ t^2 & y_2'(t) \end{vmatrix}}{W(y_1,y_2)(t)} = (t^2 + 2t + 2)e^{-t},$$

$$p_2'(t) = \frac{\begin{vmatrix} y_1(t) & 0 \\ y_1'(t) & t^2 \end{vmatrix}}{W(y_1,y_2)(t)} = -1.$$

(Is it anything more than a fortunate coincidence that $W(y_1,y_2)(t) \neq 0$?)   Integration gives

$$p_1(t) = -t^2 e^{-t} - 4te^{-t} - 6e^{-t} + c_1 \quad \text{and} \quad p_2(t) = -t + c_2,$$

so that

$$x(t) = -t^3 - 3t^2 - 6t - 6 + c_1 e^t + c_2(t^2 + 2t + 2),$$

or, more compactly, $x(t) = -t^3 + c_1 e^t + c_3(t^2 + 2t + 2)$.

The general theorem for which Example 3 is a prototype is as follows.

Theorem A (Variation of Parameters).   Let $y_1,\ldots,y_n$ be linearly independent solutions of Eq. (3) on J.   Then the general solution of Eq. (1) is given by

$$x(t) = \sum_{j=1}^{n} y_j(t) \int^t \frac{(-1)^{n+j} W_j(s)h(s)}{W(y_1,\ldots,y_n)(s)} \, ds, \tag{17}$$

where W_j is the Wronskian of $y_1,\ldots,y_{j-1},\ y_{j+1},\ldots,y_n$ (see Eq. (20) below), and each $\int^t$ indicates an indefinite integral with arbitrary constant of integration.

　　　Proof.　We seek a solution of Eq. (1)　in the form of
(14) with　k = n

$$x = y_1 p_1 + \cdots + y_n p_n \qquad (18)$$

with conditions (15) imposed for　k = n.　These　n-1　conditions plus the　n'th condition imposed by Eq. (1) itself form
a system of　n　linear algebraic equations for the unknowns
$p_1', \ldots, p_n'$:

$$y_1 p_1' + \cdots + y_n p_n' = 0$$

$$y_1' p_1' + \cdots + y_n' p_n' = 0,$$

$$\vdots \qquad\qquad (19)$$

$$y_1^{(n-2)} p_1' + \cdots + y_n^{(n-2)} p_n' = 0,$$

$$y_1^{(n-1)} p_1' + \cdots + y_n^{(n-1)} p_n' = h.$$

Verify this.　Moreover, by Theorem 11-B(ii), the determinant
of the coefficients is　$W(y_1, \ldots, y_n)(t) \neq 0$　for all　t　in
J.　So Eqs. (19) have a unique solution for　$p_1'(t), \ldots, p_n'(t)$
for each　t　in　J.　To express this, let us denote by　$W_j(t)$
(for each　j = 1, ..., n) the Wronskian of　$y_1, \ldots, y_{j-1}$,
$y_{j+1}, \ldots, y_n$,　i.e., the　(n-1) × (n-1)　determinant

$$
W_j(t) = \begin{vmatrix}
y_1 & \cdots & y_{j-1} & y_{j+1} & \cdots & y_n \\
y_1' & \cdots & y_{j-1}' & y_{j+1}' & \cdots & y_n' \\
\vdots & & & & & \\
y_1^{(n-2)} & \cdots & y_{j-1}^{(n-2)} & y_{j+1}^{(n-2)} & \cdots & y_n^{(n-2)}
\end{vmatrix} . \qquad (20)
$$

Then the unique solution of system (19) is

$$
p_j'(t) = \frac{(-1)^{n+j} W_j(t) h(t)}{W(y_1,\ldots,y_n)(t)}, \qquad (j = 1,\ldots,n).
$$

Integrating and substituting into Eq. (18), we obtain (17). And we note that if each $\int^t$ has an additive arbitrary constant of integration then, by Theorem 11-D, (17) is the general solution of Eq. (1). □

Observe that Theorem A gives a very complete result in the case of constant coefficients. For if the coefficients $a_{n-1},\ldots,a_1,$ a_0 in Eq. (1) are constants, then we know from Theorem 12-B that there do indeed exist n linearly independent solutions of Eq. (3). Thus, since h is continuous, Eq. (1) does have a general solution, which can be made to match any initial conditions of the form (2).

In solving the problems below, you are urged to use the method of variation of parameters rather than attempting to use Eq. (17) as a formula for the solution.

Problems

1. Complete Example 2 by verifying that $y_1(t) = t^{-1}$ and $y_2(t) = t^{-1} \ln |t|$ are linearly independent solutions of Eq. (11).

2. If λ_1, λ_2, and λ_3 are three distinct real numbers, prove that the functions defined by $y_1(t) = t^{\lambda_1}$, $y_2(t) = t^{\lambda_2}$, and $y_3(t) = t^{\lambda_3}$ are linearly independent on $(0,\infty)$

   (a) by computing the Wronskian, and

   (b) by adapting the method used in the proof of Theorem 12-A.

3. Find general solutions of each of the following equations and state where they are valid. The first four are examples of Euler's equation.

   (a) $4t^2x'' + 4tx' - x = 0$.

   (b) $4t^2x'' + 8tx' + x = 0$.

   (c) $t^3x''' - t^2x'' + tx' = 0$.

   (d) $t^3x''' + tx' - x = 0$.

   (e) $tx'' - (t+2)x' + 2x = t^3$ for $t > 0$. This is the equation of Example 3. But now solve it using only the observation that $y(t) = e^t$ is a solution of the associated homogeneous equation.

   (f) $tx'' - x' + t^3x = 0$. <u>Hint</u>. $x(t) = \sin(t^2/2)$ is one solution.

4. Another approach to the solution of Euler's equations begins with the introduction of new variables, $s = \ln |t|$ and $w(s) = x(e^s) = x(t)$ in case $t > 0$ or $w(s) = x(-e^s) = x(t)$ in case $t < 0$. Apply this trick to the equations of Problem 3 parts (a), (b), (c), and (d).

5. The method indicated in Problem 4 is especially appropriate in case our first method for Euler's equation leads to complex values of λ. Find the general solution of

(a) $t^2x'' + 5tx' + 5x = 0$.

(b) $t^3x''' + 3t^2x'' + 7tx' - 20x = 0$.

6.   Use the <u>method</u> of variation of parameters to find general
solutions of the following equations, and wherever possible compare your answer with that obtained using the
method of undetermined coefficients (Section 12).

(a) $x'' + 3x' + 2x = 1 + 2t + e^{-t}$.

(b) $x' + x = t$.

(c) $x'' + x = \sin t$.

(d) $x'' + x = e^t$.

(e) $x'' + x = \tan t$.

(f) $x'' + x = \sec t$.

(g) $x'' + x = \sec^2 t$.

(h) $x''' + 3x'' + 2x' = e^{-t}$.

(i) $t^2x'' - 2tx' + 2x = t^3e^t$.

(j) $t^2x'' + 3tx' + x = t^{-1}$.

*7.   Find the general solution of
$$t^5y''' + t^2y'' - 2ty' + 2y = 0$$
after observing that $y_1(t) = t$ and $y_2(t) = t^2$ give
two particular solutions.

8.   Find, by the <u>method</u> of variation of parameters, the general solution of each of the following equations when   h
is assumed to be a given continuous function.   Check your
answers by substitution.

(a) $x'' + 4x' - 5x = h(t)$.

(b) $x''' + 2x'' + 4x' + 8x = h(t)$.

(c) $x'' + x = h(t)$.

(d) $x'' + 4x' + 5x = h(t)$.

9. Let $x''(t) = h(t)$ on J where h is a continuous func-
    tion. Use the method of variation of parameters to de-
    rive the result

    $$x(t) = x(t_0) + x'(t_0)(t-t_0) + \int_{t_0}^{t} (t-s)h(s)ds,$$

    where t_0 is an arbitrary point in J. Verify the for-
    mula by differentiation, also checking the values of
    $x(t_0)$ and $x'(t_0)$. Compare this result with Taylor's
    theorem (Theorem A2-B) for the case $m = 2$. Can you
    prove that the above formula is equivalent to the one
    given by Eq. A2-(1) with $m = 2$?

*10. The most ambitious reader may want to repeat Problem 9 for
    $x^{(n)}(t) = h(t)$, obtaining

    $$x(t) = \sum_{k=0}^{n-1} \frac{x^{(k)}(t_0)}{k!}(t-t_0)^k + \frac{1}{(n-1)!} \int_{t_0}^{t} (t-s)^{n-1}h(s)ds.$$

    For an alternative easier problem, take $n = 3$.

11. Find the general solution of $mx'' + bx' + kx = 0$ when
    $b^2 = 4mk$ by first finding an exponential solution y
    $y_1(t) = e^{\lambda t}$, and then applying the method of this sec-
    tion. (Of course, the answer is already known from
    Example 10-1.)

12. It is sometimes useful when studying an equation of the
    form

    $$x'' + a_1(t)x' + a_0(t)x = h(t), \tag{21}$$

    to introduce a new unknown function z defined by

$$z(t) = x(t)e^{\frac{1}{2}\int_{t_0}^{t} a_1(s)ds},$$ (22)

using some convenient t_0. Then

$$x(t) = z(t)e^{-\frac{1}{2}\int_{t_0}^{t} a_1(s)ds}.$$

(a) Assuming that a_1 is differentiable, substitute this into Eq. (21) to find the new differential equation satisfied by z,

$$z'' + [a_0(t) - \frac{1}{4} a_1^2(t) - \frac{1}{2}a_1'(t)]z$$

$$= h(t)e^{\frac{1}{2}\int_{t_0}^{t} a_1(s)ds}.$$ (23)

Notice that, in contrast to Eq. (21), Eq. (23) contains no first-derivative term.

(b) Apply this change of variables to the "Bessel's equation of order $\frac{1}{2}$",

$$t^2 y'' + ty' + (t^2 - \frac{1}{4})y = 0 \quad \text{for} \quad t > 0.$$ (24)

You should then be able to find the general solution of Eq. (24),

$$y(t) = c_1 t^{-1/2} \cos t + c_2 t^{-1/2} \sin t.$$

Chapter IV. LINEAR ORDINARY DIFFERENTIAL SYSTEMS

The study of a system of linear differential equations (or a linear differential system) of the form

$$x_i' = \sum_{j=1}^{n} a_{ij}(t)x_j + h_i(t) \qquad (i = 1,\ldots,n)$$

is generally simplified with the aid of matrix theory. Moreover, matrix methods can often give useful information about differential systems which are not quite linear.

This chapter gives only an introduction to matrix methods for differential equations. These powerful methods are discussed more thoroughly in other texts such as Coddington and Levinson [1955], Hartman [1964], and various books on matrix theory.

In this text the reader is assumed to be familiar with only the most basic properties of matrices, including the concepts of the inverse, eigenvalues, and eigenvectors, plus the procedure for diagonalizing an $n \times n$ matrix having n linearly independent eigenvectors.

This chapter culminates with a proof of the existence of solutions for linear systems of ordinary differential equations with continuous coefficients.

14.   SOME GENERAL PROPERTIES

Consider the system of n equations in n unknown functions

$$x_i' = \sum_{j=1}^{n} a_{ij}(t)x_j + h_i(t) \quad (i = 1,\ldots,n)$$

on an interval $J = (\alpha,\beta)$. With this system we shall often associate initial conditions of the form

$$x_i(t_0) = x_{0i} \quad (i = 1,\ldots,n),$$

where $t_0 \in J$.

Let us introduce the $n \times n$ -__matrix__-__valued__ __function__, $A: J \to \mathbb{R}^n$ (or $\mathbb{C}^n$),

$$A(t) = \begin{pmatrix} a_{11}(t) & \cdots & a_{1n}(t) \\ \vdots & & \vdots \\ a_{n1}(t) & \cdots & a_{nn}(t) \end{pmatrix} \quad \text{for } t \text{ in } J,$$

where each a_{ij} is a function mapping $J \to \mathbb{R}$ (or $\mathbb{C}$).

With this notation, our system becomes

$$x' = A(t)x + h(t) \tag{1}$$

on J , and the initial conditions are written as

$$x(t_0) = x_0. \tag{2}$$

Here $x(t) = \text{col }(x_1(t),\ldots,x_n(t))$, $h(t) = \text{col }(h_1(t),\ldots,h_n(t))$, and $x_0 = \text{col }(x_{01},\ldots,x_{0n})$. Together with (1), we shall also consider the associated linear __homo__-__geneous__ system

$$y' = A(t)y, \tag{3}$$

where y is again a vector-valued function.

Proceeding as in Section 5 for vector-valued functions, we shall say a matrix-valued function A is <u>continuous</u> if each of its elements a_{ij} is continuous (i, j = 1,...,n). We shall say A is <u>differentiable</u> (or <u>integrable</u>) if each a_{ij} is differentiable (or integrable), and then the <u>derivative</u> (or <u>integral</u>) of A will be the matrix obtained by differentiating (or integrating) each element of A.

The theorems and corollaries A through E to be given in this section for system (1) are analogs of propositions 11-A through 11-E respectively for the n'th order, linear, scalar equation. However, a minor adjustment in notation is needed now. Since y_i will now stand for the i'th component of the vector-valued function y, we shall denote several different vector-valued functions by $y_{(1)}, \ldots, y_{(\ell)}$; and, when needed, we shall denote the j'th component of $y_{(i)}$ by $y_{(i)j}$.

<u>Theorem A</u> (<u>Superposition</u>). Let $h(t) = k_1 h_{(1)}(t) + \cdots$
$+ k_m h_{(m)}(t)$ where $k_1, \ldots,$ and k_m are (complex) constants, and $h_{(1)}, \ldots, h_{(m)}$ are given n-vector-valued functions on J. For j = 1,...,m, let $x_{(j)}$ be a (particular) solution of $x' = A(t)x + h_{(j)}(t)$ and let $y_{(1)}, \ldots, y_{(\ell)}$ be solutions of (3), y' = A(t)y, on J.   Then

$$x = k_1 x_{(1)} + \cdots + k_m x_{(m)} + c_1 y_{(1)} + \cdots + c_\ell y_{(\ell)}$$

is a solution of Eq. (1) for every choice of the (complex) constants $c_1, \ldots, c_\ell$.

<u>Proof</u>. Using the linearity of the differentiation operator and the linearity of multiplication by the matrix

$A(t)$,

$$x' = \sum_{j=1}^{m} k_j x'_{(j)} + \sum_{i=1}^{\ell} c_i y'_{(i)}$$

$$= \sum_{j=1}^{m} k_j [A(t)x_{(j)} + h_{(j)}(t)] + \sum_{i=1}^{\ell} c_i A(t) y_{(i)}$$

$$= A(t) \sum_{j=1}^{m} k_j x_{(j)} + h(t) + A(t) \sum_{i=1}^{\ell} c_i y_{(i)}$$

$$= A(t)x + h(t). \quad \square$$

Let us now extend to vector functions the concepts of "linearly dependence" and "linear independence", and seek analogs of the other theorems in Section 11.

Definitions. A set of vector-valued functions, $v_{(1)}, \ldots, v_{(k)}$ on an interval J is said to be linearly dependent if there exist constants $c_1, \ldots, c_k$ in $\mathbb{C}$, at least one of which is not zero such that

$$c_1 v_{(1)}(t) + \cdots + c_k v_{(k)}(t) = 0 \quad \text{for all } t \text{ in } J. \quad (4)$$

Otherwise the functions are said to be linearly independent.

Note that linear dependence of the vector-valued functions requires that each of the n equations represented by (4) hold for all t in J. Fortunately, however, in the case of functions which happen to be solutions of a linear homogeneous system (3) with a continuous coefficient matrix it suffices to examine Eq. (4) at just one value of t. This is made precise in the next theorem.

Theorem B.

(i)   Let $y_{(1)}, \ldots, y_{(k)}$ be any n-vector-valued

functions on J.   Then if the constant vectors
$y_{(1)}(t_1),\ldots,y_{(k)}(t_1)$ are linearly independent for some t_1
in J, it follows that $y_{(1)},\ldots,y_{(k)}$ are linearly independent functions on J.   (The converse is false.)

(ii)   Let A be a continuous, matrix-valued function on
the interval J, and let $y_{(1)},\ldots,y_{(k)}$ be solutions of the
linear, homogeneous equation (3) on J.   Then $y_{(1)},\ldots,y_{(k)}$
are linearly independent functions if and only if the constant vectors $y_{(1)}(t),\ldots,y_{(k)}(t)$ are linearly independent
for every t in J.

Proof.   (i)   Clearly if $y_{(1)}(t_1),\ldots,y_{(k)}(t_1)$ are
linearly independent for some t_1, then condition (4) can
only hold if $c_1 = c_2 = \cdots = c_k = 0$.   Thus $y_{(1)},\ldots,y_{(k)}$
are independent on J.

To see that the converse of (i) is false it suffices to
consider the pair of scalar-valued functions (n = 1, k = 2)
$y_{(1)}(t) = t$ and $y_{(2)}(t) = 1$ on $(-1,1)$.   Then $y_{(1)}$ and
$y_{(2)}$ are linearly independent functions on $(-1,1)$.   Why?
But $y_{(1)}(0)$ and $y_{(2)}(0)$ are linearly dependent.

(ii)   Now let $y_{(1)},\ldots,y_{(k)}$ be solutions of Eq. (3) on
J.   Then the "if" part of assertion (ii) follows from (i).
To prove the "only if" part, assume that $y_{(1)}(t_1),\ldots,y_{(k)}(t_1)$
are linearly dependent for some t_1.   Then there must exist
constants $c_1,\ldots,c_k$, not all zero, such that $c_1 y_{(1)}(t_1) +$
$\cdots + c_k y_{(k)}(t_1) = 0$.   Using these values of $c_1,\ldots,c_k$, define a new function y on J by

$$y(t) \equiv c_1 y_{(1)}(t) + \cdots + c_k y_{(k)}(t).$$

Thus y is a solution of Eq. (3) and $y(t_1) = 0$. Hence, by uniqueness, $y(t) \equiv 0$. In other words, the functions $y_{(1)}, \ldots, y_{(k)}$ are linearly dependent on J. This means that $y_{(1)}, \ldots, y_{(k)}$ can be linearly independent functions on J only if $y_{(1)}(t_1), \ldots, y_{(k)}(t_1)$ are linearly independent vectors for each t_1 in J. $\square$

Remark.   Actually Theorem B is slightly more general than its counterpart in Section 11, Theorem 11-B, in as much as we now allow $k \neq n$. However, Theorem 11-B could have been made entirely analogous to the above if it had been phrased in terms of the linear independence of the vectors col $(y_{(j)}, y'_{(j)}, \ldots, y_{(j)}^{(n-1)})$ for $j = 1, \ldots, k$ instead of the nonvanishing of the Wronskian. The Wronskian, being a determinant, can only be defined for a square array, i.e., for n columns each having n components.

Corollary C.   Let A be a continuous matrix-valued function on J. If $y_{(1)}, \ldots, y_{(n)}$ are solutions of Eq. (3), then the vectors $y_{(1)}(t), \ldots, y_{(n)}(t)$ are either linearly independent for each t in J or linearly dependent for each t in J. Thus the determinant

$$W(t) = W(y_{(1)}, \ldots, y_{(n)})(t) \equiv \begin{vmatrix} y_{(1)1}(t) & \cdots & y_{(n)1}(t) \\ \vdots & & \vdots \\ y_{(1)n}(t) & \cdots & y_{(n)n}(t) \end{vmatrix} \qquad (5)$$

is either zero for each t in J or different from zero for each t in J, according to whether $y_{(1)}, \ldots, y_{(n)}$ are linearly dependent or linearly independent.

The proof is left as Problem 2.

The scalar-valued function $W = W(y_{(1)}, \ldots, y_{(n)})$ defined on J by (5) will be called the Wronskian of the vector functions $y_{(1)}, \ldots, y_{(n)}$.

We can now describe the "general solution" of Eq. (1).

Theorem D. Let A be a continuous matrix-valued function on J. If Eq. (3) has n linearly independent solutions, $y_{(1)}, \ldots, y_{(n)}$, and if $\tilde{x}$ is any (particular) solution of Eq. (1) on J, then the general solution of Eq. (1) is defined by

$$x(t) = \tilde{x}(t) + c_1 y_{(1)}(t) + \cdots + c_n y_{(n)}(t) \quad \text{for} \quad t \text{ in J.} \quad (6)$$

By this we mean that every solution of Eq. (1) can be obtained from (6) by an appropriate choice of the (constant) scalars $c_1, \ldots, c_n$.

Moreover for each t_0 in J and each vector x_0, there does exist a (unique) solution of Eq. (1) which satisfies the initial condition (2).

Proof. Theorem A asserts that (6) defines a solution of Eq. (1) for every choice of the constants $c_1, \ldots, c_n$.

Now let $\hat{x}$ be any given solution of Eq. (1) on J. Select any t_0 in J and consider the n equations represented by

$$c_1 y_{(1)}(t_0) + \cdots + c_n y_{(n)}(t_0) = \hat{x}(t_0) - \tilde{x}(t_0). \quad (7)$$

If we regard (7) as a system of n linear, algebraic equations in the n unknowns, $c_1, \ldots, c_n$, then it follows from Corollary C that the determinant of the coefficients is

$W(t_0) \neq 0$.   Thus a solution $c_1, \ldots, c_n$ exists (and is unique). Using these values of $c_1, \ldots, c_n$, we define a function

$$x(t) \equiv \tilde{x}(t) + c_1 y_{(1)}(t) + \cdots + c_n y_{(n)}(t) \quad \text{for all} \quad t \quad \text{in} \quad J.$$

This function x is a solution of (1) and moreover $x(t_0) = \hat{x}(t_0)$.   It follows from the uniqueness theorem that $\hat{x}(t) = x(t)$ for all t in J.

If t_0 in J and a vector x_0 are given, the existence of a (unique) solution of (1) and (2) is established as above using x_0 in place of $\hat{x}(t_0)$. $\square$

Nothing we have said thus far proves that Eq. (3) with a continuous coefficient matrix, A, will have n linearly independent solutions.   This will eventually be proved (in Sec. 20) so that the hypothesis in Theorem D that Eq. (3) have n linearly independent solutions will then be seen to be unnecessary.   In the meantime, to give an application of Theorem D, we must use an example for which it is possible to explicitly find n linearly independent solutions of the homogeneous equation.   The most natural examples are those in which A is a constant matrix.

Example 1.   Find the general solution on R of the linear system with constant coefficients,

$$x_1' = -2x_1 + 2x_2 + 5 \sin 2t$$

$$x_2' = 2x_1 - 5x_2$$

(8)

(This is the system of Example 5-1.)

By analogy to the method used for scalar equations, let

us begin by trying to find solutions of the associated homogeneous system

$$y' = Ay, \quad \text{where} \quad A = \begin{pmatrix} -2 & 2 \\ 2 & -5 \end{pmatrix}, \tag{9}$$

in the form $y(t) = e^{\lambda t}\xi$, i.e.,

$$y(t) = \begin{pmatrix} \xi_1 e^{\lambda t} \\ \xi_2 e^{\lambda t} \end{pmatrix},$$

where λ, ξ_1, and ξ_2 are constants. (We have no guarantee,
at present, that this will be successful.) Substitution of
$y(t) = e^{\lambda t}\xi$ into Eq. (9) leads to the vector equation

$$\lambda e^{\lambda t}\xi = A e^{\lambda t}\xi.$$

But, since $e^{\lambda t}$ is never zero, this is equivalent to

$$(A - \lambda I)\xi = 0, \tag{10}$$

where I is the identity matrix. Now the problem of solving
Eq. (10) is exactly the problem of finding the eigenvalues,
λ, and associated eigenvectors, ξ, of the constant matrix A.

We ask the reader to compute (in Problem 3) the eigen-
values, $\lambda_1 = -1$ and $\lambda_2 = -6$, and respective eigenvectors
col $(2,1)$ and col $(1,-2)$. Solutions of (9) are therefore
given by

$$y_{(1)}(t) = \begin{pmatrix} 2e^{-t} \\ e^{-t} \end{pmatrix} \quad \text{and} \quad y_{(2)}(t) = \begin{pmatrix} e^{-6t} \\ -2e^{-6t} \end{pmatrix}. \tag{11}$$

Once the reader verifies that these two solutions, $y_{(1)}$ and
$y_{(2)}$, are linearly independent, Theorem D provides the general
solution of the homogeneous system (9), $y = c_1 y_{(1)} + c_2 y_{(2)}$.

To now find the general solution of the original non-homogeneous system (8), we require, in addition to $y_{(1)}$ and $y_{(2)}$, some particular solution of (8). Because of the nature of that system we might reasonably hope to find a solution of the form

$$\tilde{x}(t) = \begin{pmatrix} B_1 \cos 2t + C_1 \sin 2t \\ B_2 \cos 2t + C_2 \sin 2t \end{pmatrix},$$

where B_1, C_1, B_2, and C_2 are undetermined constants. Substituting this expression for $\tilde{x}$ into (8) we find

$$-2B_1 \sin 2t + 2C_1 \cos 2t$$
$$= -2B_1 \cos 2t - 2C_1 \sin 2t + 2B_2 \cos 2t + 2C_2 \sin 2t + 5 \sin 2t,$$
$$-2B_2 \sin 2t + 2C_2 \cos 2t$$
$$= 2B_1 \cos 2t + 2C_1 \sin 2t - 5B_2 \cos 2t - 5C_2 \sin 2t.$$

Let us now equate the coefficients of $\sin 2t$ and $\cos 2t$ on the two sides of each of these two equations. The result is four equations in four unknowns:

$$-2B_1 + 2C_1 \qquad\qquad - 2C_2 = 5,$$
$$2B_1 + 2C_1 - 2B_2 \qquad\quad = 0,$$
$$- 2C_1 - 2B_2 + 5C_2 = 0,$$
$$-2B_1 \qquad\quad + 5B_2 + 2C_2 = 0.$$

These yield

$$B_1 = -\frac{33}{20}, \quad C_1 = \frac{19}{20}, \quad B_2 = -\frac{7}{10}, \quad C_2 = \frac{1}{10}.$$

Thus, by virtue of Theorem D, the general solution of system (8) is

$$x(t) = \frac{1}{20}\begin{pmatrix} -33 \cos 2t + 19 \sin 2t \\ -14 \cos 2t + 2 \sin 2t \end{pmatrix} + c_1\begin{pmatrix} 2e^{-t} \\ e^{-t} \end{pmatrix} + c_2\begin{pmatrix} e^{-6t} \\ -2e^{-6t} \end{pmatrix}.$$

The reader should realize that "real" problems seldom have answers with simple numerical coefficients. The above example was constructed (with some effort) so that the answer would look fairly simple.

The following is a straightforward consequence of Theorem D. Its proof is Problem 4.

Corollary E.   If   A   is a continuous matrix-valued function on J, then Eq. (3) can have at most   n   linearly independent solutions.

Problems

1.   Show that the equation in Problem 11-8 is equivalent to the system

$$y_1' = y_2$$
$$y_2' = y_3$$
$$y_3' = 64y_1 - 40ty_2 + 8t^2y_3.$$

Let $y_{(1)}$ and $y_{(2)}$ be the vector-valued solutions corresponding to the solutions given in Problem 11-8.   Now test Theorem B on these solutions, taking $t_0 = 0$.

2.   Prove Corollary C.

3.   Complete the derivation of the solutions $y_{(1)}$ and $y_{(2)}$ of system (9), and verify that these vector functions $y_{(1)}$ and $y_{(2)}$, given in (11), are linearly independent.

4.   Prove Corollary E.

5.   Following the procedure used in Example 1, find the gen-
     eral solutions of the following systems.

(a) $\begin{aligned} y_1' &= y_1 + 4y_2 \\ y_2' &= 3y_1 + 2y_2. \end{aligned}$ (b) $x' = \begin{pmatrix} 1 & 4 \\ 3 & 2 \end{pmatrix} x + \begin{pmatrix} 2 \\ \cos t \end{pmatrix}$

(c) $x' = \begin{pmatrix} -1 & 0 & 1 \\ -1 & -2 & -1 \\ -2 & -2 & -3 \end{pmatrix} x + \begin{pmatrix} 1 + t \\ 1 \\ t \end{pmatrix}$

6.   Find the particular solutions for the first two systems
     of Problem 5 which satisfy the initial conditions

     (a) $y(0) = \text{col } (1,-1)$,   (b) $x(0) = \text{col } (1,-1)$.

*7.   If $y_{(1)}$ and $y_{(2)}$ are solutions of Eq. (3) with $n = 2$
     prove that

$$W(y_{(1)}, y_{(2)})(t) = W(y_{(1)}, y_{(2)})(t_0) \exp\{ \int_{t_0}^{t} [a_{11}(s) + a_{22}(s)] ds \}$$

     for all t, t_0 in J.   Cf. Problem 11-10.   (This gives
     an alternate proof of Corollary C.)   Can you generalize
     this to arbitrary n?

15.   CONSTANT COEFFICIENTS

     Just as Section 14 provided analogs to the theorems of
Section 11, the present section will provide analogs for much
of Section 12.

     The coefficient matrix   A   will now be a constant and
the systems considered will be

$$x' = Ax + h(t) \tag{1}$$

with initial condition

$$x(t_0) = x_0, \qquad (2)$$

plus the associated homogeneous system

$$y' = Ay. \qquad (3)$$

The interval J can be considered to be $\mathbb{R}$.

In case the matrix A can be diagonalized, then solution of Eq. (1) is an easy matter.

Example 1. Solve the system

$$
\begin{aligned}
x_1' &= -2x_1 + 2x_2 + 5 \sin 2t, \\
x_2' &= 2x_1 - 5x_2.
\end{aligned}
\qquad (4)
$$

This is system (1) with

$$
A = \begin{pmatrix} -2 & 2 \\ 2 & -5 \end{pmatrix}
\quad \text{and} \quad
h(t) = \begin{pmatrix} 5 \sin 2t \\ 0 \end{pmatrix}.
$$

When solving this system in Example 14-1 we first sought solutions of the associated homogeneous equation in the form $y(t) = e^{\lambda t}\xi$, where λ is a constant scalar and ξ is a constant vector. Substitution into $y' = Ay$ led to the equation $(A - \lambda I)\xi = 0$, the familiar eigenvalue-eigenvector problem for the matrix A. We found two distinct eigenvalues $\lambda_1 = -1$ and $\lambda_2 = -6$ together with associated linearly independent eigenvectors $\xi_{(1)} = \text{col } (2,1)$ and $\xi_{(2)} = \text{col } (1,-2)$. Then, by Theorem 14-D, the general solution of system (5) was seen to be

$$
x(t) = \tilde{x}(t) + c_1 e^{-t}\begin{pmatrix} 2 \\ 1 \end{pmatrix} + c_2 e^{-6t}\begin{pmatrix} 1 \\ -2 \end{pmatrix}.
$$

where $\tilde{x}$ was any particular solution of (4).

Another approach to system (4) involves diagonalizing the matrix A. Using the linearly independent eigenvectors

$\xi_{(1)}$ and $\xi_{(2)}$ as columns, define the matrix

$$P = \begin{pmatrix} 2 & 1 \\ 1 & -2 \end{pmatrix}, \quad \text{and compute} \quad P^{-1} = \frac{1}{5}\begin{pmatrix} 2 & 1 \\ 1 & -2 \end{pmatrix}.$$

Then the matrix A is "diagonalized" by computing

$$P^{-1}AP = \begin{pmatrix} -1 & 0 \\ 0 & -6 \end{pmatrix} = \Lambda.$$

Now let us introduce a new unknown function z defined by

$$z(t) = P^{-1}x(t).$$

Then it follows that $x(t) = Pz(t)$, and

$$z' = P^{-1}x' = P^{-1}Ax + P^{-1}h(t) = P^{-1}APz + P^{-1}h(t)$$

or

$$z' = \Lambda z + P^{-1}h(t).$$

Thus z satisfies a much simpler system of equations, namely

$$\begin{aligned} z_1' &= -z_1 + 2 \sin 2t, \\ z_2' &= -6z_2 + \sin 2t. \end{aligned} \tag{5}$$

These equations are said to be _uncoupled_ since each equation involves only one component of z, and hence can be solved quite independently of the rest of the system. We can also say we have _decoupled_ the original system (4). The general solutions of the uncoupled equations (5) are now found by the methods of Chapter III (or even of Section 2). We find

$$z(t) = \begin{pmatrix} z_1(t) \\ z_2(t) \end{pmatrix} = \frac{1}{20}\begin{pmatrix} -16 \cos 2t + 8 \sin 2t \\ -\cos 2t + 3 \sin 2t \end{pmatrix} + \begin{pmatrix} c_1 e^{-t} \\ c_2 e^{-6t} \end{pmatrix}, \tag{6}$$

where c_1 and c_2 are arbitrary constants. But, since

$$x(t) = Pz(t) = \begin{pmatrix} 2 & 1 \\ 1 & -2 \end{pmatrix} \begin{pmatrix} z_1(t) \\ z_2(t) \end{pmatrix} .$$

we have

$$x(t) = \frac{1}{20} \begin{pmatrix} -33 \cos 2t + 19 \sin 2t \\ -14 \cos 2t + 2 \sin 2t \end{pmatrix} + \begin{pmatrix} 2c_1 e^{-t} + c_2 e^{-6t} \\ c_1 e^{-t} - 2c_2 e^{-6t} \end{pmatrix}, \quad (7)$$

the same result as in Example 14-1.

This decoupling method is sometimes advantageous when initial conditions are also specified. For if $x(t_0) = x_0$ then $z(t_0) = P^{-1}x_0$.

For example, if together with system (4) we require

$$x(0) = x_0 = \text{col} (2,-1), \quad (8)$$

then $z(0) = P^{-1}x_0$. In the present case, using (6),

$$z(0) = \frac{1}{20} \begin{pmatrix} -16 \\ -1 \end{pmatrix} + \begin{pmatrix} c_1 \\ c_2 \end{pmatrix} = \frac{1}{5} \begin{pmatrix} 3 \\ 4 \end{pmatrix}.$$

This yields

$$c_1 = \frac{7}{5}, \quad c_2 = \frac{17}{20} .$$

Then $x(t) = Pz(t)$ becomes

$$x(t) = \begin{pmatrix} 2 & 1 \\ 1 & -2 \end{pmatrix} \frac{1}{20} \begin{pmatrix} -16 \cos 2t + 8 \sin 2t + 28e^{-t} \\ -\cos 2t + 3 \sin 2t + 17e^{-6t} \end{pmatrix}$$
$$(9)$$
$$= \frac{1}{20} \begin{pmatrix} -33 \cos 2t + 19 \sin 2t + 56e^{-t} + 17e^{-6t} \\ -14 \cos 2t + 2 \sin 2t + 28e^{-t} - 34e^{-6t} \end{pmatrix}$$

In Problem 1 you are asked to also obtain the solution of (4) with the initial conditions (8) directly from the general solution given by Eq. (7).

The decoupling method illustrated in Example 1 is

described in general as follows.

Theorem A. Let A have n linearly independent eigenvectors
so that A is diagonalizable, i.e.,

$$
P^{-1}AP = \Lambda = \begin{pmatrix} \lambda_1 & 0 & \cdots & 0 \\ 0 & \lambda_2 & \cdots & 0 \\ \vdots & \vdots & \ddots & \vdots \\ 0 & 0 & \cdots & \lambda_n \end{pmatrix}
$$

for some constant $n \times n$ matrix P. Then x is a solution
of Eq. (1) if and only if z defined by

$$
z(t) = P^{-1}x(t) \tag{10}
$$

is a solution of the uncoupled system

$$
z' = \Lambda z + P^{-1}h(t). \tag{11}
$$

And then $x(t) = Pz(t)$.

Moreover, the particular solution of Eq. (1) which also
satisfies the initial condition (2) is $x(t) = Pz(t)$ where
z is that solution of (11) which satisfies

$$
z(t_0) = P^{-1}x_0. \tag{12}
$$

The proof of this theorem involves practically nothing
that was not already done in Example 1. Thus we shall consider it clear.

In the case treated in Theorem A, the general solution
of (1) can also be written in the form

$$
x(t) = \tilde{x}(t) + c_1 e^{\lambda_1 t}\xi_{(1)} + \cdots + c_n e^{\lambda_n t}\xi_{(n)}, \tag{13}
$$

where $\xi_{(1)}, \ldots, \xi_{(n)}$ are linearly independent eigenvectors of A corresponding to the eigenvalues $\lambda_1, \ldots, \lambda_n$ respectively, and $\tilde{x}$ is any particular solution of (1). This follows from Theorem 14-D together with the fact that the functions defined by $y_{(j)}(t) = e^{\lambda_j t} \xi_{(j)}$ $(j = 1, \ldots, n)$ are linearly independent if $\xi_{(1)}, \ldots, \xi_{(n)}$ are linearly independent vectors. Verify this last assertion.

The natural question left by Example 1 and Theorem A is: How can we find the general solution of Eq. (1) or (3) if A has _fewer_ than n linearly independent eigenvectors? We shall concentrate now on homogeneous systems.

Example 2. Find the general solution of

$$y' = Ay = \begin{pmatrix} -1 & -1 \\ 1 & -3 \end{pmatrix} y. \tag{14}$$

It is easily discovered that the coefficient matrix A has only one eigenvalue, $\lambda = -2$. Thus any eigenvector ξ must satisfy

$$(A + 2I)\xi = \begin{pmatrix} 1 & -1 \\ 1 & -1 \end{pmatrix} \xi = 0.$$

This requires that $\xi = \text{col } (1,1)$, or some scalar multiple of that vector. So there do not exist two linearly independent eigenvectors.

Of course

$$y_{(1)}(t) = e^{-2t}\xi = \begin{pmatrix} e^{-2t} \\ e^{-2t} \end{pmatrix}$$

defines a solution of system (14). But how can we find a second linearly independent solution?

Perhaps the first idea that comes to mind, by analogy

with Section 12, is to try a function of the form $y(t) =$ $te^{-2t}\eta$, where η is a constant vector. But this will not work. For substitution into Eq. (14) gives $e^{-2t}\eta - 2te^{-2t}\eta =$ $Ate^{-2t}\eta$, or

$$(1 - 2t)\eta = A t\eta; \tag{15}$$

and the only way Eq. (15) can hold for various different values of t is if $\eta = 0$.

It will be found, however, that the general solution of system (14) <u>can</u> be obtained if we seek solutions of the form

$$y(t) = e^{-2t}\eta + te^{-2t}\zeta. \tag{16}$$

Indeed, substitution of (16) into (14) gives

$$-2e^{-2t}\eta + e^{-2t}\zeta - 2te^{-2t}\zeta = e^{-2t}A\eta + te^{-2t}A\zeta,$$

or, dividing out the common factor of e^{-2t},

$$-2\eta + \zeta - 2t\zeta = A\eta + tA\zeta.$$

Now equating terms involving like powers of t we find

$$(A + 2I)\zeta = 0, \tag{17}$$
$$(A + 2I)\eta = \zeta. \tag{18}$$

The solutions ζ of Eq. (17) are multiples of the eigenvector ξ. Solutions η of (18) (if there are any), are called <u>gen-eralized eigenvectors</u>. But, since $\det (A + 2I) = 0$, it is not at all clear that Eq. (18) has any solution when $\zeta \neq 0$.

If $\zeta = \xi$ then (18) becomes

$$\begin{pmatrix} 1 & -1 \\ 1 & -1 \end{pmatrix} \eta = \begin{pmatrix} 1 \\ 1 \end{pmatrix}$$

which, we find, does have solutions

$$\eta = \text{col } (1,0) + c\xi \tag{19}$$

for all values of the scalar c.   Thus, for example, we can
get two linearly independent solutions of (14) by putting
into (16)

$$\zeta = 0, \quad \eta = \xi,$$

and then

$$\zeta = \xi, \quad \eta = \text{col } (1,0)$$

These give, respectively

$$y_{(1)}(t) = \begin{pmatrix} e^{-2t} \\ e^{-2t} \end{pmatrix} \quad \text{and} \quad y_{(2)}(t) = \begin{pmatrix} (1+t)e^{-2t} \\ te^{-2t} \end{pmatrix}. \tag{20}$$

The reader should verify that $y_{(1)}$ and $y_{(2)}$ are linearly
independent.   Thus the general solution of (14) is

$$y = c_1 y_{(1)} + c_2 y_{(2)}.$$

A natural extension of this last procedure will handle
the case of a system of three equations having only one or two
linearly independent eigenvectors.   Two examples are presented
to illustrate the method.   It will be noted in Problem 4 that
because of the special form of these systems, (21) and (26),
there is actually an alternative easier way of solving them.
Nevertheless, the reader is asked to follow these simple
examples, as presented, to illustrate a general method of
solution.

Example 3.   Find three linearly independent solutions of

$$y' = \begin{pmatrix} \mu & 1 & 0 \\ 0 & \mu & 1 \\ 0 & 0 & \mu \end{pmatrix} y \tag{21}$$

The equation $\det (A-\lambda I) = 0$ becomes $(\mu-\lambda)^3 = 0$.   Thus
$\lambda = \mu$ is the only eigenvalue of   A.   Or, we might say, μ is
an eigenvalue of multiplicity three.   To find the eigenvectors
we solve $(A-\mu I)\xi = 0$, i.e.,

$$\begin{pmatrix} 0 & 1 & 0 \\ 0 & 0 & 1 \\ 0 & 0 & 0 \end{pmatrix} \xi = 0.$$

A solution is

$$\xi = \text{col} (1,0,0),$$

or any multiple of this.   But there are no other linearly independent eigenvectors.   Thus there will be a deficiency of
two linearly independent solutions of system (21).   All we
have so far is

$$y_{(1)}(t) = e^{\mu t}\xi = \text{col} (e^{\mu t},0,0).$$

Let us then   seek solutions of system (21)   in the form

$$y(t) = e^{\mu t}\eta + te^{\mu t}\zeta + t^2 e^{\mu t}\theta. \tag{22}$$

Substitution into (21) gives

$$\mu e^{\mu t}\eta + \mu te^{\mu t}\zeta + e^{\mu t}\zeta + \mu t^2 e^{\mu t}\theta + 2te^{\mu t}\theta$$

$$= e^{\mu t}A\eta + te^{\mu t}A\zeta + t^2 e^{\mu t}A\theta.$$

Now divide out the common factor $e^{\mu t}$ and equate coefficients of like powers of   t   to get

$$(A - \mu I)\theta = 0, \tag{23}$$

$$(A - \mu I)\zeta = 2\theta, \qquad (24)$$

$$(A - \mu I)\eta = \zeta. \qquad (25)$$

For θ we can take any multiple of ξ. However, since det $(A - \mu I) = 0$, it is again unclear that Eq. (24) has a solution when $\theta \neq 0$. A similar doubt arises for Eq. (25). Let us try anyway. If we take $\theta = c_1\xi$ then Eq. (24) becomes

$$\begin{pmatrix} 0 & 1 & 0 \\ 0 & 0 & 1 \\ 0 & 0 & 0 \end{pmatrix}\zeta = \begin{pmatrix} 2c_1 \\ 0 \\ 0 \end{pmatrix}$$

which does have solutions, namely the "generalized eigenvectors"

$$\zeta = \begin{pmatrix} c_2 \\ 2c_1 \\ 0 \end{pmatrix} = \begin{pmatrix} 0 \\ 2c_1 \\ 0 \end{pmatrix} + c_2\xi,$$

for any scalar constant, c_2. This, in turn, gives for Eq. (25),

$$\begin{pmatrix} 0 & 1 & 0 \\ 0 & 0 & 1 \\ 0 & 0 & 0 \end{pmatrix}\eta = \begin{pmatrix} c_2 \\ 2c_1 \\ 0 \end{pmatrix} .$$

Solutions are the "generalized eigenvectors"

$$\eta = \begin{pmatrix} c_3 \\ c_2 \\ 2c_1 \end{pmatrix} = \begin{pmatrix} 0 \\ c_2 \\ 2c_1 \end{pmatrix} + c_3\xi.$$

Putting these expressions for η, ζ, and θ into (22) we get

$$y(t) = \begin{pmatrix} (c_3+c_2t+c_1t^2)e^{\mu t} \\ (c_2+2c_1t)e^{\mu t} \\ 2c_1e^{\mu t} \end{pmatrix}.$$

Now using various different choices of the arbitrary con-

stants c_1, c_2, and c_3 we obtain three solutions,

$$y_{(1)}(t) = \begin{pmatrix} e^{\mu t} \\ 0 \\ 0 \end{pmatrix}, \quad y_{(2)}(t) = \begin{pmatrix} te^{\mu t} \\ e^{\mu t} \\ 0 \end{pmatrix}, \quad y_{(3)}(t) = \begin{pmatrix} t^2e^{\mu t} \\ 2te^{\mu t} \\ 2e^{\mu t} \end{pmatrix}.$$

The reader should verify that these are linearly independent.

The situation can actually become more complicated when

an eigenvalue of multiplicity m yields more than one but

fewer than m linearly independent eigenvectors, as in the

next example.

Example 4.   Find the general solution of

$$y' = \begin{pmatrix} \mu & 1 & 0 \\ 0 & \mu & 0 \\ 0 & 0 & \mu \end{pmatrix} y. \tag{26}$$

Again one easily finds that $\lambda = \mu$ is an eigenvalue of A

of multiplicity m = 3. The equation for the eigenvectors,

$(A - \mu I)\xi = 0$, becomes

$$\begin{pmatrix} 0 & 1 & 0 \\ 0 & 0 & 0 \\ 0 & 0 & 0 \end{pmatrix} \xi = 0.$$

This gives two linearly independent eigenvectors, for example

$$\xi_{(1)} = \text{col} \ (1,0,1) \quad \text{and} \quad \xi_{(2)} = \text{col} \ (0,0,1),$$

but no more than two.   Thus we immediately obtain two linearly independent solutions of system (26), defined by

$$y_{(1)}(t) = e^{\mu t}\xi_{(1)} \quad \text{and} \quad y_{(2)}(t) = e^{\mu t}\xi_{(2)}.$$

There is now a deficiency of only one linearly independent solution, and we might reasonably seek a third linearly independent solution in the form

$$y(t) = e^{\mu t}\eta + te^{\mu t}\zeta. \tag{27}$$

Substitution of (27) into (26) leads to

$$(A - \mu I)\zeta = 0, \tag{28}$$

$$(A - \mu I)\eta = \zeta. \tag{29}$$

(Putting $\zeta = 0$ gives again the two solutions, $y_{(1)}$ and $y_{(2)}$, already obtained.) To find a third linearly independent solution, let us put $\zeta = c\xi_{(1)}$ and try to solve Eq. (29) for η:

$$(A - \mu I)\eta = \begin{pmatrix} 0 & 1 & 0 \\ 0 & 0 & 0 \\ 0 & 0 & 0 \end{pmatrix}\eta = \begin{pmatrix} c \\ 0 \\ c \end{pmatrix}.$$

But this system has no solution when $c \neq 0$!

We try again, taking $\zeta = c\xi_{(2)}$. Then (29) becomes

$$\begin{pmatrix} 0 & 1 & 0 \\ 0 & 0 & 0 \\ 0 & 0 & 0 \end{pmatrix}\eta = \begin{pmatrix} 0 \\ 0 \\ c \end{pmatrix},$$

and again, there is no solution if $c \neq 0$.

What next?

We have not exhausted all possibilities, because we can use for ζ any eigenvector of A. In other words, we can try to satisfy Eq. (29) by using for ζ any linear combination of $\xi_{(1)}$ and $\xi_{(2)}$, say

$$\zeta = c_1 \xi_{(1)} + c_2 \xi_{(2)} = \text{col } (c_1, 0, c_1 + c_2).$$

Then (29) becomes

$$\begin{pmatrix} 0 & 1 & 0 \\ 0 & 0 & 0 \\ 0 & 0 & 0 \end{pmatrix} \eta = \begin{pmatrix} c_1 \\ 0 \\ c_1 + c_2 \end{pmatrix}.$$

A little study of this equation shows that a solution exists if and only if $c_1 + c_2 = 0$. Thus we can take $c_1 = 1$ and $c_2 = -1$ to get

$$\zeta = \text{col } (1,0,0) \quad \text{and} \quad \eta = \text{col } (0,1,0).$$

This gives, as a third solution of system (26),

$$y_{(3)}(t) = e^{\mu t} \eta + t e^{\mu t} \zeta = \begin{pmatrix} t e^{\mu t} \\ e^{\mu t} \\ 0 \end{pmatrix}.$$

The reader should verify the linear independence of $y_{(1)}$, $y_{(2)}$, and $y_{(3)}$.

Remark. The fact that every linear combination of eigenvectors of A in Example 4 is again an eigenvector is a special case of the following theorem: Let $\xi_{(1)}, \ldots, \xi_{(k)}$ be eigenvectors of a matrix A, each associated with the same eigenvalue, λ. Then any linear combination of $\xi_{(1)}, \ldots, \xi_{(k)}$ is also an eigenvector of A (associated with λ). If this is not clear, prove it.

The following theorem is presented to complete a pre-scription for solving the homogeneous system (3) with con-stant coefficients. We give the proof only for the case $n = 2$.

Theorem B.   Let the characteristic equation   det $(A - \lambda I) = 0$
have a root λ of multiplicity   m.   Let   k   be the number of
linearly independent eigenvectors associated with the eigenvalue λ.   Then one can find   m   linearly independent solutions of (3), of the form

$$y(t) = e^{\lambda t}\eta + te^{\lambda t}\zeta + \cdots + t^{m-k}e^{\lambda t}\theta$$

for some appropriate vectors $\eta, \zeta, \ldots, \theta$.   Moreover if these
solutions are found for each distinct eigenvalue, the result
will be a total of   n   linearly independent solutions of (3).

Proof for   n = 2.   If the 2×2 matrix, A, has two
linearly independent eigenvectors, $\xi_{(1)}$ and $\xi_{(2)}$, we know
that the general solution of (3) has the form

$$y(t) = c_1 e^{\lambda_1 t}\xi_{(1)} + c_2 e^{\lambda_2 t}\xi_{(2)}.$$

In case   A   has only one linearly independent eigenvec-
tor ξ, and hence a double eigenvalue λ, proceed as follows.
(This will explain why the method worked in Example 2.)   One
solution is given by $e^{\lambda t}\xi$.   We seek a second linearly independent solution in the form

$$e^{\lambda t}\eta + te^{\lambda t}\zeta, \tag{30}$$

as in Eq. (16).   Of course, for linear independence, we must
have $\zeta \neq 0$.   As in Example 2, the expression (30) satisfies
Eq. (3) if and only if

$$(A - \lambda I)\zeta = 0 \tag{31}$$

and

$$(A - \lambda I)\eta = \zeta. \tag{32}$$

To construct a nontrivial solution of (31) and (32),
let η be any nonzero vector which is not an eigenvector.
Then define ζ by Eq. (32). It follows that $\zeta \neq 0$. Why?
We will find that ζ is an eigenvector and hence satisfies
Eq. (31).

We first show that ζ and η are linearly independent.
Suppose they were not. Then we would have $\zeta = c\eta$ for some
$c \neq 0$. Why? Hence, from (32),

$$(A - \lambda I - cI)\eta = 0.$$

But this would say that η was an eigenvector and $\lambda + c$ an
eigenvalue for A, which is a contradiction.

The fact that ζ and η are linearly independent means
that

$$\begin{vmatrix} \zeta_1 & \eta_1 \\ \zeta_2 & \eta_2 \end{vmatrix} \neq 0.$$

Hence we can solve, for c_1 and c_2, the equations

$$c_1\zeta_1 + c_2\eta_1 = \xi_1,$$
$$c_1\zeta_2 + c_2\eta_2 = \xi_2,$$

where ξ is our original given eigenvector. Then

$$\xi = c_1\zeta + c_2\eta. \tag{33}$$

Since η is not an eigenvector, it follows that $c_1 \neq 0$ in
(33). Multiplying Eq. (33) by $A - \lambda I$ one now finds

$$0 = c_1(A - \lambda I)\zeta + c_2\zeta$$

or

$$(A - \lambda I + \frac{c_2}{c_1} I)\zeta = 0.$$

This says that $\lambda - c_2/c_1$ is an eigenvalue of A. Hence
$c_2 = 0$ and, from (33), $\xi = c_1\zeta$ so that ζ is an eigenvector. Thus (31) and (32) are both satisfied. $\square$

Deeper results of matrix analysis, which are beyond the
scope of this text, would be of great value if we wanted to
study systems of high order. But most systems encountered in
applications are of second or third order (or occasionally
fourth). And then the methods indicated in the examples of
this section should be adequate for the homogeneous system
with constant coefficients.

For a nonhomogeneous system (1), in case A is diagonalizable, Theorem A shows how to decouple the system into n
independent linear equations of first order. Thus in that
case the problem is completely solved, in principle.

When A is not diagonalizable we have concentrated on
the homogeneous system (3). Having solved the homogeneous
system we can then solve certain nonhomogeneous systems by
the "method of undetermined coefficients". This procedure
for finding a particular solution was illustrated in Example
14-1 and Problem 14-5 and will be needed again in Problem
5(c) below. It depends upon successfully guessing an appropriate form for a particular solution of the nonhomogeneous
system. This will usually be possible if the vector function
h in Eq. (1) is a simple combination of polynomials, exponentials, sines, and cosines.

A very general procedure for handling nonhomogeneous
systems will be presented in Section 17 under the heading
"variation of parameters".

Problems.

1. Use the general solution (7) of system (4) to find that
    solution which also satisfies the initial condition (8),
    i.e., find the appropriate values of c_1 and c_2. Com-
    pare your answer with Eq. (9).

2. If $A = \begin{pmatrix} -1 & -1 \\ 1 & -3 \end{pmatrix}$, as in Example 2, prove that these can-

    not be any matrix P such that

    $$P^{-1}AP = \begin{pmatrix} -2 & 0 \\ 0 & -2 \end{pmatrix}.$$

*3. If A is as in Problem 2, show that there cannot be any
     matrix P such that

     $$P^{-1}AP = \begin{pmatrix} a & 0 \\ 0 & b \end{pmatrix}, \quad \text{any diagonal matrix.}$$

     Hint: First show that the eigenvalues of any such matrix
     $P^{-1}AP$ are the same as those of A. This requires know-
     ledge of a theorem from matrix theory which asserts that
     det (AB) = (det A)(det B) for any two n × n matrices
     A and B.

4. (a) Solve system (21) by first solving its third equa-
             tion, which is "uncoupled", for y_3. Then solve the
             second equation for y_2, and finally the first equa-
             tion for y_1.
    (b) Same for system (26).

5. Find the general solutions of the following systems.

    (a) $x' = \begin{pmatrix} 2 & -1 \\ -2 & 3 \end{pmatrix} x + \begin{pmatrix} 4e^{4t} + 2t \\ e^{4t} + 2t \end{pmatrix}.$

(b) $y' = \begin{pmatrix} 1 & 2 \\ -1 & -1 \end{pmatrix} y$. (c) $x' = \begin{pmatrix} -1 & -1 \\ 1 & -3 \end{pmatrix} x + \begin{pmatrix} t-2e^{-t} \\ 1+e^{-t} \end{pmatrix}$.

You may use the result of Example 2.

(d) $y' = \begin{pmatrix} -1 & 0 & 3 \\ -8 & 1 & 12 \\ -2 & 0 & 4 \end{pmatrix} y$. (e) $y' = \begin{pmatrix} -1 & 0 & 3 \\ -8 & 1 & 11 \\ -2 & 0 & 4 \end{pmatrix} y$.

(f) $y' = \begin{pmatrix} -1 & 1 & 0 \\ -1 & -4 & 1 \\ -1 & -2 & -1 \end{pmatrix} y$. (g) $y' = \begin{pmatrix} -1 & 2 & -1 \\ -1 & -4 & 1 \\ -1 & -2 & -1 \end{pmatrix} y$.

(h) $y' = \begin{pmatrix} 0 & 1 & 2 \\ 1 & 0 & 3 \\ -1 & -1 & -3 \end{pmatrix} y$.

6. Find the particular solutions of Problem 5(a)-(e) which satisfy respectively the initial conditions

   (a) $x(0) = \text{col} (2,-1)$.   (b) $y(\pi) = \text{col} (-1,1)$.

   (c) $x(0) = \text{col} (2,0)$.   (d) and (e) $y(0) = \text{col} (1,2,3)$.

7. Solve each of the following scalar equations first by the method of Chapter III and then by converting to an equivalent system through $y_1 = z$ and $y_2 = z'$. Compare the results.

   (a) $z'' + 4z' + 3z = 0$.
   (b) $z'' + 4z' + 4z = 0$.

8. The given 3×3 matrix in Problem 5(d) has only two (distinct) eigenvalues, and yet has three linearly independent eigenvectors. Prove that it is impossible for any nondiagonal $n \times n$ matrix to have n linearly independent eigenvectors if it has only one (distinct) eigenvalue.

16.   OSCILLATIONS AND DAMPING IN APPLICATIONS

Systems of linear equations with constant coefficients arise naturally in the analysis of mechanical systems and electrical circuits.   However, the resulting differential equations may be rather cumbersome to solve exactly, even though such solution is actually possible.   Thus it would be useful to have ways of gaining _some_ information without carrying out the complete solution.

In the analysis of a system

$$x' = Ax + h(t) \tag{1}$$

with initial conditions

$$x(t_0) = x_0, \tag{2}$$

it is useful to first study the associated homogeneous system

$$y' = Ay. \tag{3}$$

We shall continue in this section to concentrate our attention on system (3) where A is a constant $n \times n$ matrix with real elements.   Various conclusions about the behavior of solutions of (3) can be obtained by examining the eigenvalues of A.   Again we take $J = \mathbb{R}$.

Theorem A.   Let ρ be any number such that

$$\text{Re } \lambda < \rho$$

for every eigenvalue λ of A.   Then for each solution y of (3) there exists some constant $K > 0$ such that

$$||y(t)|| \le Ke^{\rho t} \quad \text{for} \quad t \ge 0.$$

Proof. Choose $\delta > 0$ such that

$$\text{Re } \lambda < \rho - \delta$$

for every eigenvalue λ. Then if p is any polynomial we
will have $|p(t)| \leq Be^{\delta t}$ for all $t \geq 0$ for some $B > 0$.
Why? Thus for any eigenvalue

$$|e^{\lambda t}p(t)| \leq e^{\text{Re } \lambda t}Be^{\delta t} < Be^{\rho t} \quad \text{for } t \geq 0.$$

The assertion of the theorem then follows from Theorem 15-B.

Of course, since we proved Theorem 15-B only for the
case $n = 2$, the present proof is complete only for the case
$n = 2$. □

Corollary B. Every solution of system (3) tends to zero as
$t \to \infty$ if and only if every eigenvalue of A has negative
real part, i.e., Re $\lambda < 0$ for every eigenvalue λ. (In this
case A is called a stable matrix.)

The proof is left to the reader as Problem 1.

Corollary C. If A has n linearly independent eigenvectors and if Re $\lambda \leq 0$ for every eigenvalue λ of A, then
every solution of (3) remains bounded as $t \to \infty$.

The proof is Problem 2.

In order to apply Corollaries B and C we require some
way of determining the algebraic sign of the real parts of
the eigenvalues. Since the eigenvalue of a matrix A are
solutions of the polynomial equation det $(A - \lambda I) = 0$, our
problem is to study the zeros of a polynomial. If we are

interested in systems of 2, 3, or 4 linear, first order
equations the following Lemmas D, E, and F will be useful.

Lemma D.   The roots of the quadratic equation

$$\lambda^2 + a_1 \lambda + a_0 = 0 \tag{4}$$

with real coefficients (a_1 and a_0) both have negative real
parts if and only if

$$a_1 > 0 \quad \text{and} \quad a_0 > 0. \tag{5}$$

Proof.   Let λ_1 and λ_2 be the roots of Eq. (4).
Then $\lambda^2 + a_1 \lambda + a_0 = (\lambda - \lambda_1)(\lambda - \lambda_2)$, and so

$$\lambda_1 + \lambda_2 = -a_1 \quad \text{and} \quad \lambda_1 \lambda_2 = a_0.$$

If λ_1 and λ_2 are both real (possibly identical),
then it is easily seen that they are both negative if and
only if $a_1 > 0$ and $a_0 > 0$.

If λ_1 and λ_2 are not both real, then they must be
complex conjugates, say $\lambda_2 = \bar{\lambda}_1$. Thus

$$2 \operatorname{Re} \lambda_1 = -a_1 \quad \text{and} \quad |\lambda_1|^2 = a_0.$$

So it follows that $\operatorname{Re} \lambda_1 < 0$ if and only if $a_1 > 0$. And
in this case $a_0 > 0$ also. $\square$

Lemma E.   The roots of the cubic equation

$$\lambda^3 + a_2 \lambda^2 + a_1 \lambda + a_0 = 0 \tag{6}$$

with real coefficients all have negative real parts if and
only if

$$a_2 > 0, \quad a_1 > 0, \quad a_0 > 0, \quad \text{and} \quad a_2 a_1 > a_0. \qquad (7)$$

Proof.   Since the coefficients in Eq. (6) are all real, it follows that any complex roots must occur in conjugate pairs.   Thus there must be at least one real root, call it λ_1.   Hence we can factor the polynomial in Eq. (6) and write

$$(\lambda^2 + b_1 \lambda + b_0)(\lambda + c_0) = 0 \qquad (8)$$

where b_1, b_0, and $c_0 = -\lambda_1$ are real.   It follows that

$$a_2 = b_1 + c_0, \quad a_1 = b_0 + b_1 c_0, \quad \text{and} \quad a_0 = b_0 c_0. \quad (9)$$

From these we compute

$$
\begin{aligned}
a_2 a_1 - a_0 &= b_1 a_1 + c_0 a_1 - b_0 c_0 \\
&= b_1 (a_1 + c_0^2).
\end{aligned}
\qquad (10)
$$

Now let conditions (7) hold.   Then it follows from Eq. (10) that $b_1 > 0$.   Also, from (9), the condition $a_0 = b_0 c_0$ shows that b_0 and c_0 have like signs.   Now we could not have $b_0 < 0$ and $c_0 < 0$, or we would violate $b_0 + b_1 c_0 = a_1 > 0$.   Thus we conclude that $b_0 > 0$ and $c_0 > 0$.

The positivity of c_0 implies $\lambda_1 < 0$.   And, by Lemma D, the positivity of b_1 and b_0 guarantees that the remaining two roots of (6), alias (8), have negative real parts.

The proof of the converse part of the lemma is Problem 3.   □

Lemma F.   The roots of the quartic equation

$$\lambda^4 + a_3\lambda^3 + a_2\lambda^2 + a_1\lambda + a_0 = 0 \tag{11}$$

with real coefficients all have negative real parts if and only if

$$a_3 > 0, \quad a_2 > 0, \quad a_1 > 0, \quad a_0 > 0, \quad \text{and} \tag{12}$$

$$a_3 a_2 a_1 > a_3^2 a_0 + a_1^2.$$

Proof. Since any complex roots of Eq. (11) must occur in conjugate pairs, we can factor the left hand side of (11) and get

$$(\lambda^2 + b_1\lambda + b_0)(\lambda^2 + c_1\lambda + c_0) = 0, \tag{13}$$

where b_1, b_0, c_1, and c_0 are all real. It follows that

$$a_3 = b_1 + c_1, \qquad a_2 = b_1 c_1 + b_0 + c_0,$$

$$a_1 = b_0 c_1 + b_1 c_0, \quad \text{and} \quad a_0 = b_0 c_0. \tag{14}$$

From these we can compute

$$(a_3 a_2 - a_1)a_1 - a_3^2 a_0$$

$$= (b_1^2 c_1 + b_1 b_0 + b_1 c_1^2 + c_1 c_0)(b_0 c_1 + b_1 c_0)$$

$$\quad - (b_1^2 + 2b_1 c_1 + c_1^2)b_0 c_0$$

$$= b_1 c_1(b_0^2 - 2b_0 c_0 + c_0^2 + b_1 b_0 c_1 + b_0 c_1^2 + b_1^2 c_0 + b_1 c_1 c_0),$$

or

$$a_3 a_2 a_1 - a_1^2 - a_3^2 a_0 = b_1 c_1[(b_0 - c_0)^2 + a_1 a_3]. \tag{15}$$

Now let conditions (12) hold. Then it follows from (15) that $b_1 c_1 > 0$ and from (14) that $b_1 + c_1 = a_3 > 0$. These can both hold only if $b_1 > 0$ and $c_1 > 0$. It also follows from (14) that $b_0 c_0 = a_0 > 0$ which means b_0 and c_0 must have the same sign. But b_0 and c_0 cannot both be negative since $b_0 c_1 + b_1 c_0 = a_1 > 0$. Thus we must have $b_0 > 0$ and $c_0 > 0$. It now follows from Lemma D that all the roots of (13), and hence of (11), have negative real parts.

Again the proof of the converse is left to the reader. Problem 4. □

The unmotivated and mysterious looking criteria given in Lemmas D, E, and F are actually special cases of a systematic test which applies to polynomials of arbitrary order. If you ever need to know the nature of the zeros of a polynomial of degree $n \geq 5$, the general criteria, which includes Lemmas D, E, and F as special cases, can be found in books on matrix theory under the heading "Routh-Hurwitz" criteria.

Imagine that the homogeneous system (3) represents some real physical problem. Then we should expect all solutions to be "damped" out, i.e. to approach zero as $t \to \infty$, because of loss of energy due to friction. Such damping was indeed observed in the equations for the simple mechanical and electrical problems of Examples 10-1 and 12-4. Let us determine whether solutions also tend to zero in a more complicated case.

Example 1. Let two masses, m_1 and m_2, be suspended (and bouncing) on idealized weightless springs hanging from a fixed support as shown in Figure 1. Let $z_1(t)$ and $z_2(t)$ be their respective distances, at time t, below their equilibrium positions. Let k_1 and k_2 be the two spring

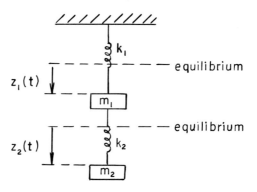

Figure 1

constants and let b_1 and b_2 be coefficients of frictional resistance for the two masses. These are all positive numbers. Then, by an argument like that in Example 6-1, one finds

$$m_1 z_1'' = - k_1 z_1 - k_2 z_1 + k_2 z_2 - b_1 z_1'$$
$$m_2 z_2'' = k_2 z_1 - k_2 z_2 - b_2 z_2'. \tag{16}$$

To analyze this system using the methods of Section 15 we should change it into an equivalent system of first order equations.

Let

$$y_1 = z_1, \quad y_2 = z_1', \quad y_3 = z_2, \quad y_4 = z_2'.$$

Then

$$y_1' = y_2$$

$$y_2' = -\frac{k_1 + k_2}{m_1}y_1 - \frac{b_1}{m_1}y_2 + \frac{k_2}{m_1}y_3$$

$$y_3' = y_4$$

$$y_4' = \frac{k_2}{m_2}y_1 - \frac{k_2}{m_2}y_3 - \frac{b_2}{m_2}y_4.$$

(17)

Computation of the general solution of this system would be straightforward but tedious.   Let us be satisfied with an examination of the eigenvalues.

The characteristic equation for system (17) is

$$-\begin{vmatrix} -\lambda & 1 & 0 & 0 \\ \dfrac{k_1 + k_2}{m_1} & -\dfrac{b_1}{m_1} - \lambda & \dfrac{k_2}{m_1} & 0 \\ 0 & 0 & -\lambda & 1 \\ \dfrac{k_2}{m_2} & 0 & -\dfrac{k_2}{m_2} & -\dfrac{b_2}{m_2} - \lambda \end{vmatrix} = 0 \quad (18)$$

which gives (Problem 5)

$$\lambda^4 + (\frac{b_1}{m_1} + \frac{b_2}{m_2})\lambda^3 + (\frac{k_1+k_2}{m_1} + \frac{k_2}{m_2} + \frac{b_1b_2}{m_1m_2})\lambda^2$$

$$+ (\frac{b_1}{m_1}\frac{k_2}{m_2} + \frac{k_1+k_2}{m_1}\frac{b_2}{m_2})\lambda + \frac{k_1k_2}{m_1m_2} = 0.$$

(19)

Now if friction means what we want it to mean then, according to Corollary B, we would expect all solutions of Eq. (19) to have negative real parts.   Let us apply Lemma F to check this.

The condition that a_3, a_2, a_1, and a_0 all be positive is clearly satisfied by Eq. (19).   To meet the last requirement of (12) the reader should verify (Problem 6) that

$$(a_3 a_2 - a_1)a_1 - a_3^2 a_0$$

$$= (\frac{b_1 k_1}{m_1^2} + \frac{b_1 k_2}{m_1^2} + \frac{b_1^2 b_2}{m_1^2 m_2} + \frac{b_2 k_2}{m_2^2} + \frac{b_1 b_2^2}{m_1 m_2^2})(\frac{b_1 k_2}{m_1 m_2} + \frac{k_1 + k_2}{m_1} \frac{b_2}{m_2})$$

$$- (\frac{b_1^2}{m_1^2} + \frac{2 b_1 b_2}{m_1 m_2} + \frac{b_2^2}{m_2^2})\frac{k_1 k_2}{m_1 m_2}$$

$$= \frac{b_1 b_2}{m_1 m_2}\frac{k_1}{m_1} - \frac{k_2}{m_2})^2 + \frac{b_1 k_2}{m_1^3 m_2}(2 b_2 k_1 + b_1 k_2 + b_2 k_2)$$

$$+ \frac{b_1 b_2}{m_1^3 m_2^2}(b_1^2 k_2 + b_1 b_2 k_1 + b_1 b_2 k_2) + \frac{b_2^2 k_2^2}{m_1 m_2^3}$$

$$+ \frac{b_1 b_2^2}{m_1^2 m_2^3}(b_1 k_2 + b_2 k_1 + b_2 k_2), \tag{20}$$

which is clearly positive.

An alternative, possibly easier, method for treating
such a system will be presented in Section 32.

Example 2.   Sometimes one considers the idealized case of
Example 1 without friction, i.e., with $b_1 = 0$ and $b_2 = 0$.
Then we might expect, on the basis of "physical intuition",
that solutions would be undamped oscillations of some sort.
Again this can be checked by considering the characteristic
equation (19) which now reduces to

$$\lambda^4 + (\frac{k_1 + k_2}{m_1} + \frac{k_2}{m_2})\lambda^2 + \frac{k_1 k_2}{m_1 m_2} = 0. \tag{21}$$

This is a quadratic equation in λ^2 and the discriminant is

$$(\frac{k_1 + k_2}{m_1} + \frac{k_2}{m_2})^2 - 4\frac{k_1 k_2}{m_1 m_2} = (\frac{k_1}{m_1} - \frac{k_2}{m_2})^2 + 2\frac{k_1 k_2}{m_1^2} + \frac{k_2^2}{m_1^2} + 2\frac{k_2^2}{m_1 m_2} > 0.$$

Hence the values of λ^2 satisfying Eq. (21) must be real and distinct. But λ^2 cannot be positive since the coefficients of Eq. (21) are all positive. We conclude that there are two negative values of λ^2 which satisfy Eq. (21). Thus the roots, λ, are distinct pairs of conjugate pure imaginary numbers,

$$\lambda = \pm i\omega_1, \quad \pm i\omega_2.$$

Corollary C then asserts that the solutions of system (16) or (17) are bounded.

Example 3. The electrical currents $x_1(t)$ and $x_2(t)$ flowing in the network of Figure 2 satisfy the differential

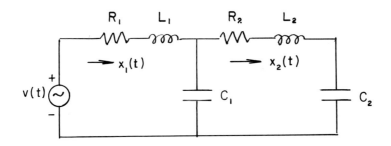

Figure 2

equations

$$L_1 x_1'' + R_1 x_1' + \frac{1}{C_1} x_1 - \frac{1}{C_1} x_2 = v'(t)$$

$$L_2 x_2'' + R_2 x_2' + (\frac{1}{C_1} + \frac{1}{C_2}) x_2 - \frac{1}{C_1} x_1 = 0,$$

(22)

where L_1, L_2, R_1, R_2, C_1, and C_2 are given positive constants. But in the case $v(t) \equiv 0$ this system is equivalent to (16). To see this, one merely sets $x_1 = z_2$, $x_2 = z_1$, $L_1 = m_2$, $L_2 = m_1$, $R_1 = b_2$, $R_2 = b_1$, $1/C_1 = k_2$, and $1/C_2 = k_1$.

Thus, without any further discussion, we conclude that all solutions of system (22) with $v(t) \equiv 0$ tend to zero as $t \to \infty$.

Example 4.   Consider the two electrical circuits of Figure 3 which are coupled only by the "mutual inductance" M between

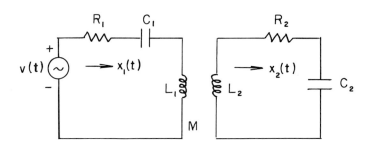

Figure 3

their two coils.   The currents $x_1(t)$ and $x_2(t)$ flowing in these circuits satisfy

$$L_1 x_1'' + M x_2'' + R_1 x_1' + \frac{1}{C_1} x_1 = v'$$

$$M x_1'' + L_2 x_2'' + R_2 x_2' + \frac{1}{C_2} x_2 = 0,$$

(23)

where M, L_1, L_2, R_1, R_2, C_1, and C_2 are given positive constants with $M^2 < L_1 L_2$.   Note that the equations of system (23) are not "solved" for x_1'' and x_2''.   It is left as Problem 7, for the reader to transform (23) into a system of four equations of first order, and then show that if $v(t) \equiv 0$ all solutions tend to zero as $t \to \infty$.

In conclusion let us add the following observation about the role of the imaginary parts of the eigenvalues of A.

Theorem G. If Im $\lambda \neq 0$ for some eigenvalue λ of A, then system (3) has oscillatory solutions. This means that some component of some solution changes sign infinitely often as $t \to \infty$.

The proof is Problem 8.

This last theorem serves in particular to assure the existence of oscillatory solutions in Example 2. The reader may be able to do better than this. It is not too difficult to show in that case that every component of every nontrivial solution changes sign infinitely often as $t \to \infty$.

Problems

1. Prove Corollary B.

2. Prove Corollary C.

3. Complete the proof of Lemma E by showing that conditions (7) are necessary conditions for the roots of Eq. (6) to have negative real parts.

4. Complete the proof of Lemma F by showing the necessity of conditions (12).

5. Verify that Eqs. (19) and (18) are equivalent.

6. Verify Eqs. (20).

7. Transform (23) into a system of four equations of first order. Then prove that if $v(t) \equiv 0$ all eigenvalues for this system have negative real parts.

8. Prove Theorem G.

17.   VARIATION OF PARAMETERS

Let us now return to the general linear system

$$x' = A(t)x + h(t),$$ (1)

where A is a given continuous n × n-matrix-valued function
and h is a given continuous n-vector-valued function on an
interval J = (α,β).   Let us assume an initial condition of
the form

$$x(t_0) = x_0,$$ (2)

where $t_0 \in J$ and x_0 is an n-vector.

In Section 13 we studied the corresponding problem for
a scalar equation of order n.   The important method introduced
there was the method of variation of parameters (or variation
of constants).   This provided a systematic procedure for solving a nonhomogeneous equation, assuming we could first find
the general solution of the associated homogeneous equation.
The procedure was summarized in Theorem 13-A.

We will soon obtain an analogous method for the system
(1) of n first order equations.   More precisely, we will express the solutions of (1) in terms of the solutions of the
associated homogeneous system

$$y' = A(t)y,$$ (3)

provided we can find the general solution of (3).   Thus we
need n linearly independent solutions of (3).

If A is a constant n × n matrix, the methods of Section 15 will provide n linearly independent solutions of (3).

It will be proved in Section 20 that whenever A is
continuous, then (3) does have n linearly independent

solutions.   It may, however, be difficult to find these solutions.   For the purposes of this section, let us assume that
A is continuous and that system (3) has n linearly independent solutions on J.

    Before presenting the general solution of Eqs. (1) and
(2) by "variation of parameters", let us preview the result
by reconsidering Example 13-1.   That example treated the
scalar linear equation

$$x' = -a(t)x + h(t),\tag{4}$$

with initial condition

$$x(t_0) = x_0.\tag{5}$$

The homogeneous equation associated with (4) is

$$y' = -a(t)y.\tag{6}$$

Let us now define

$$y(t;s,c) \equiv ce^{-\int_s^t a(u)\,du}.$$

Thus $y(\cdot;s,c)$ is that solution of Eq. (6) which satisfies
the initial condition $y(s) = y(s;s,c) = c$.   With this notation the solution of (4) and (5), given by Eq. 13-(9), is

$$x(t) = y(t;t_0,x_0) + \int_{t_0}^t y(t;s,h(s))\,ds.\tag{7}$$

    We can interpret (7) as follows.   To obtain the solu-
tion of the nonhomogeneous Eq. (4) with initial condition
$x(t_0) = x_0$, one begins with the solution of the homogeneous
equation with the same initial condition, $y(\cdot;t_0,x_0)$.   To this
one adds the infinite "sum" (i.e., integral) of the solutions
of the homogeneous equation with initial values $h(s)\,ds$ at

s   for all values of   s   from t_0 up to   t.   Note that for
any number Δs, $y(t;s,h(s)\Delta s) = y(t;s,h(s))\Delta s$.

In other words, the forcing function, h, plays the role
of many little boosts or additional "initial conditions",
h(s)ds   at time   s, to the solution of the homogeneous equation.

As a matter of fact, this last observation could be
used as the basis for a proof that (7) is the solution of Eqs.
(4) and (5).   However such a proof would be much more complicated than the one given in Example 13-1.

Let us now proceed for systems by first conjecturing
(and then proving) that the solution of (1) and (2) is analogous to that given by (7) for the scalar case.   As indicated
following Eq. (3), throughout this section we assume the
existence of   n   linearly independent solutions of Eq. (3) on
J.   Thus for any given point $(s,\xi) \in J \times \mathbb{R}^n$, we can find the
unique solution   y   of Eq. (3) on   J   which satisfies
$y(s) = \xi$.   We shall denote this solution by $y(\cdot;s,\xi)$.

Theorem A   (Variation of Parameters).   The (unique) solution
of Eqs. (1) and (2) on   J   is given by

$$x(t) = y(t;t_0,x_0) + \int_{t_0}^{t} y(t;s,h(s))ds. \tag{8}$$

Proof.   Let us introduce

$$z(t) \equiv \int_{t_0}^{t} y(t;s,h(s))ds. \tag{9}$$

It will suffice to show that   z   is a solution of Eq. (1).
For then it will follow from superposition (Theorem 14-A) that
Eq. (8) defines another solution of (1), and, from (8),

$$x(t_0) = y(t_0;t_0,x_0) + 0 = x_0.$$

Hence (8) will define <u>the</u> solution of (1) and (2).

To show that z satisfies Eq. (1) we could either substitute directly (which would involve differentiating under
the integral sign in (9)), or we could try to show that z
satisfies the integral equation

$$z(t) = \int_{t_0}^{t} [A(u)z(u) + h(u)]du, \tag{10}$$

from which $z'(t) = A(t)z(t) + h(t)$ follows. We elect the
latter procedure because it will more easily generalize to
delay differential equations later. The calculation involves
interchanging the order of an iterated integral. The appropriate theorem can be found for scalar functions (such as the
components of a vector function) in Corollary A2-N.

To verify Eq. (10) we use (9), the definition of z,
to compute

$$\int_{t_0}^{t} [A(u)z(u) + h(u)]du$$

$$= \int_{t_0}^{t}\int_{t_0}^{u} A(u)y(u;s,h(s))ds\,du + \int_{t_0}^{t} h(u)du$$

$$= \int_{t_0}^{t}\int_{s}^{t} y'(u;s,h(s))du\,ds + \int_{t_0}^{t} h(u)du$$

$$= \int_{t_0}^{t} [y(t;s,h(s)) - h(s)]ds + \int_{t_0}^{t} h(u)du$$

$$= z(t). \quad \square$$

Apart from the analogy between Eqs. (8) and (7), it remains unclear why Theorem A is called variation of parameters.
In Problem 1 the reader is asked to complete an alternate

proof of Theorem A which resembles the method used in Section
13.   This should provide a better understanding for the name
of Theorem A.

Now if we intend to use Eq. (8) to explicitly find solutions of Eqs. (1) and (2) we should have some convenient representation for $y(t;s,\xi)$.

Let us define $I_{(j)} = \text{col }(0,\ldots,0,1,0,\ldots,0)$, the
column n-vector with each component 0 except the j'th
which is 1.   In other words $I_{(j)}$ is the j'th column of
the $n \times n$ identity matrix, I.   Now suppose we have solved
Eq. (3) with the initial condition $y(s) = I_{(j)}$ for each
$j = 1,\ldots,n$, obtaining the n vector functions

$$y(\cdot;s,I_{(1)}), \quad \ldots, \quad y(\cdot;s,I_{(n)}).$$

Let us define the <u>transition matrix</u> $T(\cdot;s)$ to be the $n \times n$
matrix whose columns are the vector functions listed above,
i.e.,

$$T(t,s) \equiv \left(y(t;s,I_{(1)}) \quad \cdots \quad y(t;s,I_{(n)})\right) \tag{11}$$

for all t and s in J.

The usefulness of the transition matrix, as we show
next, is that for each $\xi \in R^n$

$$y(t;s,\xi) = T(t;s)\xi \tag{12}$$

for all t and s in J.   To prove this, note that

$$T(t;s)\xi = \xi_1 y(t;s,I_{(1)}) + \cdots + \xi_n y(t;s,I_{(n)}).$$

This is a linear combination of solutions of Eq. (3), and
hence is also a solution of (3) itself.   Moreover $T(s;s)\xi =$
$I\xi = \xi$.   So the unique solution of Eq. (3) with $y(s) = \xi$

is given by $T(t;s)\xi$, which proves (12).

We can now rewrite (8), the solution of Eqs. (1) and (2), in the equivalent form

$$x(t) = T(t;t_0)x_0 + \int_{t_0}^{t} T(t;s)h(s)ds. \tag{13}$$

Example 1. Use Eq. (13) to re-solve the system of Example 15-1,

$$x' = \begin{pmatrix} -2 & 2 \\ 2 & -5 \end{pmatrix} x + \begin{pmatrix} 5 \sin 2t \\ 0 \end{pmatrix} \tag{14}$$

on $\mathbb{R}$, with

$$x(0) = x_0 = \text{col } (2,-1). \tag{15}$$

Proceeding as in Example 14-1, we obtain the general solution of the associated homogeneous system, $y' = Ay$, in the form

$$y(t) = c_1 \begin{pmatrix} 2e^{-t} \\ e^{-t} \end{pmatrix} + c_2 \begin{pmatrix} e^{-6t} \\ -2e^{-6t} \end{pmatrix} .$$

To obtain $y(t;s,I_{(1)})$ we must choose c_1 and c_2 so that $y(s) = \text{col } (1,0)$, i.e., so that

$$2e^{-s}c_1 + e^{-6s}c_2 = 1$$

$$e^{-s}c_1 - 2e^{-6s}c_2 = 0.$$

This gives $c_1 = (2/5)e^{s}$ and $c_2 = (1/5)e^{6s}$. Thus

$$y(t;s,I_{(1)}) = \frac{1}{5} \begin{pmatrix} 4e^{-(t-s)} + e^{-6(t-s)} \\ 2e^{-(t-s)} - 2e^{-6(t-s)} \end{pmatrix} .$$

By a similar computation, the reader should find $y(t;s,I_{(2)})$ and hence

$$T(t;s) = \frac{1}{5} \begin{pmatrix} 4e^{-(t-s)} + e^{-6(t-s)} & 2e^{-(t-s)} - 2e^{-6(t-s)} \\ 2e^{-(t-s)} - 2e^{-6(t-s)} & e^{-(t-s)} + 4e^{-6(t-s)} \end{pmatrix}.$$

Substituting this into Eq. (13) we have

$$x(t) = \frac{1}{5} \begin{bmatrix} 6e^{-t} + 4e^{-6t} \\ 3e^{-t} - 8e^{-6t} \end{bmatrix}$$

$$+ \begin{pmatrix} \displaystyle\int_{t_0}^{t} [4e^{-(t-s)} + e^{-6(t-s)}]\sin 2s\, ds \\ \displaystyle\int_{t_0}^{t} [2e^{-(t-s)} - 2e^{-6(t-s)}]\sin 2s\, ds \end{pmatrix} \qquad (16)$$

as the solution of (14) and (15).   The reader should perform the indicated integrations in (16) and compare this with the solution obtained previously, namely Eq. 15-(9).   Problem 2.

For a linear system with constant coefficients, such as Eq. (14), we make the following, small labor-saving observation.   If $A(t) = A$ is constant then, as we shall prove below, the transition matrix for Eq. (1) or (3) has the property

$$T(t;s) = T(t-s;0). \qquad (17)$$

So to find $T(t;s)$, in this case, it is simplest to first compute $T(t,0)$, and then apply (17).

Equation (17) is valid when A is constant because, for any $\xi \in \mathbb{R}^n$, both $y(t;s,\xi)$ and $y(t-s;0,\xi)$ are solutions of (3) which take the value ξ when $t = s$.   Hence, by uniqueness, $y(t;s,\xi) = y(t-s;0,\xi)$.   In particular, $y(t;s,I_{(j)}) = y(t-s;0,I_{(j)})$ for each $j = 1,...,n$.   This implies (17).

Example 2.   Find the transition matrix $T(t;s)$ for the system of Example 15-2,

$$y' = \begin{pmatrix} -1 & -1 \\ 1 & -3 \end{pmatrix} y. \tag{18}$$

Recall that this matrix, A, has only one eigenvalue, $\lambda = -2$ and only one linearly independent eigenvector. Nevertheless, we found in Example 15-2 that the general solution is given by

$$y(t) = c_1 \begin{pmatrix} e^{-2t} \\ e^{-2t} \end{pmatrix} + c_2 \begin{pmatrix} (1+t)e^{-2t} \\ te^{-2t} \end{pmatrix}. \tag{19}$$

We shall proceed by first finding $T(t;0)$. The condition $y(0) = I_{(1)}$ gives $c_1 = 0$ and $c_2 = 1$. Substitution of these values into Eq. (19) yields $y(t;0,I_{(1)})$, the first column of $T(t,0)$. Similarly, to get $y(t;0,I_{(2)})$ we require $c_1 = 1$ and $c_2 = -1$ in Eq. (19). Thus

$$T(t;0) = \begin{pmatrix} (1+t)e^{-2t} & -te^{-2t} \\ te^{-2t} & (1-t)e^{-2t} \end{pmatrix}.$$

Finally we apply Eq. (17) to find

$$T(t;s) = \begin{pmatrix} (1+t-s)e^{-2(t-s)} & -(t-s)e^{-2(t-s)} \\ (t-s)e^{-2(t-s)} & (1-t+s)e^{-2(t-s)} \end{pmatrix} \tag{20}$$

for all t and s in $\mathbb{R}$.

Problems.

1.   A "variation-of-parameters" derivation of Theorem A: Let $y_{(1)}, \ldots, y_{(n)}$ be any n linearly independent solutions of (3), $y' = A(t)y$, on J; and form a matrix-valued function Y by using $y_{(1)}, \ldots, y_{(n)}$ as columns: $Y(t) \equiv (y_{(1)}(t) \cdots y_{(n)}(t))$. Any such Y is called a fundamental matrix for Eq. (3). For example, $T(t;s)$ is a

fundamental matrix for each choice of s.

(a) Verify that $Y(t)^{-1}$ exists for all t in J.

(b) Verify that y is a solution of Eq. (3) if and only
     if $y(t) = Y(t)\xi$ for some ξ in $\mathbb{R}^n$.

(c) Show that for each t and s in J and each ξ
     in $\mathbb{R}^n$, $y(t;s,\xi) = Y(t)Y(s)^{-1}\xi$; and from this con-
     clude that $T(t;s) = Y(t)Y(s)^{-1}$.

(d) By analogy with Section 13, seek a solution of the
     nonhomogeneous Eq. (1) in the form $x(t) = Y(t)p(t)$,
     where p is an unknown vector-valued function.
     Substituting this expression for x into Eq. (1),
     you should find $p'(t) = Y(t)^{-1}h(t)$.   (See Eq. 19-(13).)
     Hence conclude that, for any t_0 in J and any n-vector,
     ξ,
     $$x(t) = Y(t)[\xi + \int_{t_0}^{t} Y(s)^{-1}h(s)ds]$$
     defines a solution of Eq. (1).

(e) Use parts (c) and (d) to show that the solution of
     Eq. (1) with $x(t_0) = x_0$ is given by (13).

2. Evaluate the integrals in (16) and compare with the solu-
    tion found in Example 15-1. Hint.

$$\int e^{as}\sin bs\ ds = \frac{1}{a^2+b^2}(ae^{as}\sin bs - be^{as}\cos bs) + C.$$

3. Find the transition matrix $T(t,s)$ for the systems in
    Problem 15-5(a), (b), (d), (e), and (f).

4. Use the transition matrices found above for Problem 15-
    5(b) and for Example 2 to solve the following equations
    with $x(0) = x_0$.

(a) $x' = \begin{pmatrix} 1 & 2 \\ -1 & -1 \end{pmatrix} x + \begin{pmatrix} 1 \\ -2 \end{pmatrix}$, $J = \mathbb{R}$.

(b) $x' = \begin{pmatrix} 1 & 2 \\ -1 & -1 \end{pmatrix} x + \begin{pmatrix} 2\cos t \\ 0 \end{pmatrix}$, $J = \mathbb{R}$.

(c) $x' = \begin{pmatrix} 1 & 2 \\ -1 & -1 \end{pmatrix} x + \begin{pmatrix} 0 \\ \tan t \end{pmatrix}$, $J = (-\frac{\pi}{2}, \frac{\pi}{2})$.

(d) $x' = \begin{pmatrix} -1 & -1 \\ 1 & -3 \end{pmatrix} x + \begin{pmatrix} 0 \\ e^{-2t} \end{pmatrix}$, $J = \mathbb{R}$.

5.   Verify that the transition matrix for

$$y' = \begin{pmatrix} 0 & t \\ -t & 0 \end{pmatrix} y \quad \text{is} \quad T(t,s) = \begin{pmatrix} \cos\frac{t^2-s^2}{2} & \sin\frac{t^2-s^2}{2} \\ -\sin\frac{t^2-s^2}{2} & \cos\frac{t^2-s^2}{2} \end{pmatrix}.$$

6.   Solve on $J = \mathbb{R}$,

$$x' = \begin{pmatrix} 0 & t \\ -t & 0 \end{pmatrix} x + \begin{pmatrix} 0 \\ 2t \end{pmatrix} , \quad x(0) = \begin{pmatrix} x_{01} \\ x_{02} \end{pmatrix}.$$

18.   A MATRIX NORM

In Chapter II the norm of a vector was introduced and found to be a useful tool in the study of systems of equations. When, as in this chapter, the system is linear,

$$x' = A(t)x + h(t) \tag{1}$$

on $J = (\alpha, \beta)$, the matrix $A(t)$ plays a central role; and it will be useful to have a "norm" for matrices.   We now define such a norm, and illustrate its usefulness by re-proving uniqueness and continuous dependence of the solutions of (1) with the usual initial conditions,

$$x(t_0) = x_0. \tag{2}$$

Although it is not needed in this section, let us preserve our numbering system for Eqs. (1), (2), and (3) by writing

$$y' = A(t)y. \tag{3}$$

Definition. The norm of a matrix A with elements a_{ij} will be defined as

$$||A|| = \max_{j=1,\ldots,n} \sum_{i=1}^{n} |a_{ij}|, \tag{4}$$

i.e., the maximum of the sums of the absolute values of the elements in each column of A.

Many other definitions of norm are also available. One is given in Problem 5.

Recall that, in Section 5, the norm of a vector ξ with components $\xi_1,\ldots,\xi_n$ was defined by

$$||\xi|| = \sum_{j=1}^{n} |\xi_j|. \tag{5}$$

We shall use the same symbol $||\cdot||$ for both the matrix norm and the vector norm.

Note that for an n × n matrix A

$$\max_{i,j} |a_{ij}| \le ||A|| \le n \cdot \max_{i,j} |a_{ij}|. \tag{6}$$

Observe also that $||I|| = 1$.

Other important properties of the matrix norm and its relation to the vector norm are given in the following. Note that (i), (ii), (iii), and (iv) below correspond to the properties found in Section 5 for the norm of a vector.

Theorem A (Properties of the Norm). Let A and B be any

$n \times n$ matrices and let ξ be any column n-vector. Then

(i) $||A|| \geq 0$,

(ii) $||A|| = 0$ if and only if $A = 0$ (the zero matrix)

(iii) $||cA|| = |c| \cdot ||A||$ for every scalar c.

(iv) $||A + B|| \leq ||A|| + ||B||$ (the triangle inequality),

(v) $||A\xi|| \leq ||A|| \cdot ||\xi||$,

(vi) $||AB|| \leq ||A|| \cdot ||B||$.

Proof. The proofs of (i), (ii), (iii), and (iv) are straightforward and will be left to the reader.

The proof of (v) is as follows

$$||A\xi|| = \sum_{i=1}^{n} |\sum_{j=1}^{n} a_{ij}\xi_j| \leq \sum_{i=1}^{n} \sum_{j=1}^{n} |a_{ij}| \cdot |\xi_j|$$

$$= \sum_{j=1}^{n} \sum_{i=1}^{n} |a_{ij}| \cdot |\xi_j| \leq ||A|| \sum_{j=1}^{n} |\xi_j| = ||A|| \cdot ||\xi||.$$

The proof of (vi) is similar:

$$||AB|| = \max_{j} \sum_{i=1}^{n} |\sum_{k=1}^{n} a_{ik}b_{kj}| \leq \max_{j} \sum_{i=1}^{n} \sum_{k=1}^{n} |a_{ik}| \cdot |b_{kj}|$$

$$= \max_{j} \sum_{k=1}^{n} |b_{kj}| \sum_{i=1}^{n} |a_{ik}| \leq ||B|| \cdot ||A||.$$

Corollary B. If A and B are any two $n \times n$ matrices, then

$$\Big| ||A|| - ||B|| \Big| \leq ||A - B||. \tag{7}$$

The proof, using Theorem A(iv), is the same as for vector norms in Problem 5-5, and so is left to the reader again.

Sometimes a different but equivalent definition of $||A||$ is useful. This definition and a proof of its equivalence to (4) are as follows.

Lemma C.   For an   n × n   matrix, A,

$$||A|| = \sup\{||A\xi||: ||\xi|| \le 1 \}.$$

Proof.

$$||A|| = \max_{j} ||AI_{(j)}||$$

$$\le \sup\{||A\xi||: ||\xi|| \le 1\}$$

$$\le \sup\{||A||\cdot||\xi||: ||\xi|| \le 1\} \le ||A||. \quad \square$$

Using the matrix norm notation, it is possible to re-phrase the definitions given in Section 14 for the continuity and differentiability of matrix-valued functions.

In Problem 3(a) the reader is asked to show that a matrix-valued function   A   is continuous at a point t_1 if and only if

$$\lim_{t \to t_1} ||A(t) - A(t_1)|| = 0. \tag{8}$$

And in Problem 3(b) you should show that   A   is differentiable at t_1 if and only if there is a matrix $A'(t_1)$, the deriva-tive of   A   at t_1, such that

$$\lim_{\Delta t \to 0} ||\frac{A(t_1+\Delta t) - A(t_1) - \Delta t A'(t_1)}{\Delta t}|| = 0. \tag{9}$$

Corollary D.   If   A(t)   is continuous at t_1, then so is $||A(t)||$.

Proof.   From Corollary B

$$\Big| ||A(t)|| - ||A(t_1)|| \Big| \le ||A(t) - A(t_1)||.$$

But the latter tends to zero as $t \to t_1$ by (8). $\square$

To further illustrate the use of the norm notation for matrices, we will prove a minor generalization of Theorem 8-C. Assuming the continuity of A, the following theorem asserts that the solution of (1) and (2) is unique and depends continuously on the initial conditions.

<u>Theorem E</u>. Let A be a continuous matrix-valued function and let h be any vector-valued function on J. Let $t_0 \in J$ and let x_0 and $\tilde{x}_0$ be any two (constant) n-vectors. Then if x and $\tilde{x}$ are solutions of Eq. (1) on J which satisfy $x(t_0) = x_0$ and $\tilde{x}(t_0) = \tilde{x}_0$, it follows that for all t in J

$$||x(t) - \tilde{x}(t)|| \leq ||x_0 - \tilde{x}_0|| e^{|\int_{t_0}^{t} ||A(s)|| ds|}. \tag{10}$$

Proof. It is given that

$$x'(t) = A(t)x(t) + h(t) \quad \text{and} \quad \tilde{x}'(t) = A(t)\tilde{x}(t) + h(t)$$

Thus, for all s in J,

$$\frac{d}{ds}[x(s) - \tilde{x}(s)] = A(s)[x(s) - \tilde{x}(s)].$$

Now integrate from t_0 to t, using the initial condition $x(t_0) - \tilde{x}(t_0) = x_0 - \tilde{x}_0$, to find

$$x(t) - \tilde{x}(t) = x_0 - \tilde{x}_0 + \int_{t_0}^{t} A(s)[x(s) - \tilde{x}(s)]ds.$$

Applying 5-(9) and Theorem A-(v), one finds

$$||x(t) - \tilde{x}(t)|| \leq ||x_0 - \tilde{x}_0|| + ||\int_{t_0}^{t} A(s)[x(s) - \tilde{x}(s)]ds||$$

$$\leq ||x_0 - \tilde{x}_0|| + \left|\int_{t_0}^{t} ||A(s)[x(s) - \tilde{x}(s)]|| ds\right|$$

$$\leq ||x_0 - \tilde{x}_0|| + \left| \int_{t_0}^{t} ||A(s)|| \cdot ||x(s) - \tilde{x}(s)|| ds \right|.$$

The assertion of the theorem now follows from Lemma 8-A.   □

Note that if $||A(t)|| \leq K$, then from Theorem E we can obtain the weaker estimate of Theorem 8-C,

$$||x(t) - \tilde{x}(t)|| \leq ||x_0 - \tilde{x}_0|| e^{K|t - t_0|}. \qquad (11)$$

Example 3.   Consider the differential system

$$x_1' = -(\sin t)x_2 + 4,$$

$$x_2' = -x_1 + 2tx_2 - x_3 + e^t,$$

$$x_3' = 3(\cos t)x_1 + x_2 + \frac{1}{t} x_3 - 5t^2.$$

Assuming vector solutions x and $\tilde{x}$ exist on the interval (1, 3) with

$$x(2) = \text{col} (7, 3, -2) \quad \text{and} \quad \tilde{x}(2) = \text{col} (6.7, 3.2, -1.9)$$

let us estimate the "error", $||x(t) - \tilde{x}(t)||$, for $1 < t < 3$.

In order to apply Theorem E we must first compute the norm of the matrix

$$A(t) = \begin{pmatrix} 0 & -\sin t & 0 \\ -1 & 2t & -1 \\ 3 \cos t & 1 & 1/t \end{pmatrix},$$

namely,

$$||A(t)|| = \max \{1 + 3|\cos t|, |\sin t| + 2t + 1, 1 + 1/t\}.$$

From this we can easily obtain an upper bound for $||A(t)||$ when $1 < t < 3$, namely

$$||A(t)|| \leq \max \{4, 2+2t, 2\} = 2 + 2t \leq 8,$$

It therefore follows from either (10) or (11) that

$$||x(t) - \tilde{x}(t)|| \leq ||x(2) - \tilde{x}(2)||e^{\left|\int_2^t ||A(s)||ds\right|}$$

$$\leq (0.3 + 0.2 + 0.1)e^{\left|\int_2^t 8ds\right|}$$

$$= 0.6e^{8|t-2|} < 0.6e^8$$

for all t in (1, 3).

A sharper estimate results if we use $||A(t)|| \leq 2 + 2t$

in (10):

$$||x(t) - \tilde{x}(t)|| \leq 0.6e^{\left|\int_2^t (2+2s)ds\right|} \leq 0.6e^7.$$

In addition to the estimate of the growth rate of errors the following estimate for the growth of a solution itself is often useful.

Theorem F.   Let   A   be a continuous matrix-valued function on J with $||A(t)|| \leq K$ for some constant K > 0, and let h be a continuous vector-valued function on J. Then, if x is a solution of (1) and (2),

$$||x(t)|| \leq [||x_0|| + \left|\int_{t_0}^t ||h(s)||ds\right|]e^{K|t-t_0|}$$

for t in J as far as the solution exists.

Proof. The solution of (1) and (2) satisfies

$$x(t) = x_0 + \int_{t_0}^t [A(s)x(s) + h(s)]ds.$$

Thus

$$||x(t)|| \leq ||x_0|| + \left|\int_{t_0}^t ||h(s)||ds\right| + \left|\int_{t_0}^t K||x(s)||ds\right|.$$

The assertion of the theorem now follows from Corollary 8-D. ▫

Problems

1. Let $A = \begin{bmatrix} 0 & 2 \\ -4 & 1 \end{bmatrix}$ and $B = \begin{bmatrix} -1 & 3 \\ 1 & 2 \end{bmatrix}$. Then verify that,

    as asserted by Theorem A, $||A + B|| \leq ||A|| + ||B||$,
    $||AB|| \leq ||A|| \cdot ||B||$, and $||BA|| \leq ||B|| \cdot ||A||$.

2. If A is an $n \times n$ invertible matrix, prove that
    $||A|| > 0$ and $||A^{-1}|| \geq \dfrac{1}{||A||}$.

3. (a) Show that condition (8) is equivalent to the defini-
        tion given in Section 14 for the continuity of a
        matrix-valued function.

    (b) Show that condition (9) is equivalent to the defini-
        tion given in Section 14 for the differentiability
        of a matrix-valued function.

4. Let A be an $n \times n$ matrix with $||A|| < 1$. For each
    integer $k = 1, 2, \ldots$ let $a_{ij}^{(k)}$ be the ij'th element of
    the matrix A^k. Then prove that $\lim\limits_{k \to \infty} a_{ij}^{(k)} = 0$ for each

    $i, j = 1, \ldots, n$. (We will then say $\lim\limits_{k \to \infty} A^k = 0$, the zero
    matrix.)

5. Another possible way of defining the norm of an $n \times n$
    matrix A is
    $$||A||_S = \sum_{j=1}^{n} \sum_{i=1}^{n} |a_{ij}|.$$
    Prove that with this definition Theorem A remains true.

6. Compute $||I||_S$ if the norm is defined as in Problem 5.

7. Assume the hypothesis of Theorem F.   Then

(a) Show that $||T(t,s)|| \le e^{K|t-s|}$ where $T(t,s)$ is the transition matrix of Section 17.

(b) Using variation of parameters Eq. 17-(13), show that if $||h(t)|| \le M$, then

$$||x(t)|| \le ||x_0||e^{K|t-t_0|} + \frac{M}{K}(e^{K|t-t_0|} - 1).$$

(c) Show that the latter is sharper than the estimate

$$||x(t)|| \le [||x_0|| + M|t-t_0|]e^{K|t-t_0|}$$

given by Theorem F when $||h(t)|| \le M$.

8. Similarly to Lemma C, prove that the norm defined by Eq. (4) satisfies

(a) $||A|| = \sup \{||A\xi||: ||\xi|| = 1\}$.

(b) $||A|| = \sup \{\frac{||A\xi||}{||\xi||}: \xi \ne 0\}$.

*19. MATRIX EXPONENTIAL

Section 15 provided methods for solving linear homogeneous systems of differential equations with constant coefficients. And, once the homogeneous system is completely solved, the method of variation of parameters gives the solution of a nonhomogeneous system

$$x' = Ax + h(t) \tag{1}$$

with initial condition

$$x(t_0) = x_0. \tag{2}$$

It is the purpose of the present section to define the concept of a "matrix exponential". We shall then use this to give an expression for the transition matrix $T(t,s)$ of system

(1) or its associated homogeneous system

$$y' = Ay,$$ (3)

plus a new method of solution for Eqs. (1) and (2).

By analogy to the meaning of e^a when a is a real number, we intend to define

$$e^A = I + \frac{1}{1!} A + \frac{1}{2!} A^2 + \cdots$$

where A is a constant $n \times n$ matrix. Thus we must first discuss infinite sequences and series of matrices.

<u>Definition</u>. Let S and $S(k)$ for $k = 1,2,\ldots$ be $n \times n$ matrices with elements s_{ij} and $s_{ij}(k)$ respectively. We shall say the <u>sequence</u> $\{S(k)\}$ <u>converges</u> to S and write

$$\lim_{k \to \infty} S(k) = S$$

if

$$\lim_{k \to \infty} s_{ij}(k) = s_{ij} \quad \text{for each} \quad i, j = 1,\ldots,n.$$

The matrix S is called the <u>limit</u> of the sequences $\{S(k)\}$.

Note how this definition of convergence follows the pattern set by the definitions of continuity, differentiability, and integrability of matrices. We extend a familiar concept from scalars to matrices by requiring that the individual elements of the matrices satisfy the old definition.

It will also be useful to have the equivalent definition of convergence given in the following lemma.

<u>Lemma A</u>. If $S, S(1), S(2),\ldots$ are $n \times n$ matrices then

$$\lim_{k \to \infty} S(k) = S \quad \text{if and only if} \quad \lim_{k \to \infty} ||S(k) - S|| = 0.$$

Proof.   From 18-(6),

$$\max_{i,j} |s_{ij}(k) - s_{ij}| \le ||S(k) - S|| \le n \max_{i,j} |s_{ij}(k) - s_{ij}|,$$

and the assertion of the lemma follows. $\square$

Definition.   If $B(k)$ is an $n \times n$ matrix for each $k = 0$, 1, 2,... we shall say the series $\sum_{k=0}^{\infty} B(k)$ converges to an $n \times n$ matrix S if

$$\lim_{\ell \to \infty} S(\ell) = S,$$

where

$$S(\ell) \equiv \sum_{k=0}^{\ell} B(k)$$

is a partial sum of the given series.   In case the series converges to S we shall call S the sum of the series and write

$$\sum_{k=0}^{\infty} B(k) = S.$$

Remark.   Note that

$$\sum_{k=0}^{\infty} B(k) = S \quad \text{if and only if} \quad \sum_{k=0}^{\infty} b_{ij}(k) = s_{ij}$$

for each $i, j = 1,...,n$, where $b_{ij}(k)$ and s_{ij} are the ij'th elements of $B(k)$ and S respectively.

Example 1.   For $k = 0, 1, 2,...$, let

$$B(k) = \begin{pmatrix} 1/k! & 1/3^k \\ 0 & 2^k/k! \end{pmatrix}.$$

Then

$$\sum_{k=0}^{\infty} B(k) = \begin{pmatrix} \sum_{k=0}^{\infty} \dfrac{1}{k!} & \sum_{k=0}^{\infty} \dfrac{1}{3^k} \\ 0 & \sum_{k=0}^{\infty} \dfrac{2^k}{k!} \end{pmatrix} = \begin{pmatrix} e & \dfrac{3}{2} \\ 0 & e^2 \end{pmatrix}$$

Theorem B.   If

$$\sum_{k=0}^{\infty} B(k) = S \quad \text{and} \quad \sum_{k=0}^{\infty} C(k) = T$$

are two convergent series of n × n matrices, and if A is
another n × n matrix, then

$$\sum_{k=0}^{\infty} [B(k) + C(k)] = S + T, \tag{4}$$

$$\sum_{k=0}^{\infty} AB(k) = AS, \tag{5}$$

and

$$\sum_{k=0}^{\infty} B(k)A = SA. \tag{6}$$

(Note that, since we do not assume commutativity of multiplication, it is important to distinguish between AS and SA).

Proof.   By Lemma A,

$$\lim_{\ell \to \infty} \| \sum_{k=0}^{\ell} B(k) - S \| = 0 \quad \text{and} \quad \lim_{\ell \to \infty} \| \sum_{k=0}^{\ell} C(k) - T \| = 0.$$

Thus, as $\ell \to \infty$,

$$\| \sum_{k=0}^{\ell} [B(k) + C(k)] - [S + T] \|$$

$$\leq \| \sum_{k=0}^{\ell} B(k) - S \| + \| \sum_{k=0}^{\ell} C(k) - T \| \to 0$$

This proves (4).

To prove (5) we note that, as $\ell \to \infty$,

$$\| \sum_{k=0}^{\ell} AB(k) - AS \| = \| A(\sum_{k=0}^{\ell} B(k) - S) \|$$

$$\leq \| A \| \cdot \| \sum_{k=0}^{\ell} B(k) - S \| \to 0$$

The proof of (6) is analogous.   □

The matrix series with which we shall have the most concern is

$$\sum_{k=0}^{\infty} \frac{1}{k!} A^k = I + \frac{1}{1!} A + \frac{1}{2!} A^2 + \cdots, \tag{7}$$

where A is a given $n \times n$ matrix. Observe that the notational convention $a^0 \equiv 1$, used for series of numbers, has now been extended to matrices as

$$A^0 \equiv I.$$

When the series (7) converges, we will define its sum to be e^A.

Example 2. Consider the series (7) when

$$A = \begin{pmatrix} 1 & 0 & 0 \\ 0 & -3 & 0 \\ 0 & 0 & 0 \end{pmatrix}.$$

For this simple matrix it is easy to compute the required powers, A^k, and to prove that the series (7) converges with the following results:

$$\sum_{k=0}^{\infty} \frac{1}{k!} A^k = I + \sum_{k=1}^{\infty} \frac{1}{k!} \begin{pmatrix} 1 & 0 & 0 \\ 0 & (-3)^k & 0 \\ 0 & 0 & 0 \end{pmatrix}$$

$$= \begin{pmatrix} \sum_{k=0}^{\infty} \frac{1}{k!} & 0 & 0 \\ 0 & \sum_{k=0}^{\infty} \frac{1}{k!}(-3)^k & 0 \\ 0 & 0 & 1 \end{pmatrix} = \begin{pmatrix} e & 0 & 0 \\ 0 & e^{-3} & 0 \\ 0 & 0 & 1 \end{pmatrix}.$$

With the aid of the following theorem, we will be able to prove that the series (7) always converges for any $n \times n$ matrix, A.

Theorem C.   If $B(0)$, $B(1)$, $B(2)$,... are $n \times n$ matrices such that the series of numbers $\Sigma_{k=0}^{\infty} \|B(k)\| < \infty$, then the series of matrices

$$\sum_{k=0}^{\infty} B(k)$$

converges to some matrix S and one obtains the convergence estimate

$$\|S - \sum_{k=0}^{\ell} B(k)\| \le \sum_{k=\ell+1}^{\infty} \|B(k)\| \quad \text{for each} \quad \ell = 0,1,... \quad (8)$$

Proof.   From the definition of norm,

$$\sum_{k=0}^{\infty} |b_{ij}(k)| \le \sum_{k=0}^{\infty} \|B(k)\| < \infty \quad \text{for each} \quad i,j = 1,...,n.$$

By the comparison test for scalar series, this assures the convergence of $\Sigma_{k=0}^{\infty} b_{ij}(k)$ for each $i,j = 1,...,n$, and hence implies the convergence of $\Sigma_{k=0}^{\infty} B(k) = S$.

Now, by application of the triangle inequality, we find for $m > \ell$

$$\|S - \sum_{k=0}^{\ell} B(k)\| = \|S - \sum_{k=0}^{m} B(k) + \sum_{k=\ell+1}^{m} B(k)\|$$

$$\le \|S - \sum_{k=0}^{m} B(k)\| + \sum_{k=\ell+1}^{\infty} \|B(k)\|$$

$$\to \sum_{k=\ell+1}^{\infty} \|B(k)\| \quad \text{as} \quad m \to \infty. \quad \square$$

Corollary D and a Definition.   For any $n \times n$ matrix A the series (7) converges. We shall define the exponential of A

$$e^{A} \equiv \sum_{k=0}^{\infty} \frac{1}{k!} A^{k}. \quad (9)$$

(Note, from this, that $e^{0} = I$.)

Proof. By Theorem C, the convergence of the series (7) follows from the fact that

$$\sum_{k=0}^{\infty} \left|\left|\frac{1}{k!} A^k\right|\right| \leq \sum_{k=0}^{\infty} \frac{1}{k!} ||A||^k = e^{||A||} < \infty. \quad \square$$

Even though we have proved that the series for e^A always converges, it would clearly be impractical in most cases to try to compute e^A directly from Eq. (9). However, if A can diagonalized the computation of e^A becomes quite simple.

Theorem E. If A is an $n \times n$ matrix and P is a matrix with inverse such that

$$P^{-1}AP = \Lambda = \begin{pmatrix} \lambda_1 & 0 & \cdots & 0 \\ 0 & \lambda_2 & \cdots & 0 \\ \vdots & \vdots & & \vdots \\ 0 & 0 & & \lambda_n \end{pmatrix}$$

then

$$e^A = Pe^{\Lambda}P^{-1} = P \begin{pmatrix} e^{\lambda_1} & 0 & \cdots & 0 \\ 0 & e^{\lambda_2} & \cdots & 0 \\ \vdots & \vdots & & \vdots \\ 0 & 0 & \cdots & e^{\lambda_n} \end{pmatrix} P^{-1}. \quad (10)$$

Proof. From $A = P\Lambda P^{-1}$, it follows easily that $A^k = P\Lambda^k P^{-1}$ for each $k = 0,1,2,\ldots$. Thus

$$e^A = \sum_{k=0}^{\infty} \frac{1}{k!} P\Lambda^k P^{-1}.$$

But now we have only to invoke Eqs. (5) and (6) to find

$$e^A = \sum_{k=0}^{\infty} P \frac{1}{k!} \Lambda^k P^{-1} = P(\sum_{k=0}^{\infty} \frac{1}{k!} \Lambda^k) P^{-1} = Pe^{\Lambda}P^{-1}.$$

The second equality in (10), evaluating e^{Λ}, follows just as in Example 2. $\square$

Example 3. Let us compute e^A when $A = \begin{pmatrix} 1 & 4 \\ 3 & 2 \end{pmatrix}$. For this matrix, one finds eigenvalues $\lambda_1 = 5$ and $\lambda_2 = -2$ with corresponding (linearly independent) eigenvectors $\xi_{(1)} =$ col $(1,1)$ and $\xi_{(2)} =$ col $(4, -3)$. Using $\xi_{(1)}$ and $\xi_{(2)}$ as columns, we form the matrix

$$P = \begin{pmatrix} 1 & 4 \\ 1 & -3 \end{pmatrix} \quad \text{and compute} \quad P^{-1} = \frac{1}{7}\begin{pmatrix} 3 & 4 \\ 1 & -1 \end{pmatrix}.$$

Then $P^{-1}AP = \begin{pmatrix} 5 & 0 \\ 0 & -2 \end{pmatrix} = \Lambda$. Hence, by Theorem E,

$$e^A = P\begin{pmatrix} e^5 & 0 \\ 0 & e^{-2} \end{pmatrix}P^{-1} = \frac{1}{7}\begin{pmatrix} 3e^5+4e^{-2} & 4e^5-4e^{-2} \\ 3e^5-3e^{-2} & 4e^5+3e^{-2} \end{pmatrix}.$$

The solutions of Eq. (1) will eventually be expressed in terms of the matrix _exponential_ _function_ defined by

$$e^{(t-t_0)A} = \sum_{k=0}^{\infty} \frac{1}{k!} (t-t_0)^k A^k. \tag{11}$$

This function is well defined for all t in $\mathbb{R}$ since, for each t, $(t - t_0)A$ is merely another $n \times n$ matrix to which we can apply Corollary D.

Theorem F. The matrix exponential function is differentiable, and for all t

$$\frac{d}{dt} e^{(t-t_0)A} = Ae^{(t-t_0)A} = e^{(t-t_0)A}A \tag{12}$$

-- exactly the result obtained by differentiating the series in Eq. (11) termwise.

Proof. The second equality in (12) follows from (5) and (6) and the fact that A commutes with each term in the series (11).

According to Eq. 18-(9) it will now suffice to prove that, for each fixed t, $||f(\Delta t)|| \to 0$ as $\Delta t \to 0$, where

$$f(\Delta t) \equiv \frac{1}{\Delta t}[e^{(t+\Delta t-t_0)A} - e^{(t-t_0)A}] - Ae^{(t-t_0)A}.$$

Thus we compute for $\Delta t \neq 0$ (with the aid of Theorem B)

$$f(\Delta t) = \sum_{j=1}^{\infty} \frac{1}{j!} \frac{(t+\Delta t-t_0)^j - (t-t_0)^j}{\Delta t} A^j - \sum_{k=0}^{\infty} \frac{1}{k!}(t-t_0)^k A^{k+1}$$

$$= \sum_{k=0}^{\infty} \frac{1}{k!} [\frac{(t+\Delta t-t_0)^{k+1} - (t-t_0)^{k+1}}{(k+1)\Delta t} - (t-t_0)^k] A^{k+1}.$$

But, since $(d/dt)(t-t_0)^{k+1} = (k+1)(t-t_0)^k$, it follows from the mean value theorem that there exist numbers s_k between t and t+Δt such that

$$f(\Delta t) = \sum_{k=0}^{\infty} \frac{1}{k!}[(s_k-t_0)^k - (t-t_0)^k] A^{k+1}.$$

In order to show that $\lim_{\Delta t \to 0} ||f(\Delta t)|| = 0$ we now invoke a standard argument which is used when "uniform convergence" is available. Let any $\varepsilon > 0$ be given, and choose any $R > |t-t_0|$. Then, since the power series for $e^{R||A||}$ converges, there exists N such that

$$\sum_{k=N+1}^{\infty} \frac{1}{k!} R^k ||A||^k < \frac{\varepsilon}{2||A||+1}.$$

If we now require $0 < |\Delta t| < R - |t-t_0|$, it follows that $|t+\Delta t-t_0| < R$ and hence $|s_k-t_0| < R$ for each $k = 0,1,2,...$ Thus, by (8) with $\ell = N$,

$$||f(\Delta t)|| \leq ||\sum_{k=0}^{N} \frac{1}{k!}[(s_k-t_0)^k - (t-t_0)^k] A^{k+1}||$$

$$+ ||\sum_{k=N+1}^{\infty} \frac{1}{k!}(s_k-t_0)^k A^{k+1}|| + ||\sum_{k=N+1}^{\infty} \frac{1}{k!}(t-t_0)^k A^{k+1}||$$

$$\leq \sum_{k=1}^{N} \frac{1}{k!} |(s_k-t_0)^k - (t-t_0)^k| \cdot ||A||^{k+1}$$

$$+ 2||A|| \frac{\varepsilon}{2||A||+1} \, .$$

Finally we can choose $\delta > 0$ sufficiently small so that the finite sum, $\Sigma_{k=1}^{N}$, in this last expression is less than $\varepsilon/2$ provided each $|s_k-t| < \delta$. Thus it follows that

$$||f(\Delta t)|| < \varepsilon \quad \text{whenever} \quad 0 < |\Delta t| < \min (\delta, R-|t-t_0|).$$

Since $\varepsilon > 0$ was arbitrary, we have proved that

$$\lim_{\Delta t \to 0} ||f(\Delta t)|| = 0. \quad \square$$

The following is a useful consequence of Eq. (12). It is, perhaps, surprising that differentiation should be of any use in proving this result.

Corollary G. If A is any $n \times n$ matrix and a and b are any two real numbers, then

$$e^{aA} e^{bA} = e^{(a+b)A} = e^{bA} e^{aA}.$$

In particular, this shows that e^A always has an inverse and that

$$(e^A)^{-1} = e^{-A}.$$

Proof. For arbitrary real numbers, a and t, let us consider the matrices

$$B(t) \equiv e^{aA} e^{tA} \quad \text{and} \quad C(t) \equiv e^{(a+t)A}.$$

Then, by Eq. (12),

$$B'(t) = e^{aA} A e^{tA} = AB(t)$$

and

$$C'(t) = Ae^{(a+t)A} = AC(t).$$

Moreover

$$B(0) = e^{aA} = C(0).$$

If $b_{(j)}$ and $c_{(j)}$ are the j'th columns of B and C respectively, then we have proved that $b_{(j)}$ and $c_{(j)}$ are both solutions of the differential system $y' = Ay$, and that $b_{(j)}(0) = c_{(j)}(0)$. It follows from uniqueness that $b_{(j)}(t) = c_{(j)}(t)$ for all t. This means that

$$B(t) = C(t) \quad \text{for all} \quad t.$$

In particular

$$e^{aA}e^{bA} = e^{(a+b)A}.$$

Interchanging the roles of a and b, we also have

$$e^{bA}e^{aA} = e^{(b+a)A} = e^{(a+b)A}.$$

The assertions about $(e^A)^{-1}$ follow from the fact that

$$e^A e^{-A} = e^0 = I = e^{-A}e^A. \quad \square$$

Lemma H.   If B is a differentiable matrix-valued function and v is a differentiable vector-valued function, then

$$\frac{d}{dt}[B(t)v(t)] = B'(t)v(t) + B(t)v'(t), \tag{13}$$

If B is constant and v is continuous, then

$$\int_{t_0}^{t} Bv(s)ds = B \int_{t_0}^{t} v(s)ds, \tag{14}$$

and if B is continuous and v is constant, then

$$\int_{t_0}^{t} B(s)vds = [\int_{t_0}^{t} B(s)ds]v. \tag{15}$$

Proof.   To prove (13), we consider the i'th component of the left hand side:

$$\frac{d}{dt} \sum_{j=1}^{n} b_{ij}(t)v_j(t) = \sum_{j=1}^{n} b'_{ij}(t)v_j(t) + \sum_{j=1}^{n} h_{ij}(t)v'_j(t).$$

The result is the i'th component of the right hand side of (13).   The proofs of (14) and (15) are similar. $\square$

Using the machinery assembled in this section, we are ready to solve Eqs. (1) and (2) in a manner entirely analogous to the method used for scalar, linear, first order equations in Section 2.

Assuming (at first) that there exists a solution, subtract $Ax(t)$ from both sides of Eq. (1) and then multiply (from the left) by the "integrating factor" e^{-tA} to obtain

$$e^{-tA}x'(t) - e^{-tA}Ax(t) = e^{-tA}h(t).$$

By Eq. (13), this is   equivalent to

$$\frac{d}{dt} [e^{-tA}x(t)] = e^{-tA}h(t).$$

Now integrate this equation (componentwise) to get

$$e^{-tA}x(t) - e^{-t_0 A}x(t_0) = \int_{t_0}^{t} e^{-sA}h(s)ds.$$

Multiply (from the left) by e^{tA} to find

$$x(t) = e^{(t-t_0)A}x_0 + e^{tA}\int_{t_0}^{t} e^{-sA}h(s)ds. \tag{16}$$

Since e^{tA} is a constant matrix as far as the integration is concerned, Eq. (14) asserts that this can also be written as

$$x(t) = e^{(t-t_0)A}x_0 + \int_{t_0}^{t} e^{(t-s)A}h(s)ds. \tag{16a}$$

As in Section 2, all we have shown thus far is that no

function can be a solution of Eqs. (1) and (2) except the vector function defined by Eq. (16). It remains for the reader to prove (Problem 4) that the solution candidate defined by Eq. (16) or (16a) really is a solution.

But Theorem 17-A gave another representation for the solution of Eqs. (1) and (2). Comparing Eq. (16) and Eq. 17-(13) in the case. h = 0 we conclude that for all t and t_0

$$T(t,t_0)x_0 = e^{(t-t_0)A}x_0$$

for all x_0. Hence (by Problem 5) it follows that

$$T(t,t_0) = e^{(t-t_0)A}. \tag{17}$$

Thus we now have an explicit expression for the fundamental matrix whenever A is constant.

Example 4. Compute e^{tA} when $A = \begin{pmatrix} -1 & -1 \\ 1 & -3 \end{pmatrix}$. This matrix, which occurred in Problems 15-2 and 3, cannot be diagonalized. However, using Eq. (17) and the result of Example 17-2, we easily find

$$e^{tA} = T(t,0) = \begin{pmatrix} (1+t)e^{-2t} & -te^{-2t} \\ te^{-2t} & (1-t)e^{-2t} \end{pmatrix}.$$

Problems

1. Compute e^{tA} (using Theorem E when possible), for the matrices

(a) $A = \begin{pmatrix} 2 & -1 \\ -2 & 3 \end{pmatrix}$,   (b) $A = \begin{pmatrix} 1 & 2 \\ -1 & -1 \end{pmatrix}$,

(c) $A = \begin{pmatrix} -1 & 0 & 3 \\ -8 & 1 & 12 \\ -2 & 0 & 4 \end{pmatrix}$,   (d) $A = \begin{pmatrix} -1 & 2 & -1 \\ -1 & -4 & 1 \\ -1 & -2 & -1 \end{pmatrix}$.

Hint.   You may wish to use some of your previous work on Problem 15-5.

2. Convince yourself, by writing out the first few terms of the various series involved, that

$$e^A e^B = e^{A+B} = e^B e^A \qquad \text{if} \qquad AB = BA.$$

(Corollary G is a special case of this.)

3. Let $A = \begin{bmatrix} 1 & 0 \\ 0 & 0 \end{bmatrix}$ and $B = \begin{bmatrix} 0 & 1 \\ 0 & 0 \end{bmatrix}$, and compute e^A, e^B, and e^{A+B}. (This is easily done directly from the definition in these cases.) Then show that $e^A e^B \neq e^{A+B} \neq e^B e^A$. Why does this not contradict Problem 2?

4. Verify, by direct substitution, that (16) defines a solution of Eqs. (1) and (2).   Hint.   You will need to recall Eq. 5-(5).

5. If, for some $n \times n$ matrix A, $A\xi = 0$ for every n-vector ξ prove that $A = 0$.   Hint.   Consider, in turn, $\xi = I_{(1)}$, $\xi = I_{(2)}, \ldots,$ $\xi = I_{(n)}$.

6. It might appear (by analogy with Section 2) that the solution of Eq. (1) could be found in terms of some matrix exponential even when A is a nonconstant function of t. In the case h = 0, and A continuous, would we not merely get

$$x(t) = e^{\int_{t_0}^{t} A(s)ds} x_0 \qquad \text{instead of} \qquad x(t) = e^{(t-t_0)A} x_0?$$

Show that this is not true, in general, because

$$\frac{d}{dt} e^{\int_{t_0}^{t} A(s)ds} \neq A(t) e^{\int_{t_0}^{t} A(s)ds}.$$

Hint: Try $A(t) = \begin{pmatrix} 1 & 2t \\ 0 & 0 \end{pmatrix}$ with $t_0 = 0$.

7. (a) If B and C are two differentiable $n \times n$-matrix-
        valued functions prove that

$$\frac{d}{dt}[B(t)C(t)] = B'(t)C(t) + B(t)C'(t).$$

   (b) Why can one not prove that $\frac{d}{dt}B^2(t) = 2B(t)B'(t)$?

        If $B(t) = \begin{pmatrix} 1 & t \\ 0 & 0 \end{pmatrix}$, verify that

$$\frac{d}{dt}B^2(t) = B'(t)B(t) + B(t)B'(t) \neq 2B(t)B'(t)$$

*8. If $||A|| < 1$, prove that

$$\sum_{k=0}^{\infty} A^k = (I - A)^{-1}.$$

9. Invent a reasonable definition for $\sin A$ where A is an
    $n \times n$ matrix. Then compute $\sin A$ for

    (a) $A = \begin{pmatrix} 1 & 0 \\ 0 & 2 \end{pmatrix}$,   (b) $A = \begin{pmatrix} 1 & 4 \\ 3 & 2 \end{pmatrix}$ (as in Example 3).

10. If A is any constant $n \times n$ matrix, prove that for any
     constant vector ξ, $y(t) = (\sin tA)\xi$ is a solution of the
     system of second order equations $y'' + A^2 y = 0$. Use
     your definition of $\sin tA$ from Problem 9.

<u>20</u>.   <u>EXISTENCE</u> <u>OF</u> <u>SOLUTIONS</u> (<u>SUCCESSIVE</u> <u>APPROXIMATIONS</u>)

To this point we have demonstrated the existence of solutions only in the case of linear systems with constant coefficients (and certain other very special cases).

Now we shall consider the general linear system with variable coefficients

$$x' = A(t)x + h(t),\tag{1}$$

where the given $n \times n$ matrix function A and vector function h are continuous on $J = (\alpha,\beta)$. We shall then prove that this equation together with an initial condition

$$x(t_0) = x_0\tag{2}$$

has a (unique) solution. As a corollary it will follow that the homogeneous equation

$$y' = A(t)y\tag{3}$$

has n linearly independent solutions on J. And finally, we shall also prove the existence of solutions of the general n'th order linear equations discussed in Chapter III.

The proof (of existence for (1) and (2)) is based on the equivalent system of integral equations

$$x(t) = x_0 + \int_{t_0}^{t} [A(s)x(s) + h(s)]ds,\tag{4}$$

for t in J. We shall use a simple and intuitively appealing method of "successive approximations". This method was used by J. Liouville and others in the early 1800's. But the method is often referred to as "Picard iteration", after E. Picard who further developed it in 1893.

Before proceeding with a general proof let us introduce the method of successive approximations by treating a
simple and familiar example.

Example 1.   Consider the scalar equation

$$x'(t) = cx(t) \quad \text{with} \quad x(0) = 1, \tag{5}$$

where c is a given constant and J = $\mathbb{R}$. (Of course we already know how to solve this elementary problem.) The problem
represented by Eqs. (5) is equivalent to

$$x(t) = 1 + \int_0^t cx(s)ds. \tag{6}$$

Let us now pick some continuous function $x_{(0)}$ to be
considered as a "first approximation" or "first guess" for the
solution. For convenience in computation, let

$$x_{(0)}(t) \equiv 1 \quad \text{for all} \quad t \quad \text{in} \quad \mathbb{R}.$$

Now, to generate a "next approximation", let us substitute
$x_{(0)}$ into the right hand side of Eq. (6) and define

$$x_{(1)}(t) = 1 + \int_0^t cx_{(0)}(s)ds = 1 + ct.$$

Similarly, put $x_{(1)}$ into the right hand side of (6) and define

$$x_{(2)}(t) = 1 + \int_0^t cx_{(1)}(s)ds$$

$$= 1 + c\int_0^t (1 + cs)ds = 1 + ct + \frac{(ct)^2}{2}.$$

Continuing in this manner, we find

$$x_{(3)}(t) = 1 + \int_0^t c x_{(2)}(s)ds$$

$$= 1 + ct + \frac{(ct)^2}{2!} + \frac{(ct)^3}{3!} ,$$

and in general, the "successive approximations" are

$$x_{(\ell)}(t) = 1 + ct + \frac{(ct)^2}{2!} + \cdots + \frac{(ct)^\ell}{\ell!} ,$$

for $\ell = 1,2,\ldots$.   Thus, for each t,

$$\lim_{\ell \to \infty} x_{(\ell)}(t) = e^{ct}.$$

And as we well know, $x(t) = e^{ct}$ is the unique solution of (5).

This nice result is much more than a coincidence. We shall now show that this same method provides a solution of Eqs. (1) and (2). In other words, successive approximations can be defined as in the example; they form a convergent sequence; and the limit of that sequence is a solution of (1) and (2).

<u>Theorem A</u>. Let A and h be continuous on J and let $(t_0, x_0) \in J \times \mathbb{R}^n$. Then Eqs. (1) and (2) have a (unique) solution on the entire interval $J = (\alpha, \beta)$.

<u>Proof</u>. As in Example 1, we are going to try to use the right hand side of Eq. (4) to generate a sequence of "successive approximations" for a solution of Eqs. (1) and (2). Thus we will begin with an arbitrary first guess for the continuous function $x_{(0)}: J \to \mathbb{R}^n$. Then we hope to define successively better approximations $x_{(1)}, x_{(2)}, \ldots,$ by

$$x_{(1)}(t) = x_0 + \int_{t_0}^{t} [A(s)x_{(0)}(s) + h(s)]ds, \qquad (7)$$

$$x_{(2)}(t) = x_0 + \int_{t_0}^{t} [A(s)x_{(1)}(s) + h(s)]ds, \qquad (8)$$

$$\vdots$$

$$x_{(\ell+1)}(t) = x_0 + \int_{t_0}^{t} [A(s)x_{(\ell)}(s) + h(s)]ds, \qquad (9)$$

But can we do this? Since the functions A and h are unspecified, we cannot actually evaluate the integrals in Eqs. (7), (8), and (9) as we did in Example 1. So we must ask: Will the integrands in the right hand sides of these equations always be well defined and continuous? And if so will the resulting sequence of functions $\{x_{(\ell)}\}$ converge? And if it does converge, say to x, will x be a solution of Eqs. (1) and (2)?

Since A, $x_{(0)}$, and h are continuous on J it follows that the integrand in Eq. (7) is continuous. Thus $x_{(1)}$ exists and is continuous on J. This in turn shows that $x_{(2)}$, defined by Eq. (8), exists and is continuous on J. The existence of $x_{(\ell)}$ for all $\ell = 1,2,\ldots$ follows by induction.

Our next objective is to show that the sequence $\{x_{(\ell)}\}$ of "successive approximations" does indeed converge. Let α_0 and β_0 be any two numbers such that

$$\alpha < \alpha_0 < t_0 < \beta_0 < \beta.$$

Then the continuity of A on the closed bounded interval $[\alpha_0, \beta_0]$ assures the continuity of $\|A(t)\|$ (by Corollary 18-D). Hence there exists a constant $K > 0$ such that

$$||A(t)|| \leq K \quad \text{for} \quad \alpha_0 \leq t \leq \beta_0.$$

Now consider the magnitude of the difference between two consecutive functions of the sequence $\{x_{(\ell)}\}$, namely

$$||x_{(\ell+1)}(t) - x_{(\ell)}(t)|| \quad \text{for} \quad \alpha_0 \leq t \leq \beta_0.$$

We shall obtain a recursive relation for these differences. For $\alpha_0 \leq t \leq \beta_0$ and $\ell = 0,1,2,\ldots$ one easily finds

$$||x_{(\ell+2)}(t) - x_{(\ell+1)}(t)|| = ||\int_{t_0}^{t} A(s)[x_{(\ell+1)}(s) - x_{(\ell)}(s)]ds||.$$

Thus

$$||x_{(\ell+2)}(t) - x_{(\ell+1)}(t)|| \leq K\left|\int_{t_0}^{t} ||x_{(\ell+1)}(s) - x_{(\ell)}(s)||ds\right|. \quad (10)$$

By the continuity of $x_{(1)}$ and $x_{(0)}$ it follows that $||x_{(1)}(t) - x_{(0)}(t)||$ is continuous. Thus, since $[\alpha_0,\beta_0]$ is closed and bounded, there exists a constant M such that

$$||x_{(1)}(t) - x_{(0)}(t)|| \leq M \quad \text{for} \quad \alpha_0 \leq t \leq \beta_0.$$

Putting this into (10) with $\ell = 0$ gives, for $\alpha_0 \leq t \leq \beta_0$,

$$||x_{(2)}(t) - x_{(1)}(t)|| \leq K\left|\int_{t_0}^{t} M ds\right| = MK|t-t_0|.$$

Now put this last estimate into (10) with $\ell = 1$ to find

$$||x_{(3)}(t) - x_{(2)}(t)|| \leq K\left|\int_{t_0}^{t} MK|s-t_0|ds\right| = M\frac{K^2|t-t_0|^2}{2}.$$

You should verify the evaluation of this integral by considering two cases, $t \geq t_0$ and $t \leq t_0$. Then carry out the next similar computation to show that

$$||x_{(4)}(t) - x_{(3)}(t)|| \leq M\frac{K^3|t-t_0|^3}{3!}.$$

At about this point one might suspect that a pattern is

emerging. It appears likely that

$$||x_{(\ell+1)}(t) - x_{(\ell)}(t)|| \leq M \frac{K^\ell |t-t_0|^\ell}{\ell!} \tag{11}$$

$$\text{for } \alpha_0 \leq t \leq \beta_0,$$

for $\ell = 0,1,2,\ldots$ This conjecture is verified by mathematical induction as follows. Assuming (11) holds for some $\ell = 0$, or 1, or 2, ... we compute from (10)

$$||x_{(\ell+2)}(t) - x_{(\ell+1)}(t)|| \leq K \left| \int_{t_0}^t M \frac{K^\ell |s-t_0|^\ell}{\ell!} \, ds \right|$$

$$= M \frac{K^{\ell+1} |t-t_0|^{\ell+1}}{(\ell + 1)!} \quad .$$

(Again one evaluates the integral for the two cases $t \geq t_0$ and $t \leq t_0$). Since this estimate is just (11) with ℓ replaced by $\ell+1$ the verification of (11) is complete.

Now, for each $\ell = 1,2,\ldots$ and each t in $[\alpha_0,\beta_0]$, we can write

$$x_{(\ell)}(t) = x_{(0)}(t) + \sum_{p=0}^{\ell-1} [x_{(p+1)}(t) - x_{(p)}(t)]. \tag{12}$$

Why?

Inequality (11) gives

$$||x_{(p+1)}(t) - x_{(p)}(t)|| \leq M \frac{(Ka)^p}{p!} \quad \text{for } \alpha_0 \leq t \leq \beta_0,$$

where $a = \max \{\beta_0-t_0, t_0-\alpha_0\}$. Hence for each $k = 1,\ldots,n$

$$|x_{(p+1)k}(t) - x_{(p)k}(t)| \leq M \frac{(Ka)^p}{p!} \quad \text{for } \alpha_0 \leq t \leq \beta_0.$$

Applying this and the comparison test, we conclude that each scalar series

$$\sum_{p=0}^{\infty} [x_{(p+1)k}(t) - x_{(p)k}(t)] \quad (k = 1,\ldots,n) \tag{13}$$

converges for all t in $[\alpha_0, \beta_0]$, because of the convergence
of $\sum_{p=0}^{\infty} M(Ka)^p/p!$, the well-known series for Me^{Ka}. This
means that the sequence of "partial sums" of (13) involved in
(12) must converge on $[\alpha_0, \beta_0]$. But by the arbitrariness of
α_0 and β_0, this sequence must converge for all t in J.
Let

$$x(t) = \lim_{\ell \to \infty} x_{(\ell)}(t) \quad \text{for t in J.}$$

We note further that for all t in $[\alpha_0, \beta_0]$ and for
any integer $m > \ell \geq 0$

$$||x_{(m)}(t) - x_{(\ell)}(t)|| = ||\sum_{p=\ell}^{m-1} [x_{(p+1)}(t) - x_{(p)}(t)]||$$

$$\leq M \sum_{p=\ell}^{\ell} \frac{(Ka)^p}{p!} .$$

Hence, letting $m \to \infty$, we find

$$||x(t) - x_{(\ell)}(t)|| \leq M \sum_{p=\ell}^{\infty} \frac{(Ka)^p}{p!} \quad \text{for all t in } [\alpha_0, \beta_0]. \quad (14)$$

This proves that $x_{(\ell)k}$ converges to x_k uniformly on
$[\alpha_0, \beta_0]$ as $\ell \to \infty$ for each $k = 1, \ldots, n$. See Appendix 2
for a discussion of uniform convergence. Thus, since each
$x_{(\ell)k}$ is continuous it follows that the limit x_k is continuous on $[\alpha_0, \beta_0]$ by Theorem A2-K. Hence the vector-valued
function x is also continuous on $[\alpha_0, \beta_0]$.

Now we should like to take limits as $\ell \to \infty$ in Eq. (9)
and conclude that x satisfies Eq. (4). To justify this we
note that, for $\alpha_0 \leq t \leq \beta_0$

$$||\int_{t_0}^{t} A(s)x(s)ds - \int_{t_0}^{t} A(s)x_{(\ell)}(s)ds||$$

$$\leq K \left| \int_{t_0}^{t} ||x(s) - x_{(\ell)}(s)||ds \right| .$$

But, by (14), the scalar sequence $\{||x(s) - x_{(\ell)}(s)||\}$ con-
verges uniformly to zero for s in $[\alpha_0,\beta_0]$. So, by part
(ii) of Theorem A2-K, the above integral $\to 0$ as $\ell \to \infty$. It
follows that x satisfies Eq. (4) on $[\alpha_0,\beta_0]$. This means
x is a solution of (1) and (2) on (α_0,β_0). Hence, invoking
the arbitrariness of α_0 and β_0 once more, x is a solution
on J.

     And, of course, we have known since Chapter II that the
solution is unique. □

Example 2. Let A be a constant n × n matrix, and let us
apply the method of successive approximations to solve (once
again) the system

$$y' = Ay \quad \text{with} \quad y(t_0) = y_0.$$

Let $y_{(0)}(t) \equiv y_0$. Then one finds on $\mathbb{R}$

$$y_{(1)}(t) = y_0 + \int_{t_0}^{t} Ay_0 ds = y_0 + (t-t_0)Ay_0,$$

$$y_{(2)}(t) = y_0 + \int_{t_0}^{t} A[y_0 + (s-t_0)Ay_0]ds$$

$$= y_0 + (t-t_0)Ay_0 + \frac{1}{2!}(t-t_0)^2 A^2 y_0,$$

and, by induction,

$$y_{(\ell)}(t) = y_0 + \frac{1}{1!}(t-t_0)Ay_0 + \frac{1}{2!}(t-t_0)^2 A^2 y_0$$

$$+ \cdots + \frac{1}{\ell!}(t-t_0)^\ell A^\ell y_0.$$

Now, by Theorem A, we know that this process converges to the
solution. Hence

$$y(t) = \sum_{k=0}^{\infty} \frac{1}{k!}(t-t_0)^k A^k y_0 = (\sum_{k=0}^{\infty} \frac{1}{k!}(t-t_0)^k A^k) y_0.$$

If you have read Section 19, you should recognize this as $y(t) = e^{(t-t_0)A}$ once again.

The proof of Theorem A shows that, starting with any continuous function $x_{(0)}: J \to \mathbb{R}^n$, the successive approximations defined by (9) will always converge to _some_ solution of (1) and (2). But we know that there is at most one solution on J. Thus, no matter what the "first approximation" $x_{(0)}$, the successive approximations will always converge to the one and only solution.

Note further that inequality (14) gives a specific error estimate for the successive approximations.   In other words, even though we may not know the solution x, we can say that $x_{(\ell)}$ lies within a certain small distance of x for all sufficiently large ℓ. As a matter of fact we can obtain the following simpler and possibly sharper error estimate, provided we define $x_{(0)}$ by $x_{(0)}(t) = x_0$ for t in J.

Corollary B.   Under the same hypotheses and using the same notation as in Theorem A and its proof except that $x_{(0)}(t) \equiv x_0$, we have for $\ell = 0,1,\ldots$

$$||x(t) - x_{(\ell)}(t)|| \leq \frac{Be^{K|t-t_0|}K^\ell|t-t_0|^{\ell+1}}{(\ell+1)!} \quad \text{for} \quad \alpha_0 \leq t \leq \beta_0.$$

(15)

Here   K   is an upper bound for $||A(t)||$ on $[\alpha_0,\beta_0]$ and

$$B \geq \max_{\alpha_0 \leq t \leq \beta_0} ||A(t)x_0 + h(t)||.$$

Proof.   For $\alpha_0 \leq t \leq \beta_0$

$$||x(t) - x_0|| = ||\int_{t_0}^{t} [A(s)x(s) + h(s)]ds||$$

$$\leq ||\int_{t_0}^{t} A(s)[x(s) - x_0]ds|| + ||\int_{t_0}^{t} [A(s)x_0 + h(s)]ds||$$

$$\leq K\left|\int_{t_0}^{t} ||x(s) - x_0||ds\right| + B|t-t_0|.$$

From this, Corollary 8-D gives

$$||x(t) - x_0|| \leq B|t-t_0|e^{K|t-t_0|},$$

which is equivalent to (15) with $\ell = 0$. The remainder of the proof -- a straightforward induction on (15) -- is left as Problem 2. ☐

In Section 17, when we solved Eqs. (1) and (2) by variation of parameters, we assumed that the associated homogeneous equation (3) had n linearly independent solutions. The following important corollary of Theorem A asserts that such an assumption is superfluous.

Corollary C. Let A be a continuous n × n-matrix-valued function on J. Then system (3) has n linearly independent solutions on J.

Proof. Choose any t_0 in J and any n linearly independent vectors $\xi_{(1)}, \ldots, \xi_{(n)}$. For example, the latter could be $I_{(1)}, \ldots, I_{(n)}$. Then for $j = 1, \ldots, n$, let $y_{(j)}$ be the unique solution of (3) on J which satisfies $y_{(j)}(t_0) = \xi_{(j)}$. The solutions $y_{(j)}$ exist by Theorem A, and they are linearly independent by Corollary 14-C. ☐

Chapter III treated the linear scalar differential equation of order n

$$D^n x + a_{n-1}(t)D^{n-1}x + \cdots + a_1(t)Dx + a_0(t)x = h(t). \qquad (16)$$

But most of the results were restricted to the case of constant coefficients.   Now, at last, we are ready to prove a general existence theorem for Eq. (16) in case of variable continuous coefficients and with arbitrary initial conditions of the form

$$x(t_0) = b_0, \quad Dx(t_0) = b_1, \quad \ldots, \quad D^{n-1}x(t_0) = b_{n-1}. \qquad (17)$$

<u>Theorem</u> <u>D</u>.   Let $a_{n-1}, \ldots, a_0$, and h be continuous (real-valued) functions on J and let $(t_0, b_0, b_1, \ldots, b_{n-1})$ be any point in $J \times \mathbb{R}^n$.   Then Eqs. (16) and (17) have a (unique) solution on J.

Moreover, the associated homogeneous equation (Eq. (16) with $h(t) \equiv 0$) has n linearly independent solutions $y_1, \ldots, y_n$ on J.

<u>Remark</u>.   This theorem shows that it was unnecessary to hypothesize, in Theorem 11-D, the existence of a particular solution of Eq. (16) and n linearly independent solutions of the associated homogeneous equation.

<u>Proof</u> <u>of</u> <u>Theorem</u> <u>D</u>.   Defining $z_1 = x$, $z_2 = Dx, \ldots$, $z_n = D^{n-1}x$, we can convert Eq. (16) into the equivalent system

$$
\begin{aligned}
z_1' &= z_2 \\
z_2' &= z_3 \\
&\;\;\vdots
\end{aligned}
\qquad (18)
$$

$$z_n' = -a_0(t)z_1 - a_1(t)z_2 - \cdots - a_{n-1}(t)z_n + h(t)$$

with initial condition

$$z(t_0) = \mathrm{col}\ (b_0, \ b_1, \ \ldots, \ b_{n-1}). \qquad (19)$$

By Theorem A, Eqs. (18) and (19) have a unique solution z on J. Thus, taking $x = z_1$ we have a (unique) solution of (16) and (17) on J.

To prove the existence of n linearly independent solutions of the homogeneous equation (16) with $h(t) \equiv 0$ it suffices to apply the above result using, in turn,

$$\text{col} (b_0, b_1, \ldots, b_{n-1}) = I_{(1)}, I_{(2)}, \ldots, \text{ and } I_{(n)}. \quad \square$$

Problems

1. Obtain the first few successive approximations for the following problems and compare with the known solutions obtained by other methods.

    (a) $x' = cx$ with $x(0) = 1$

        using $x_{(0)}(t) = 1 - t$.

    (b) $x' = x + t$ with $x(1) = 2$

        using $x_{(0)}(t) = 2$.

    (c) $x' = \begin{bmatrix} -1 & -2 \\ 1 & -4 \end{bmatrix} x$ with $x(0) = \begin{bmatrix} 1 \\ 0 \end{bmatrix}$

        using $x_{(0)}(t) = x(0)$.

2. Complete the proof of Corollary B by giving the induction calculation for (15).

3. Choose appropriate values of α_0 and β_0 and take $x_{(0)} \equiv x(t_0)$ in the equations of Problem 1(a), (b), and (c). Then use Corollary B to estimate $|x(t) - x_{(4)}(t)|$ on (α_0, β_0).

4. How might the proof of Theorem A be slightly simplified if we had $\beta - t_0 < 1/K$ and $t_0 - \alpha < 1/K$ with $||A(t)|| \leq K$ on (α, β)? Hint: Consider inequality (10)

and the resulting simplifications.

5. Repeat the calculation of Example 2 for the equation
    $x' = Ax + h(t)$ with $x(t_0) = x_0$, where   A   is a constant
    $n \times n$ matrix and   h   is continuous.   Your answer should
    ultimately agree with Eq. 19-(16a).

Chapter V. INTRODUCTION TO DELAY DIFFERENTIAL EQUATIONS

Up to now we have considered only certain ordinary differential equations in which the unknown function and its derivatives are all evaluated at the same instant, t.

A more general type of differential equation, often called a functional differential equation, is one in which the unknown function occurs with various different arguments. For example we might have

$$x'(t) = -2x(t-1),$$

$$x'(t) = x(t) - x(t/2) + x'(t-1),$$

$$x'(t) = x(t)x(t-1) + t^2 x(t+2), \quad \text{or}$$

$$x''(t) = -x'(t) - x'(t-1) - 3\sin x(t) + \cos t.$$

In the Russian literature these are called "differential equations with deviating argument".

The simplest and perhaps most natural type of functional differential equation is a "delay differential equation" (or "differential equation with retarded argument"). This means an equation expressing some derivative of x at time t in terms of x (and its lower order derivatives if any) at t and at earlier instants. The first and fourth examples above are of this type. The other two are not, for the second example involves the highest order derivative at two different instants and the third contains an advanced argument.

This book treats only delay differential equations.

Clearly, when working with delay differential equations, one must write out the arguments of the unknown function. We cannot omit them as we have been doing in ordinary differential equations.

225

21.   EXAMPLES AND THE METHOD OF STEPS

In this section we give some examples of physical and biological systems in which the present rate of change of some unknown function(s) depends upon past values of the same function(s).

Mixing of Liquids. Consider a tank containing B gallons of salt water brine. Fresh water flows in at the top of the tank at a rate of q gallons per minute. The brine in the tank is continually stirred, and the mixed solution flows out through a hole at the bottom, also at the rate of q gallons per minute. See Figure 2-1.

Let $x(t)$ be the amount (in pounds) of salt in the brine in the tank at time t. If we assume continual, in-stantaneous, perfect mixing throughout the tank, then the brine leaving the tank contains $x(t)/B$ lbs. of salt per gallon, and hence

$$x'(t) = -qx(t)/B.$$

But, more realistically, let us agree that mixing can-not occur instantaneously throughout the tank. Thus the con-centration of the brine leaving the tank at time t will equal the average concentration at some earlier instant, say t-r. We shall assume that r is a positive constant, al-though this assumption may also be subject to improvement. The differential equation for x then becomes a delay dif-ferential equation, $x'(t) = -qx(t-r)/B$ or, setting $c = q/B$,

$$x'(t) = -cx(t-r), \tag{1}$$

where r is the "delay" or "time lag".

The first mathematical question now is, what kind of
initial conditions should one use, if any, or order to obtain
a unique solution of Eq. (1)? Various possibilities will be
considered in the problems and in the next section.

From these considerations we will find that the most
natural answer appears to be that one should specify an initial function on some interval of length r, say $[t_0-r,t_0]$,
and then try to satisfy Eq. (1) for $t \geq t_0$. Thus we set

$$x(t) = \theta(t) \quad for \quad t_0-r \leq t \leq t_0, \tag{2}$$

where θ is some given function. Then we seek a continuous
extension of θ into the future, to a function x which
satisfies Eq. (1) for $t \geq t_0$. See Figure 1. (At t_0,

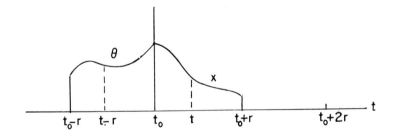

Figure 1

$x'(t_0)$ should be interpreted as a right hand derivative.)
We are assuming that the past history of the salt content of
the tank is known, and represented by θ, without regard to
whether or not that past history satisfies the delay differential equation.

For example, let us take $\theta(t) \equiv \theta_0$, a positive constant. In other words, assume that the tank contained θ_0
pounds of salt thoroughly mixed in B gallons of brine prior

to time t_0. Then at t_0 valves were opened allowing fresh
water to flow in at the top and mixed brine to flow out at
the bottom, each at the rate of q gallons per minute.

With this constant initial function, it is easy to
solve Eqs. (1) and (2) on $[t_0, t_0+r]$. For there Eq. (1) becomes simply

$$x'(t) = -c\theta_0$$

with initial condition $x(t_0) = \theta_0$. The solution, of course,
is

$$x(t) = \theta_0 - c\theta_0(t-t_0) \quad \text{for} \quad t_0 \le t \le t_0+r. \quad (3)$$

Now that x is known up to t_0+r, we consider the interval
$[t_0+r, t_0+2r]$. There Eq. (1) becomes

$$x'(t) = -c\theta_0 + c^2\theta_0(t-r-t_0)$$

with initial condition, obtained from (3), $x(t_0+r) = \theta_0 - cr\theta_0$.
This can easily be solved, and one can then consider the interval $[t_0+2r, t_0+3r]$, etc. This procedure can, in principle,
be continued as far as desired. It is called, quite naturally,
the method of steps.

But the calculations quickly become unwieldly, and it
is difficult to determine even the most essential properties
of the solution. For example, since $x(t)$ represents the
amount of salt in the tank we must have $x(t) \ge 0$ for all t.
Hence Eq. (3) shows that we should have $cr < 1$ if the model
is to be meaningful. But $cr < 1$ is not a sufficient smallness condition on r. Problem 1. How do we know that $x(t)$
does not eventually become negative regardless of how small
we make r?

Population Growth. If N(t) is the population at time t
of an isolated colony of animals, the most naive model for the
growth of the population is

$$N'(t) = kN(t), \qquad (4)$$

where k is a positive constant. This implies exponential
growth, $N(t) = N_0 e^{kt}$ where $N_0 = N(0)$.

    A somewhat more realistic model is obtained if we ad-
mit that the growth rate coefficient k will not be constant
but will diminish as N(t) grows, because of overcrowding
and shortage of food. This leads to the differential equation

$$N'(t) = k[1 - N(t)/P]N(t), \qquad (5)$$

where k and P are both positive constants. This equation
with $N(0) \equiv N_0$ can be solved by separation of variables.
The result (Problem 2) is

$$N(t) = \frac{N_0 e^{kt}}{1 + (N_0/P)(e^{kt} - 1)} .$$

If N_0/P is small compared to 1, the solution is like
$N_0 e^{kt}$ when t is close to 0, as one would expect from Eq.
(5). But as $t \to \infty$, regardless of the value of $N_0 > 0$, N(t)
approaches the equilibrium value P. The solution curves are
sketched in Figure 2 for $N_0 = P/10$, P/2, P, and 3P/2.

    Now suppose that the biological self-regulatory re-
action represented by the factor [1 - N(t)/P] in (5) is not
instantaneous, but responds only after some time lag r > 0.
Then instead of (5) we have the delay differential (or difference differential) equation

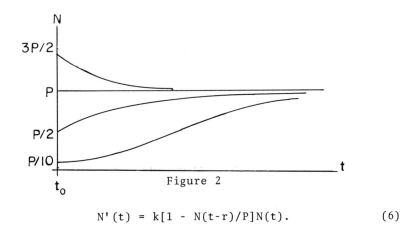

Figure 2

$$N'(t) = k[1 - N(t-r)/P]N(t). \tag{6}$$

This equation has been studied extensively by Wright [1945], [1955], Kakutani and Markus [1958], Jones [1962], Kaplan and Yorke [1975], and others. It is often rewritten, by introducing $x(s) = N(rs)/P - 1$ and $c = kr$, in the form

$$x'(s) = -cx(s-1)[1 + x(s)].$$

Having made this change of variables (Problem 3), let us now re-label s   as   t   and write

$$x'(t) = -cx(t-1)[1 + x(t)]. \tag{7}$$

It is customary to specify an initial function θ on [-1, 0] and then seek a function   x   such that

$$x(t) = \theta(t) \quad \text{for} \quad -1 \le t \le 0, \tag{8}$$

and   x   satisfies Eq. (7) for $t \ge 0$ -- as we did for Eq. (1).

If we assume θ to be continuous, then it is quite easy to establish the existence and uniqueness of a solution

of Eq. (7) subject to initial condition (8) by the method of
steps.   On [0, 1], Eq. (7) becomes a linear first order
<u>ordinary</u> differential equation.

$$x'(t) + c\theta(t-1)x(t) = -c\theta(t-1)$$

with initial condition $x(0) = \theta(0)$.   As in Section 2, the
solution of this problem is found, with the aid of an integrating factor.   We get

$$\frac{d}{dt}[x(t)e^{\int_0^t c\theta(s-1)ds}] = -c\theta(t-1)e^{\int_0^t c\theta(s-1)ds}$$

which leads to the unique solution on   [0, 1],

$$x(t) = [\theta(0)+1]e^{-\int_0^t c\theta(s-1)ds} - 1. \tag{9}$$

Having found the exact solution on   [0, 1], we can now, in a
similar manner, reduce Eq. (7) on   [1, 2]   to a linear first
order ordinary differential equation and solve on that interval.   And so on.

    While this procedure can, in principle, be continued
to obtain the exact solution of Eqs. (7) and (8) as far as
desired, the resulting integrals quickly become very cumbersome.   Thus once again it is difficult to draw any general
conclusions about the solution from this exact procedure.   It
would be of interest to determine whether a solution of Eq.
(7) is bounded, whether it oscillates, and whether like the
solutions of (5), it approaches a limit as $t \to \infty$.   Some
questions of this sort are formulated in Problem 16.

    It is not at all obvious that the type of initial con-
dition, (8) or (2), which we have used is the only reasonable
one for a delay differential equation.   Another possible type

of initial condition is considered in Problem 5.

Many other models of population growth have been proposed -- both with and without time lags. For example Cooke
and Yorke [1972] have studied the equation

$$N'(t) = g(N(t)) - g(N(t-L)),$$

where g is a given continuous, positive function and L is
the lifetime of members of the species.

Prey-Predator Population Models. Let $x(t)$ be the population
at time t of some species of animal called prey and let
$y(t)$ be the population of a predator species which lives off
these prey. We assume that $x(t)$ would increase at a rate
proportional to $x(t)$ if the prey were left alone, i.e., we
would have $x'(t) = a_1 x(t)$, where $a_1 > 0$. However the predators are hungry, and the rate at which each of them eats
prey is limited only by his ability to find prey. (This
seems like a reasonable assumption as long as there are not
too many prey available.) Thus we shall assume that the ac-
tivities of the predators reduce the growth rate of $x(t)$ by
an amount proportional to the product $x(t)y(t)$, i.e.,

$$x'(t) = a_1 x(t) - b_1 x(t) y(t),$$

where b_1 is another positive constant.

Now let us also assume that the predators are completely dependent on the prey as their food supply. If there
were no prey, we assume $y'(t) = -a_2 y(t)$, where $a_2 > 0$, i.e.,
the predator species would die out exponentially. However,
given food the predators breed at a rate proportional to their
number and to the amount of food available to them. Thus we

consider the pair of equations

$$x'(t) = a_1 x(t) - b_1 x(t) y(t)$$
$$y'(t) = -a_2 y(t) + b_2 x(t) y(t), \tag{10}$$

where a_1, a_2, b_1, and b_2 are positive constants. This well-known model was invented and studied by Lotka [1920], [1925] and Volterra [1928], [1931].

Vito Volterra was trying to understand the observed fluctuations in the sizes of populations $x(t)$ of commercially desirable fish and $y(t)$ of larger fish which fed on the smaller ones in the Adriatic Sea in the decade from 1914 to 1923. System (10) will be discussed in Section 38, and we shall see then that this model, naive as it is, does lead to the prediction of "population cycles".

Wangersky and Cunningham [1957] proposed to modify system (10) so that the birth rate of prey has a further limitation as in Eq. (5), while the birth rate of predators responds to changes in the magnitudes of x and y only after a delay $r > 0$. Thus they replaced system (10) with

$$x'(t) = a_1 [1-x(t)/P] x(t) - b_1 y(t) x(t)$$
$$y'(t) = -a_2 y(t) + b_2 x(t-r) y(t-r). \tag{11}$$

System (11) can be solved by the method of steps. Indeed, the reader will find that the solution of (11) is, in principle, more elementary than the solution of (10). Problem 11. But the problem of obtaining useful information about the solutions of (11) is much more difficult.

Volterra himself had studied several generalizations of (10) in his book [1931]. One of these was the system

$$x'(t) = [a_1 - b_1 y(t) - \int_{-r}^{0} h_1(\sigma)y(t+\sigma)d\sigma]x(t)$$

$$(12)$$

$$y'(t) = [-a_2 + b_2 x(t) + \int_{-r}^{0} h_2(\sigma)x(t+\sigma)d\sigma]y(t),$$

where h_1 and h_2 are given continuous non-negative functions. Here the past histories of x and y over an entire interval of length r enter the equations through the integrals. Such equations may be said to have infinitely many delays.

The method of steps cannot be applied to system (12). Why not? Try it. The trouble is that (12) contains arbitrarily small delays.

Control Systems. Any system involving a feedback control will almost certainly involve time delays. These arise because a finite time is required to sense information and then react to it.

In the 1930's and 1940's Callender, Hartree, and Porter [1936], Minorsky [1947], [1948], [1962], and others began studies of certain stabilization problems in which delays become significant.

Consider, as a simple example, a system whose motion is governed by a second order, linear homogeneous differential equation with positive constant coefficients

$$mx''(t) + bx'(t) + kx(t) = 0. \qquad (13)$$

The solution of this equation with arbitrarily specified initial conditions, $x(t_0)$ and $x'(t_0)$, is a function which decays exponentially toward zero. Recall that we classified the equation (in Example 10-1) in terms of the magnitude of the damping coefficient b.

The case $b^2 < 4mk$ is called "underdamped", and the
solution oscillates as it decays.

The case $b^2 > 4mk$ is called "overdamped". The solutions do not oscillate, but die out exponentially at a slower
rate the larger b becomes.

The case $b^2 = 4mk$ is called "critically damped" because, in a sense, solutions approach zero most rapidly in
this case.

Let us assume that the system is underdamped
$(b^2 < 4mk)$ and we wish to somehow increase the damping coefficient to bring the system closer to critical damping,
thereby diminishing the oscillations more rapidly. Perhaps
we would even prefer a slight overdamping so as to eliminate
oscillations.

If our physical system is a simple mass hanging from
a spring in the laboratory then it is quite simple to increase b. We might simply immerse the whole system in
molasses. Or if that makes b _too_ big we could try No. 10
motor oil instead.

However, if, as in Minorsky's case, our system is a
ship rolling in the waves and x is the angle of tilt from
the normal upright position, we must be more ingenious in
trying to increase b. We might, for example, introduce
ballast tanks, partially filled with water, in each side of
the ship. We would also have a servomechanism designed to
pump water from one tank to the other in an attempt to
counteract the roll of the ship. Hopefully, this would introduce another term proportional to $x'(t)$ in the equation,

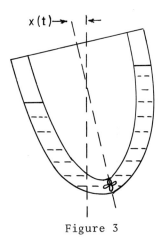

Figure 3

say   qx'(t).   Thus we consider

$$mx''(t) + bx'(t) + qx'(t) + kx(t) = 0. \tag{14}$$

But now one must recognize that the servomechanism
cannot respond instantaneously.   Thus instead of Eq. (14) we
should consider

$$mx''(t) + bx'(t) + qx'(t-r) + kx(t) = 0. \tag{15}$$

The control takes time   r > 0   to respond and thus the control term is proportional to the velocity at the earlier instant, t-r.   It seems possible that such a time lag could result in the force represented by   qx'(t-r)   being in the opposite direction to that which is desired.   Thus it is con-
ceivable that rather than helping to stabilize the system,
such a control might make matters worse, and even cause unde-
sired oscillations.   Problem 12.

This is a simplified version of a problem which actually arose during tests of systems for antirolling stabilization of a ship before World War II (Minorsky [1947]).   The

delay in that case was found to be due to "cavitation" at
the pump itself.

Distribution of Prime Numbers.   In the 1940's Cherwell (see
de Visme [1961] and Wright [1961]) attempted to find a heuristic analytical proof of the prime number theorem, which asserts that if $\pi(n)$ is the number of primes $\leq n$ then
$\lim_{n\to\infty}[n^{-1}\pi(n) \ln n] = 1$.

Proceeding very heuristically one can "derive" the
equation

$$z'(n) = -z(n)z(n^{1/2})/2n \qquad (16)$$

where $z(n)$ is the "density of primes" in the "neighborhood"
of   n   for large (positive)   n.   It is easily verified that
$z(n) = 1/(\ln n)$ is a solution of this equation, and this is
the result one would want for the prime number theorem.   However there are many other solutions as well.   The question is,
do all solutions behave like $1/(\ln n)$ for large   n?

Two-Body Problem of Electrodynamics.   It is possible to consider the interaction of two electrons (or other charged
particles) as a problem of "action at a distance".   However,
one must then take into account the fact that the interactions
between the two particles travel not instantaneously, but
with finite speed   c -- the speed of light.   Thus the influence of particle   2   on particle   1   at time   t   must have
been "generated" by   2   at some earlier instant   t-r.

For simplicity let us consider two (classical) charged
particles moving on the   x   axis.   Let $x_1(t)$ and $x_2(t)$
represent the positions of the two particles at time   t   (all
relative to some inertial reference frame).   Then the fields

reaching $x_1(t)$ at time t from particle 2 were generated
at t-r, where the delay r must satisfy the functional

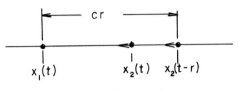

Figure 4

equation

$$cr = |x_1(t) - x_2(t-r)|.$$

Clearly the (unknown) delay r is not a constant, but depends
on t via the (unknown) trajectories x_1 and x_2. A similar equation determines the time delay for the influence of
particle 1 on particle 2. Thus we need two delays $r_{21}(t)$
and $r_{12}(t)$ which are to satisfy the functional equations

$$cr_{21}(t) = |x_1(t) - x_2(t-r_{21}(t))|$$

$$cr_{12}(t) = |x_2(t) - x_1(t-r_{12}(t))|.$$

(17)

Not only are the delays not constant but they depend upon the
unknown functions, through the functional equations. In terms
of these delays one can obtain equations of motion of the
form

$$x_1''(t) = f_1(x_1(t)-x_2(t-r_{21}(t)), x_1'(t), x_2'(t-r_{21}(t)))$$

$$x_2''(t) = f_2(x_2(t)-x_1(t-r_{12}(t)), x_2'(t), x_1'(t-r_{12}(t))),$$

(18)

where f_1 and f_2 map some open subset of $\mathbb{R}^3 \to \mathbb{R}$.

    A fair amount is known about these equations. See
Driver [1963], [1969], Driver and Norris [1967], Travis

[1975], Ždanov [1975], and Hsing [1977].

Geometrical Problems. For history's sake we should mention
that the earliest examples of functional differential equations arose in problems of geometry. Euler and others, as
far back as 1750, sought the equations of curves having certain properties. This led to some rather exotic functional
differential equations such as

$$y^2(x) + y^2(x)[y'(x)]^2 = a + y^2(x+y(x)y'(x))$$

(cf. Poisson [1806] or Babbage [1816]), and

$$y(x)y'(x) = y(y(x))$$

(cf. Barba [1930]).

Others. Hopefully one or more of the examples given above
has convinced the reader that delay differential equations are
worth studying. But if not, we could cite others which have
arisen in studies of: nuclear reactors (Ergen [1954], Levin
and Nohel [1960], Gorjačenko [1971]), neutron shielding
(Placzek [1946], Boffi and Scozzafava [1967]), electron energy
distribution in a gas discharge (Sherman [1960]), liquid fuel
rocket engines (Norkin [1965]), transistor circuits (Gumowski
[1962]), electromagnetic vibrators (Norkin [1965]), photo
emulsions (Silberstein [1940]), transmission lines (Brayton
[1967], Fox, Mayers, Ockendon and Tayler [1971]), elasticity
theory (Volterra [1909], Gurtin and Sternberg [1962]), the
spread of infectious diseases (Lotka and Sharpe [1923], Wilson
and Burke [1942], Hoppensteadt and Waltman [1970], Yorke
[1971], Cooke and Yorke [1972], London and Yorke [1973]),

neurology (Melzak [1961]), the respiratory system (Grodins,
Buell, and Bart [1967]), business cycles and economic growth
(Kalecki [1935], Goodwin [1951], Cooke and Yorke [1972]),
inventory maintenance (Arrow, Harris, and Marschak [1951]),
the production and death of red blood cells (Chow [1974]),
and metal rolling control systems (Johnson [1972]). Many of
these examples and further references are found in the books
of Bellman and Cooke [1963], Norkin [1965], and Mitropol'skiĭ
and Martynjuk [1969], and in the survey papers of Myškis
[1949], Zverkin, Kamenskiĭ, Norkin and Èl'sgol'c [1964],
Cooke [1963], [1967], Myškis and Èl'sgol'c [1967], and
Manitius [1974].

Problems

1. (a) How small must cr be in order that the solution of
         Eqs. (1) and (2) with $\theta(t) \equiv \theta_0 > 0$ remain posi-
         tive for $t_0 \leq t \leq t_0 + 2r$?

    (b) Can you find a smallness condition on r which
         would guarantee $x(t) > 0$ for all $t \geq t_0$?

    (c) In the "mixing-of-liquids" example, instead of fresh
         water, let the incoming liquid be brine containing
         a lbs of salt per gallon. Then show that,

         $$x'(t) = qa - cx(t-r).$$

         Find a change of variable which transforms this into
         Eq. (1).

2.   Solve Eq. (5) with $N(0) = N_0 \geq 0$. <u>Hint</u>. You will need
to use "partial fractions" in the case when $N_0 \neq 0$ or
P.   What should you do if $N_0 = 0$ or $N_0 = P$?

3.   Justify the transformation of Eq. (6) into Eq. (7).

4.   Apply the method of steps to solve on $[0,2]$ the equation

$$x'(t) = ax(t) + bx(t-1),\qquad\qquad(19)$$

with $x(t) = \theta(t) = 1+t$ on $[-1,0]$. Assume that a and
b are constants with $b \neq 0$. <u>Caution</u>. Most students
get this problem wrong. Check your answer carefully on
$[0,1]$ before proceeding to $[1,2]$.

5.   (a)   As another approach to Eq. (19), we might seek exponential solutions of the form $x(t) = e^{\lambda t}$, where
λ is a constant. Find appropriate values of the
coefficients a and b in Eq. (19) so that $x(t) = e^{\lambda t}$ will be a solution both for $\lambda = 0$ and $\lambda = 1$.

(b)   Using these values of a and b, show that Eq. (19)
with $x(0) = x_0$ can have infinitely many solutions
valid <u>for all</u> t in $\mathbb{R}$.

(c)   Do you suppose the solution of the equation in part
(b) would be unique if we specified both $x(0) = x_0$
and $x'(0) = x_1$?

6.   Example 3-5 showed that the equation $x'(t) = [x(t)]^{2/3}$
with $x(0) = 0$ has many solutions. What can you say
about existence and uniqueness of solutions of the equation

$$x'(t) = [x(t-r)]^{2/3}\qquad\text{for}\quad t \geq 0,$$

with $x(t) = \theta(t)$ for $-r \leq t \leq 0$, where r is a positive constant and θ is a given continuous initial function?

7.   We found in Example 3-2 that the equation $x'(t) = 1 +$
$x^2(t)$ with $x(0) = 1$ has a unique solution, but the
solution cannot be continued beyond $\pi/4$.   What can you
say about existence and uniqueness of solutions of the
equation

$$x'(t) = 1 + x^2(t-r) \quad \text{for} \quad t \geq 0,$$

with $x(t) = \theta(t)$, for $-r \leq t \leq 0$, where r is a positive constant and θ is a given continuous initial function?

8.   Prove that the equation $x'(t) = x(t)x(t-r)$ with continuous $\theta: [-r,0] \to \mathbb{R}$ has a unique solution for all
$t \geq 0$.   Compare this result, which holds for any $r > 0$
(no matter how small), with the situation when $r = 0$.

9.   State the best existence and uniqueness theorem you can
for the delay differential equation

$$x'(t) = f(t, x(t-1)),$$

where f is a given continuous function.

10. For the delay differential equation

$$x'(t) = f(t,x(t),x(t-1)),$$

find the best uniqueness theorem which follows from
Theorem 4-A.

11. (a) Show that, in principle, it is always possible to
         solve system (11) exactly for $t \geq 0$ when $x(t) =$
         $\theta_1(t)$ and $y(t) = \theta_2(t)$ for $-r \leq t \leq 0$, where
         θ_1 and θ_2 are given continuous functions. Hint.
         You will need to know how to handle a "Bernoulli
         Equation".

    (b) Can you say the same for system (10)?

    (c) Show that the solution of system (10) will be unique
         if $x(0) = x_0$ and $y(0) = y_0$ are given.

12. Show that it is possible for Eq. (15) to have undamped
    periodic solutions when m, k, b, q, and r are posi-
    tive constants. Hint. Try to find a solution of the
    form $x(t) = e^{\lambda t}$ where $\lambda = i\omega$ with ω real. Then
    verify that $\cos \omega t$ and $\sin \omega t$ are solutions. For
    simplicity set $m = 1$, $b = 1$, and $r = 1$.

13. Solve $x'(t) = [1 + x^2(t)]x(t/2)$ for $t \geq 1$ if
    $x(t) = 1$ on $[\frac{1}{2}, 1]$. How far can the solution be con-
    tinued?

*14. Consider $x'(t) = [1 + x(t)]x(t/2)$ for $t \geq 1$ with
    $x(t) = 1$ on $[\frac{1}{2}, 1]$. How far can the solution be con-
    tinued?

*15. Consider the equation (essentially that of Problem 4),

$$x'(t) = ax(t) + bx(t-r) \quad \text{for} \quad t \geq 0,$$

    with $x(t) = \theta(t)$ on $[-r,0]$, where θ is continuous.

    (a) Prove that a unique solution exists for all $t \geq 0$.

    (b) If $a < 0$ and $|b| \leq |a|$, prove that the solution

is bounded.   Outline.   Define a new function

$$v(t) = x^2(t) + |a| \int_{t-r}^{t} x^2(s)ds, \text{ where } x \text{ is the}$$

solution, and show that $v'(t) \leq 0$.   From this it
will follow that

$$|x(t)| \leq [\theta^2(0) + |a| \int_{-r}^{0} \theta^2(s)ds]^{1/2} \quad \text{for} \quad t \geq 0.$$

(c)   In the special case of part (b) when $a < 0$ and
$0 < b < |a|$, prove that the solution tends to zero
exponentially as $t \to \infty$.   Outline.   Define $\Delta(\lambda) =$
$\lambda - a - be^{-r\lambda}$, and note that $\Delta(0) > 0$ while
$\Delta(a) < 0$.   Hence for some γ in $(0, |a|)$,
$\Delta(-\gamma) = 0$.   Why?   Define $z(t) = x(t)e^{\gamma t}$, and show
that $z'(t) = (a+\gamma)z(t) - (a+\gamma)z(t-r)$ for $t \geq 0$.
Hence, by part (b), $z(t)$ is bounded.   From this
conclude that $x(t) = z(t)e^{-\gamma t} \to 0$ exponentially
as $t \to \infty$.

*16.   Consider Eq. (7) for $t > 0$ with $x(t) = \theta(t)$ for
$t \in [-1,0]$ where θ is continuous.

(a)   Prove that a unique solution exists for all $t \geq 0$
and that $x(t) > -1$ if $\theta(0) > -1$, $x(t) = -1$ if
$\theta(0) = -1$, and $x(t) < -1$ if $\theta(0) < -1$.

(b)   Prove that if $c > 0$ and $\theta(0) > -1$, then either
(i) $\lim_{t\to\infty} x(t) = 0$ or (ii) $x(t)$ oscillates about
$x = 0$, i.e., assumes both positive and negative
values for arbitrarily large   t.   In case (i) show
also that $\lim_{t\to\infty} x'(t) = 0$.   Interpret these results
for Eq. (6).

(c) What can you say about the solution of Eq. (7) in
case $c > 0$ and $\theta(0) < -1$?

17. (a) Prove that if we define $x(t) = z(e^{2^t})2^t - 1$, then
Eq. (16) is transformed into Eq. (7). What is c?

(b) Is $c > 0$? What function z corresponds to $x = 0$?

(c) Using Problem 16, what have you proved about solutions of Eq. (16)?

22. SOME DISTINGUISHING FEATURES AND SOME "WRONG" QUESTIONS

Before proving any general theorems about delay differential equations we shall illustrate some further ways in
which these equations differ from ordinary differential equations. We shall illustrate some inappropriate questions and
an invalid "method" which one might be tempted to apply to
delay differential equations.

But, after getting this out of the way, much of the
further discussion of delay differential equations in this
book will be aimed at showing how similar they are to ordinary differential equations when the "basic initial problem"
is properly formulated.

In discussing delay differential equations we adopt
some terminology from ordinary differential equations. For
example, the _order_ of a delay differential equation will mean
the order of the highest derivative involved in the equation.
Thus

$$x'(t) = ax(t) + bx(t-r) \tag{1}$$

and

$$x'(t) = -cx(t - 1)[1 + x(t)] \tag{2}$$

are of first order, while

$$mx''(t) + bx'(t) + qx'(t-r) + kx(t) = h(t) \qquad (3)$$

is of second order. Similarly, we will refer to a system
like 21-(11) or 21-(12) as a _system_ of first order equations.

    We shall say Eqs. (1) and (3) are linear and Eq. (2)
is nonlinear. The definition of "linear equation" is as
follows.

Definition. A _linear equation_ on $J = (\alpha, \beta)$ is an equation
of the form

$$L(t,x) = h(t)$$

where h is a given function and $L(t, \cdot)$ is a _linear opera-
tor_ for each t in J. That is,

$$L(t, c_1 y + c_2 z) = c_1 L(t,y) + c_2 L(t,z)$$

whenever y and z belong to some class of "admissible" functions, and c_1 and c_2 are real (or complex) numbers.
Briefly we will say L is a linear operator.

    For example, if we take as "admissible" the differen-
tiable functions, then Eq. (1) has the form $L(t,x) = 0$, where

$$L(t,x) \equiv x'(t) - ax(t) - bx(t-r).$$

The reader should verify that L is a linear operator.

    Similarly, if we take as "admissible" the twice dif-
ferentiable functions, then Eq. (3) has the form $L(t,x) =$
$h(t)$ where L is a linear operator.

    Why is Eq. (2) not linear? Problem 1.

    A linear equation of the form $L(t,x) = 0$ is said to

be a (linear) <u>homogeneous</u> equation.

Convince yourself that the above definitions of "linear" and "homogeneous" are satisfied by the ordinary differential equations and systems of equations to which we applied these terms in Chapters III and IV.

In this section we shall make observations about functional differential equations and illustrate those observations with examples. Each of the six examples will be a special case of the linear homogeneous delay differential system

$$x'(t) = A(t)x(t) + B(t)x(t-r), \qquad (4)$$

where A and B are continuous $n \times n$ matrix valued functions on $J = \mathbb{R}$ and $r > 0$ is a constant.

<u>Observation 1</u>. A first order (scalar), linear, homogeneous delay differential equation with real coefficients can have a nontrivial, oscillating solution (i.e., a solution, x, such that $x(t) = 0$ for arbitrarily large values of t, and yet $x(t) \not\equiv 0$). Show that this is impossible if $r = 0$. Problem 2.

<u>Example 1</u>. The equation

$$x'(t) = -x(t - \tfrac{\pi}{2}) \qquad (5)$$

has solutions of the form $x(t) = c_1 \cos t + c_2 \sin t$ for arbitrary constants c_1 and c_2. Verify it.

<u>Observation 2</u>. If one specifies

$$x(t) = \theta(t) \quad \text{(continuous)} \quad \text{for} \quad t_0 - r \leq t \leq t_0, \qquad (6)$$

and then tries to satisfy Eq. (4) for $t \leq t_0$, there will, in general, be no (continuous) solution. This remains true even if θ is infinitely differentiable.

Example 2.   Consider Eq. (1) with $b \neq 0$ and $r > 0$, and assume $a \neq -b$. Let $\theta(t) = k$, a nonzero constant on $[-r,0]$. Then, if x were to be a solution, we would require

$$x(t) = -ak/b \quad \text{on} \quad [-2r,-r].$$

Why? But, since $-ak/b \neq k$, we would have a discontinuity at $-r$ and so x would not be a solution.

Observation 3.   If θ is chosen so that Eqs. (4) and (6) do have a (backwards) solution for $t \leq t_0$, this solution will not, in general, be unique.

Example 3   (Winston and Yorke [1969]).   Consider the linear scalar equation

$$x'(t) = b(t)x(t-1) \quad \text{for} \quad t \geq 0, \tag{7}$$

where

$$b(t) = \begin{cases} 0 & \text{for } t \leq 0, \\ \cos 2\pi t - 1 & \text{for } 0 < t \leq 1, \\ 0 & \text{for } t > 1. \end{cases}$$

(Note that b is continuous.) Let $x(t) = k$ for $t \leq 0$, where k is any constant. Then, Eq. (7) is clearly satisfied on $(-\infty,0]$. Thus on $[0,1]$

$$x'(t) = (\cos 2\pi t - 1)k, \quad \text{with} \quad x(0) = k.$$

Hence

$$x(1) = k + k\int_0^1 (\cos 2\pi t - 1)dt = 0.$$

But then the fact that $x'(t) = 0$ for $t \geq 1$ implies $x(t) = 0$ for all $t \geq 1$.

This shows that, given $x(t) = \theta(t) \equiv 0$ for $1 \leq t \leq 2$, Eq. (7) has infinitely many continuous solutions valid for $t \leq 2$.

Note that this example also shows non-existence if $x(t) = \theta(t) \not\equiv 0$ on $[1,2]$.

<u>Observation</u> 4. If one merely specifies

$$x(t_0) = x_0, \tag{8}$$

there may still be no solution of (4) for $t \leq t_0$.

This follows from Example 3 if one sets $x(1) = x_0 \neq 0$.

However, it is of interest to show that this same phenomenon can also occur for a <u>system</u> (4) having <u>constant coefficients</u>. Several mathematicians tried for some years to prove that this could not occur. But examples of the following type were finally discovered independently and almost simultaneously by Popov [1971] in the United States and Zverkin [1971] in the Soviet Union. The following example is Popov's.

<u>Example</u> 4. Let $r = 1$ and let

$$A = \begin{bmatrix} 0 & 2 & 0 \\ 0 & 0 & -1 \\ 0 & 0 & 0 \end{bmatrix} \quad \text{and} \quad B = \begin{bmatrix} 0 & 0 & 0 \\ 1 & 0 & 0 \\ 0 & 2 & 0 \end{bmatrix}.$$

Let x satisfy system (4) for $t \geq t_1$:

$$\begin{aligned}
x_1'(t) &= 2x_2(t) \\
x_2'(t) &= -x_3(t) + x_1(t-1) \\
x_3'(t) &= 2x_2(t-1).
\end{aligned} \tag{9}$$

Then for $t \geq t_1+1$ one finds successively

$$x_1'(t-1) - x_3'(t) = 0,$$
$$x_1(t-1) - x_3(t) = c_1,$$
$$x_2(t) = c_2 + c_1 t,$$
$$x_1(t) = c_3 + 2c_2 t + c_1 t^2,$$

where c_1 , c_2 , and c_3 are arbitrary constants. Thus, for $t \geq t_1+2$

$$x_3(t) = c_3 - 2c_2 + 2(c_2-c_1)t + c_1 t^2,$$

and so,

$$x_1(t) - 2x_2(t) - x_3(t) \equiv 0.$$

Clearly then, there cannot be any solution of (4) on $(-\infty, t_0]$ with $x(t_0) = x_0 = \text{col}\ (x_{01}, x_{02}, x_{03})$ if

$$x_{01} - 2x_{02} - x_{03} \neq 0.$$

Observation 5. If a solution of (4) and (8) does exist on $(-\infty, t_0]$ it will, in general, be nonunique. In fact, even if one specifies x plus all its derivatives at t_0 ,

$$x(t_0) = x_0, \quad x'(t_0) = b_1, \quad x''(t_0) = b_2, \quad \dots, \qquad (10)$$

there may be infinitely many different solutions of (4) on $(-\infty, t_0]$ which satisfy (10).

Nonuniqueness on $(-\infty, t_0]$ for Eqs. (4) and (8) was encountered in Problem 21-5(b). It is also illustrated by Examples 1 and 3 above. How?

To show that even (10), the specification of x and all its derivatives at t_0 , does not assure uniqueness on $(-\infty, t_0]$ one could use Eq. (7) with $x(t_0) = 0$, $x'(t_0) = 0$,

$x''(t_0) = 0,\ldots$ for any $t_0 > 1$. But let us also give an example using the scalar Eq. (1) with constant coefficients. This will then answer Problem 21-5(c).

Example 5. We shall consider Eq. (1) with $b \neq 0$ and $r > 0$. For simplicity, let $t_0 = 0$. Then consider the function $\theta: [-r,0] \to \mathbb{R}$ defined by

$$\theta(t) = \begin{cases} 0 & \text{for } t = -r \\ e^{-t^{-2}} e^{-(t+r)^{-2}} & \text{for } -r < t < 0 \\ 0 & \text{for } t = 0. \end{cases} \tag{11}$$

We leave it as a nontrivial Problem 3 for the reader to show that

(i) θ has continuous derivatives of all orders on $[-r,0]$, and

(ii) θ and all its derivatives are zero at $t = -r$ and $t = 0$. (We interpret the derivatives as one-sided at the ends of the interval.)

Now let us try to use θ as an "initial function" to obtain a solution y of Eq. (1) for all t. We first consider the past. If Eq. (1) is to be satisfied for $-r \leq t \leq 0$, we must have

$$y(t-r) = \frac{\theta'(t) - a\theta(t)}{b} \quad \text{for } -r \leq t \leq 0$$

or

$$y(t) = \frac{\theta'(t+r) - a\theta(t+r)}{b} \quad \text{for } -2r \leq t \leq -r. \tag{12}$$

The important thing to note is that (12) defines a function y on $[-2r,-r]$ having properties like (i) and (ii) listed above for θ. Thus we can perform a similar computation to extend y to $[-3r,-2r]$. This construction can be continued

backwards by steps indefinitely.   (One can verify that the
graph of y on each succeeding interval to the left has more
oscillations and greater amplitudes, as suggested by Figure 1.

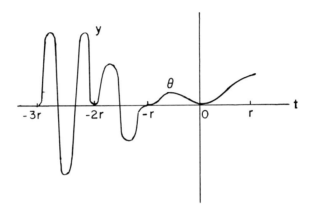

Figure 1

(Clearly y could also be continued into the future by the
method of steps.)   Thus y satisfies Eq. (1) for all t, and

$$y(0) = 0, \quad y'(0) = 0, \quad y''(0) = 0, \quad \dots \qquad (13)$$

But this solution of Eqs. (1) and (13) is not unique.   For
any multiple of y is also a solution (including the identically zero function).

Furthermore it is easily verified (Problem 4) that if
x on $(-\infty,\infty)$ is a solution of Eqs. (1) and (10) with $t_0 = 0$
and if y is a solution on $(-\infty,\infty)$ of Eqs. (1) and (13),
then, for any constant c, x+cy is another solution on
$(-\infty,\infty)$ of (1) and (10).   Thus if we have a solution of Eq.
(1) on $(-\infty,\infty)$ satisfying any given conditions of the type
(10) there must also be infinitely many other such solutions.
For instance, Eq. (5) with $x^{(2k)}(0) = (-1)^k$ and

$x^{(2k+1)}(0) = 0$ $(k = 0,1,...)$ is satisfied by $x(t) = \cos t$ and by infinitely many other functions.

Remarks. The phenomena of nonexistence and nonuniqueness described in Observations 4 and 5 could not occur for a linear ordinary differential system. For we know from Sections 5 and 20 that if A is a continuous matrix-valued function and h (not necessarily zero) is a continuous vector-valued function on $\mathbb{R}$, then the ordinary differential system

$$x'(t) = A(t)x(t) + h(t)$$

always has a unique solution on $\mathbb{R}$ satisfying any given initial condition of the form $x(t_0) = x_0$.

It should be mentioned that, in spite of Observations 4 and 5, there are some special delay differential equations arising in applications for which it <u>does</u> make sense to talk of solutions on $(-\infty, t_0]$ subject to a given initial condition $x(t_0) = x_0$. See, for example, Driver [1969], [1976a], and Ždanov [1975], and Hsing [1977].

Occasionally one encounters a published book or research paper in which the author proposes to analyze a delay differential equation by replacing $x(t-r)$ with the first few terms of a Maclaurin series, say

$$x(t) - rx'(t) + \frac{1}{2} r^2 x''(t) - \cdots + (-1)^m \frac{1}{m!} r^m x^{(m)}(t).$$

Beware of such papers!

Observation 6. The solutions of Eq. (4) or (1) may behave quite differently from the solutions of the "approximating" ordinary differential equation obtained if one replaces x(t-r) with the first few terms of the Maclaurin series.

Example 6.   Consider the scalar delay differential equation

$$x'(t) = -2x(t) + x(t-r).$$

Every solution is bounded (and tends to zero as $t \to \infty$).   (If
you have not proved this in Problem 21-15, then just accept
the statement without proof until Sections 31 and 32.)   However, the ordinary differential equation

$$x'(t) = -2x(t) + [x(t) - rx'(t) + \frac{1}{2}r^2x''(t)]$$

has exponentially increasing solutions, $ce^{\lambda t}$ with $\lambda > 0$,
regardless of how small the value of $r > 0$.   Prove this.

Moreover, the equation

$$x'(t) = -3x(t) - 2x(t-1)$$

has all its solutions bounded as $t \to \infty$; while the ordinary
differential equation obtained by replacing x(t-1) by
x(t)-x'(t) has exponentially increasing solutions x(t) =
ce^{5t}.   Check it.

It should be remarked nevertheless, that for some
equations with "sufficiently small" delay one does get useful
approximations by replacing x(t-r) by x(t) or by
x(t)-rx'(t).   See, for example, Driver, Sasser, and Slater
[1973].

Problems

1.   Show that Eq. (2) is nonlinear.

2. If a is a real valued function on $\mathbb{R}$, use the method of
    Section 2 to prove that the equation $x'(t) = a(t)x(t)$
    cannot have a nontrivial, oscillating solution.

*3. Verify properties (i) and (ii) for the function θ de-
    fined by Eq. (11).

4. Verify that if x and y are two solutions of Eq. (1)
    which satisfy conditions (10) with $t_0 = 0$ and condition
    (13) respectively, then $x + cy$ is another solution
    which satisfies (10).

23. LIPSCHITZ CONDITION AND UNIQUENESS
FOR THE BASIC INITIAL PROBLEM

The difficulties encountered in Section 22 for various
"alternative problems" associated with delay differential
equations lead us to return now to the initial-function problem described in Section 21. Thus we specify $x(t) = \theta(t)$
for $t \leq t_0$ and seek a continuation which satisfies the delay differential equation for $t \geq t_0$. We shall henceforth
refer to this type of problem as the basic initial problem
for a delay differential equation.

Our first general observation with respect to the
basic initial problem is that it is sometimes easier to prove
uniqueness or existence for a differential equation with
delay than for one without. For example, the equation

$$x'(t) = f(t, x(t-1))   \quad \text{for} t \geq 0,$$

with $x(t) = \theta(t)$ given for $-1 \leq t \leq 0$, can be treated un-
ambiguously by the method of steps. All one requires is the

integrability of $f(t,\theta(t-1))$ in order to assert the existence of a unique solution on $[0,1]$.

No comparable assertion can be made for the equation
$x'(t) = f(t,x(t))$.

We have already encountered examples in which the presence of a delay causes uniqueness (Problem 21-6) or simplifies the computation of the solution (Problem 21-11(a)).
However, in case the delay can become zero, as it does at
$t = 0$ in

$$x'(t) = x(t/2) \quad \text{for} \quad t \geq 0,$$

no such simplification results from the presence of the delay.
In fact uniqueness does not follow from any of our old
theorems.

We shall now generalize our earlier uniqueness theorems to handle delay differential equations with fairly general types of delays.

Consider a system of delay differential equations

$$x'(t) = f(t,x(g_1(t)),\ldots,x(g_m(t))), \tag{1}$$

where x and f are n-vector-valued functions, and each
$g_j(t)$ is a "retarded argument", i.e., $g_j(t) \leq t$ for
$j = 1,\ldots,m$. We will often have $g_1(t) \equiv t$.

Sometimes one writes $t - r_j(t)$ instead of $g_j(t)$.
Then the non-negative quantities $r_j(t)$ are the "delays".

We shall assume the existence of a number $\gamma \leq t_0$
such that, for each $j = 1,\ldots,m$,

$$\gamma \leq g_j(t) \leq t \quad \text{for} \quad t_0 \leq t < \beta.$$

Let J be an interval in $\mathbb{R}$ and let D be an open set in $\mathbb{R}^n$. Sometimes J will be $[t_0,\beta)$ and sometimes it will be (α,β) where $\alpha < t_0 < \beta$. Let f map $J \times D^m \rightarrow \mathbb{R}^n$. (Note that D is not the analog of the set D in Chapter II. For there D itself was the domain of f.)

We shall consider Eq. (1) for $t_0 \leq t < \beta$ together with the initial condition

$$x(t) = \theta(t) \quad \text{for} \quad \gamma \leq t \leq t_0, \tag{2}$$

where θ is a given initial function mapping $[\gamma,t_0] \rightarrow D$.

Since we shall often refer to Eq. (1), it is convenient to have a briefer notation for it. Given $t \in [t_0,\beta)$ and given any function $\chi: [\gamma,t] \rightarrow D$, define

$$F(t,\chi_t) \equiv f(t,\chi(g_1(t)),\ldots,\chi(g_m(t))).$$

With this notation, Eq. (1) takes the compact form

$$x'(t) = F(t,x_t). \tag{1'}$$

More will be made of this notation in Section 25. But for now (1') is nothing more nor less than an abbreviation for Eq. (1).

Definitions. A solution of Eqs. (1) and (2) is a continuous function $x: [\gamma,\beta_1) \rightarrow D$, for some $\beta_1 \in (t_0,\beta]$, such that

(i) $x(t) = \theta(t)$ for $\gamma \leq t \leq t_0$, and

(ii) $x'(t) = F(t,x_t)$ for $t_0 \leq t < \beta_1$.

(We understand $x'(t_0)$ to mean the right-hand derivative.)

The solution of Eqs. (1) and (2) is said to be unique if every two solutions agree with each other as far as both are defined.

We shall assume that f and $g_1, \ldots, g_m$ are continuous. Thus if x is a continuous function mapping $[\gamma, \beta_1) \to D$, for some $\beta_1 \in (t_0, \beta]$, then $F(t, x_t) = f(t, x(g_1(t)), \ldots, x(g_m(t)))$ depends continuously on t (since it is a composition of continuous functions). Hence x is a solution of Eqs. (1) and (2) if and only if

$$
x(t) = \begin{cases} \theta(t) & \text{for } \gamma \le t \le t_0 \\[2mm] \theta(t_0) + \int_{t_0}^{t} F(s, x_s) ds & \text{for } t_0 \le t < \beta_1. \end{cases} \tag{3}
$$

Equation (3) is often more convenient to work with than Eqs. (1) and (2).

You might guess that, in order to establish uniqueness, we should assume something like a Lipschitz condition on f.

Definition. We say that f satisfies a _Lipschitz condition_ with Lipschitz constant K on a set $G \subset J \times D^m$ if

$$
|| f(t, \xi_{(1)}, \ldots, \xi_{(m)}) - f(t, \tilde{\xi}_{(1)}, \ldots, \tilde{\xi}_{(m)}) ||
$$
$$
\le K \max_{j=1, \ldots, m} || \xi_{(j)} - \tilde{\xi}_{(j)} || \tag{4}
$$

for all $(t, \xi_{(1)}, \ldots, \xi_{(m)})$ and $(t, \tilde{\xi}_{(1)}, \ldots, \tilde{\xi}_{(m)})$ in G.
We shall then also say f is _Lipschitzian_ on G.

(As usual, the norm of a vector $\xi = \text{col} (\xi_1, \ldots, \xi_n)$ in R^n is defined by $||\xi|| = \Sigma_{i=1}^{n} |\xi_i|$.)

If f satisfies Lipschitz condition (4) on G and if $t \in [t_0, \beta)$ and $\chi, \tilde{\chi}: [\gamma, t] \to D$ are such that

$$
(t, \chi(g_1(t)), \ldots, \chi(g_m(t))), \quad (t, \tilde{\chi}(g_2(t)), \ldots, \tilde{\chi}(g_m(t))) \in G,
$$

then it follows that

$$||F(t,x_t) - F(t,\tilde{x}_t)|| \leq K \max_{j=1,\ldots,m} ||x(g_j(t)) - \tilde{x}(g_j(t))||$$

$$\leq K \sup_{\gamma \leq s \leq t} ||x(s) - \tilde{x}(s)||. \qquad (4')$$

If f is Lipschitzian on the entire set $J \times D^m$, it is quite easy to prove that Eqs. (1) and (2) have at most one solution. But such a global Lipschitz condition is seldom satisfied for nonlinear equations. Fortunately a weaker assumption -- a "local Lipschitz condition" -- suffices for uniqueness. So we proceed to define this condition, and then prove the correspondingly stronger uniqueness theorem.

Definition. We say that $f: J \times D^m \rightarrow \mathbb{R}^n$ satisfies a local Lipschitz condition (or f is locally Lipschitzian) if for each point $(t_1, \eta_{(1)}, \ldots, \eta_{(m)}) \in J \times D^m$ there exist numbers $a > 0$ and $b > 0$ such that each

$$A_j \equiv \{\xi \in \mathbb{R}^n : ||\xi - \eta_{(j)}|| \leq b\} \qquad (j = 1,\ldots,m) \qquad (5)$$

is a subset of D and f is Lipschitzian on the set

$$([t_1-a,t_1+a] \cap J) \times A_1 \times A_2 \times \cdots \times A_m.$$

Theorem A. Let f be continuous and locally Lipschitzian on $[t_0,\beta) \times D^m$, let each g_j be continuous with $\gamma \leq g_j(t) \leq t$ on $[t_0,\beta)$, and let θ be continuous on $[\gamma,t_0] \rightarrow D$. Then Eqs. (1) and (2) have at most one solution on any interval $[\gamma,\beta_1)$ where $t_0 < \beta_1 \leq \beta$.

Proof (Similar to that of Theorem 4-A.) Suppose (for contradiction) that there are two solutions x and $\tilde{x}$ mapping $[\gamma,\beta_1) \rightarrow D$ with $x \neq \tilde{x}$. Since $x(t) = \tilde{x}(t)$ on $[\gamma,t_0]$ it follows that $x(t) \neq \tilde{x}(t)$ for some t in (t_0,β_1). Let

$$t_1 = \inf \{t \in (t_0,\beta_1): x(t) \neq \tilde{x}(t)\}.$$

Then $t_0 < t_1 < \beta_1$ and, from the continuity of   x   and $\tilde{x}$,

$$x(t) = \tilde{x}(t) \quad \text{for} \quad \gamma \leq t \leq t_1.$$

Since $t_0 < t_1 < \beta_1$ and since each $x(g_j(t_1))$ is in the open set   D, we can choose numbers $a > 0$ and $b > 0$ such that $[t_1-a,t_1+a] \subset (t_0,\beta_1)$, each A_j as defined in (5) is a subset of   D, and   f   is Lipschitzian on $[t_1-a,t_1+a] \times A_1$ $\times \cdots \times A_m$.   Let the Lipschitz constant be   K.

Now, for some $\beta_2 \in (t_1,t_1+a)$ the points $x(g_j(t))$ and $\tilde{x}(g_j(t))$ must remain in A_j for $t_1 \leq t < \beta_2$, $j = 1,\ldots,m$.   Why?   Thus for $t_1 \leq t < \beta_2$, since   x   and $\tilde{x}$ both satisfy (3), we find

$$||x(t) - \tilde{x}(t)|| = ||\int_{t_1}^{t} [F(s,x_s) - F(s,\tilde{x}_s)]ds||$$

$$\leq K \int_{t_1}^{t} \sup_{\gamma \leq \sigma \leq s} ||x(\sigma)-\tilde{x}(\sigma)||\ ds$$

Let us now define $v(t) = \sup_{\gamma \leq s \leq t} ||x(s)-\tilde{x}(s)||$.   Then

$$||x(t) - \tilde{x}(t)|| \leq K \int_{t_1}^{t} v(s)ds \quad \text{for} \quad t_1 \leq t < \beta_2,$$

or, since $x(s) = \tilde{x}(s)$ when $\gamma \leq s \leq t_1$,

$$v(t) \leq K \int_{t_1}^{t} v(s)ds \quad \text{for} \quad t_1 \leq t < \beta_2.$$

It follows from Lemma 8-A that $v(t) = 0$, and hence $x(t) = \tilde{x}(t)$ for all   t   in $[\gamma,\beta_2)$ -- contradicting the definition of t_1. $\square$

Example 1.   The equation $x'(t) = x(t/2)$ with $x(0) = x_0$ has at most one solution on any interval $[0,\beta)$ with $\beta > 0$.

This follows at once from Theorem A because of the continuity of the functions f and g involved and the global Lipschitz condition satisfied by f.

Clearly, the requirement that f be locally Lipschitzian is weaker than the requirement that it be Lipschitzian. A simple sufficient condition for the local Lipschitz condition is as follows.

Lemma B. If $f: [t_0,\beta) \times D^m \to \mathbb{R}^n$ has continuous first partial derivatives with respect to all but its first argument, then f is locally Lipschitzian.

Proof. Let $(t_1,\eta_{(1)},\ldots,\eta_{(m)})$ be any point in $J \times D^m = [t_0,\beta) \times D^m$. Choose a > 0 sufficiently small so that $t_1+a < \beta$, and choose b > 0 sufficiently small so that whenever $||\xi-\eta_{(j)}|| \leq b$ for some j = 1,..., or m, then $\xi \in D$. (This choice of b is possible because D is an open set.)

Now $J_1 \equiv [t_1-a,t_1+a] \cap J$ is a closed bounded subinterval of J, and each

$$A_j \equiv \{\xi \in \mathbb{R}^n : ||\xi - \eta_{(j)}|| \leq b\} \subset D$$

is closed and bounded. Hence $J_1 \times A_1 \times \cdots \times A_m$ is a closed, bounded subset of $J \times D^m$ and, by Theorem A2-I, it follows that each of the continuous partial derivatives of f is bounded there. (In most cases which we will encounter it will be clear that these partial derivatives are bounded, and we will easily be able to find a bound without resorting to Theorem A2-I.) Let B be a common bound for $|D_{1+\ell}f_i|$ for all i = 1,...,n and ℓ = 1,...,mn.

The set $A_1 \times \cdots \times A_m$ is convex. For if $(\xi_{(1)}, \ldots, \xi_{(m)})$ and $(\tilde{\xi}_{(1)}, \ldots, \tilde{\xi}_{(m)})$ are any two points in $A_1 \times \cdots \times A_m$, if $0 \leq s \leq 1$, and if

$$\zeta_{(j)}(s) \equiv (1-s)\xi_{(j)} + s\tilde{\xi}_{(j)} \quad \text{for} \quad j = 1, \ldots, m,$$

then

$$||\zeta_{(j)}(s) - \eta_{(j)}|| = ||(1-s)(\xi_{(j)} - \eta_{(j)}) + s(\tilde{\xi}_{(j)} - \eta_{(j)})||$$

$$\leq (1-s)b + sb = b.$$

Hence $(\zeta_{(1)}(s), \ldots, \zeta_{(m)}(s)) \in A_1 \times \cdots \times A_m$.

Thus by the Mean Value Theorem (A2-D), if $(t, \xi_{(1)}, \ldots, \xi_{(m)})$ and $(t, \tilde{\xi}_{(1)}, \ldots, \tilde{\xi}_{(m)}) \in J_1 \times A_1 \cdots \times A_m$, then for each $i = 1, \ldots, n$

$$|f_i(t, \xi_{(1)}, \ldots, \xi_{(m)}) - f_i(t, \tilde{\xi}_{(1)}, \ldots, \tilde{\xi}_{(m)})|$$

$$\leq \sum_{j=1}^{m} \sum_{k=1}^{n} B|\xi_{(j)k} - \tilde{\xi}_{(j)k}|$$

$$\leq mB \max_{j=1, \ldots, m} ||\xi_{(j)} - \tilde{\xi}_{(j)}||.$$

Summing over $i = 1, \ldots, n$ and letting $K = nmB$, we find

$$||f(t, \xi_{(1)}, \ldots, \xi_{(m)}) - f(t, \tilde{\xi}_{(1)}, \ldots, \tilde{\xi}_{(m)})||$$

$$\leq K \max_{j=1, \ldots, m} ||\xi_{(j)} - \tilde{\xi}_{(j)}||. \quad \square$$

Example 2.   The equation

$$x'(t) = [1 + tx(t)x(t-1)]/x(t/2) \quad \text{for} \quad t > 0$$

with $x(t) = \theta(t) = \cos t$ for $-1 \leq t \leq 0$ has at most one solution on any interval $[-1, \beta_1)$. This is proved as follows. The function defined by

$$f(t, \xi_{(1)}, \xi_{(2)}, \xi_{(3)}) = [1 + t\xi_{(1)}\xi_{(2)}]/\xi_{(3)}$$

is continuous and (by Lemma B) locally Lipschitzian on $\mathbb{R} \times D^3$ where $D = (0,\infty)$. Moreover, θ maps $[-1,0] \to D$ and is continuous. Thus Theorem A applies.

   <u>Remarks</u>. The definitions we have given and Theorem A all apply to ordinary differential equations as a special case. That would be the case in which $m = 1$, $g_1(t) \equiv t$, and $\gamma = t_0$.

   But, on the other hand, Theorem A fails to assert uniqueness for some simple delay differential equations (such as $x'(t) = [x(t-1)]^{2/3}$ with $\theta(t) \equiv 0$) for which we already know that the solution <u>is</u> unique. Verify this.

   Equations of second order and higher are treated by transforming into first order systems -- as we did for ordinary differential equations. Thus we get the following corollary of Theorem A and Lemma B.

<u>Corollary C</u>. Let f be continuously differentiable on $[t_0,\beta) \times D^m \to \mathbb{R}$ for some open set $D \subset \mathbb{R}^n$, and let g_j be continuous with $\gamma \le g_j(t) \le t$ on $[t_0,\beta)$ for $j = 1,\ldots,m$. Consider the n'th order scalar delay differential equation

$$x^{(n)}(t) = f(t, x(g_1(t)), x'(g_1(t)), \ldots, x^{(n-1)}(g_1(t)),$$
$$\ldots, x(g_m(t)), x'(g_m(t)), \ldots, x^{(n-1)}(g_m(t))) \tag{6}$$

on $[t_0,\beta)$ with

$$x(t) = \theta(t) \quad \text{on} \quad [\gamma,t_0], \tag{7}$$

where θ and its first $n-1$ derivatives are continuous and $(\theta(t), \theta'(t), \ldots, \theta^{(n-1)}(t)) \in D$ on $[\gamma,t_0]$. Then there is at most one solution of (6) and (7) on any interval $[\gamma,\beta_1)$

where $t_0 < \beta_1 \leq \beta$. [A _solution_ means an $n-1$ times continuously differentiable function on $[\gamma,\beta_1) \to \mathbb{R}$ such that on
$[\gamma,t_0]$ Eq. (7) holds and on $[t_0,\beta_1)$ $(x(t),x'(t),\ldots,$
$x^{(n-1)}(t)) \in D$ and (6) is satisfied.]

   Proof. Let x be a solution of (6) and (7) on $[\gamma,\beta_1)$.
Define an n-vector-valued function by

$$y(t) = \text{col } (x(t),x'(t),\ldots,x^{(n-1)}(t))$$

for $\gamma \leq t < \beta_1$. Then y satisfies the system of first order delay differential equations

$$y_1'(t) = y_2(t)$$
$$y_2'(t) = y_3(t)$$
$$\vdots$$
$$y_{n-1}'(t) = y_n(t)$$
$$y_n'(t) = f(t,y(g_1(t)),\ldots,y(g_m(t)))$$

on $[t_0,\beta_1)$ with

$$y(t) = \text{col } (\theta(t),\theta'(t),\ldots,\theta^{(n-1)}(t))$$

on $[\gamma,t_0]$. The uniqueness of y, and hence of x, follows
from Theorem A. $\square$

   We round out this section by giving an elementary continuous dependence result. That is, we shall obtain an estimate for the change in the solution of a system of delay differential equations due to a change in the initial function
θ.

Theorem _D._ Let $f : [t_0,\beta) \times D^m \to \mathbb{R}^n$ be continuous and (globally) Lipschitzian with Lipschitz constant K, let each g_j

be continuous with $\gamma \leq g_j(t) \leq t$ on $[t_0,\beta)$, and let $\theta: [\gamma,t_0] \to D$ be continuous. Now let $\tilde{\theta}: [\gamma,t_0] \to D$ also be continuous, and let x and $\tilde{x}$ be solutions of Eq. (1) on $[\gamma,\beta_1)$ satisfying Eq. (2) with θ and $\tilde{\theta}$ respectively. Then, for $t_0 \leq t < \beta_1$

$$||x(t) - \tilde{x}(t)|| \leq \sup_{\gamma \leq s \leq t_0} ||\theta(s) - \tilde{\theta}(s)|| e^{K(t-t_0)}. \qquad (8)$$

Remark. This theorem applies, in particular, to a system of n ordinary differential equations. Compare with Theorem 8-B.

Proof of Theorem D. The two solutions x and $\tilde{x}$ satisfy Eqs. (3) and (3̃) respectively, where (3̃) means Eq. (3) with θ replaced by $\tilde{\theta}$. Subtracting we get for $t_0 \leq t < \beta$,

$$||x(t)-\tilde{x}(t)|| = ||\theta(t_0)-\tilde{\theta}(t_0) + \int_{t_0}^{t} [F(s,x_s)-F(s,\tilde{x}_s)]ds||$$

$$\leq ||\theta(t_0) - \tilde{\theta}(t_0)||$$

$$+ K\int_{t_0}^{t} \sup_{\gamma \leq \sigma \leq s} ||x(\sigma) - \tilde{x}(\sigma)||ds.$$

Once again, we define $v(t) = \sup_{\gamma \leq s \leq t} ||x(s) - \tilde{x}(s)||$. Then

$$||x(t) - \tilde{x}(t)|| \leq v(t_0) + K\int_{t_0}^{t} v(s)ds \quad \text{for } t_0 \leq t < \beta_1,$$

or, since $||\theta(t) - \tilde{\theta}(t)|| \leq v(t_0)$ for $\gamma \leq t \leq t_0$,

$$v(t) \leq v(t_0) + K\int_{t}^{t} v(s)ds \quad \text{for } t_0 \leq t < \beta_1.$$

It follows, with the aid of Lemma 8-A, that

$$||x(t) - \tilde{x}(t)|| \leq v(t_0) e^{K(t-t_0)} \quad \text{for } t_0 \leq t < \beta_1,$$

which is assertion (8). $\square$

Problems

1. Prove that each of the delay differential equations

   (a) $x'(t) = [1 + x^2(t)]x(t/2)$

   (b) $x'(t) = [1 + x(t)]x(t/2)$

   has at most one solution on $[0,\beta_1)$ for any $\beta_1 > 0$ if
   the value of $x(0) = x_0$ is specified arbitrarily.   Why
   can you not use the method of steps?

2. Without using the method of steps, prove that system
   21-(11) has at most one solution on any interval
   $[t_0-r,\beta_1)$ if $x(t) = \theta_1(t)$ and $y(t) = \theta_2(t)$ for
   $t \in [t_0-r,t_0]$, where θ_1 and θ_2 are any continuous,
   real-valued functions.

3. Without using the method of steps, prove that Eq. 21-(15)
   has at most one continuously differentiable solution on
   any interval $[t_0-r,\beta)$ if $x(t) = \theta(t)$ for $t_0-r \leq t \leq$
   t_0, where θ is given continuously differentiable.

4. Prove that Eq. 21-(16) has at most one solution on $[1,\beta_1)$
   for any $\beta_1 > 1$ if $z(1) = z_0$ is specified arbitrarily.

5. Determine whether or not the equation $x'(t) = [x(t/2)]^{2/3}$
   has a unique solution for $t \geq t_0 = 0$ if $x(0) = 0$.

6. Let x and $\tilde{x}$ be solutions of $x'(t) = ax(t) + bx(t-r)$
   on $[-r,\infty)$ corresponding to initial functions θ and $\tilde{\theta}$
   respectively on $[-r,0]$. Find an estimate for $||x(t) -$
   $\tilde{x}(t)||$ valid for $t \geq 0$.

*7.   Consider Eq. 21-(15),

$$mx''(t) + bx'(t) + qx'(t-r) + kx(t) = 0,$$

with $m \neq 0$.   If $\varepsilon > 0$ and $\beta > 0$ are considered to be
given numbers, find conditions on θ: $[-r,0] \to \mathbb{R}$ to assure that the solution with $x(t) = \theta(t)$ for $-r \leq t \leq 0$
satisfies $|x(t)| < \varepsilon$ for $-r \leq t < \beta$.

*8.   Consider the ordinary differential equation

$$x'(t) = f_1(t,x(t))\, f_2(x(t)).$$

Let f_1, mapping $(\alpha,\beta) \times (\gamma,\delta) \to \mathbb{R}$, be continuous and
locally Lipschitzian, and let f_2, mapping $(\gamma,\delta) \to \mathbb{R}$,
be <u>positive</u>.   Then prove that if $(t_0,x_0) \in (\alpha,\beta) \times (\gamma,\delta)$
the differential equation has at most one solution on
(α,β) such that $x(t_0) = x_0$.   <u>Hint</u>:   Define

$$h(\xi) \equiv \int_{x_0}^{\xi} \frac{du}{f_2(u)}$$ and then introduce $y(t) = h(x(t))$.

9.   Show that the equation $x'(t) = 1 + 3[x(t) - t]^{2/3}$ with
$x(0) = 0$ has at least two solutions, namely $x(t) = t$
and $x(t) = t + t^3$.   Why does this not contradict the
assertion of Problem 8 with $f_1(t,\xi) \equiv 1$?

*10.   Show that if g is continuous and $g(t) \leq t$ for
$t \geq t_0$, then
$$x'(t) = x(t) - x(g(t))   \text{ for }   t \geq t_0$$

with $x(t) = \theta(t) \equiv k$ for all $t \leq t_0$ has at most one
solution on $(-\infty,\beta_1)$ for any $\beta_1 > t_0$.   Find a solution.

Chapter VI. <u>EXISTENCE</u> <u>THEORY</u>

In Section 20, we proved the existence of solutions of linear systems of ordinary differential equations with variable coefficients, even though we were not able to write down the solution explicitly. In this chapter existence theorems will be proved for some rather general nonlinear differential systems, both with and without delays.

The fact that we have not given any such proofs thus far has not meant any lack of rigor. For none of the nonlinear examples or problems up to now has required any knowledge of existence of solutions beyond what was clearly true in the particular cases considered. For example we have never assumed the existence of a solution for the problem

$$x'(t) = \sin x(t) - \sin t \quad \text{with} \quad x(1) = 1.$$

Before one attempts to go much further in the study of differential equations, however, it is important to have some general existence theorems. Moreover, the proofs of existence theorems sometimes provide methods for the numerical computation of solutions in case our simple methods fail.

The theorems in this chapter will make use of Lipschitz conditions, as well as continuity, for the given functions in the differential equations. It is possible to weaken these hypotheses; but the existence theorems given here will suffice for most practical purposes.

24. ORDINARY DIFFERENTIAL SYSTEMS

The method of successive approximations, previously
used to prove existence for a linear system of ordinary differential equations, will now be applied to a nonlinear system. This requires some extra care. For, unless the values
of t and the first guess, $x_{(0)}(t)$, are suitably restricted,
the succeeding approximations may not be well defined; or, if
defined, the sequence $\{x_{(\ell)}\}$ may not converge.

The need for restrictions on t should not be surprising since we already know that the actual solutions of a nonlinear equation may not be continuable very far. For example,
the solution of

$$x'(t) = x^2(t) \quad \text{with} \quad x(0) = x_0 > 0.$$

is only valid for $t < 1/x_0$. However, we shall find that the
restrictions to be imposed on t in the existence proof are
often much stronger than actually dictated by any inherent
noncontinuability of the solution.

In this section we consider a system of n ordinary
differential equations (without delays) of the form

$$x'(t) = f(t,x(t)) \tag{1}$$

with initial conditions

$$x(t_0) = x_0. \tag{2}$$

The theorems to be proved in Section 26 for delay differential systems will essentially supercede those of the
present section. But it seems worthwhile to go through the
slightly simpler special case of ordinary differential systems

first.

Let $J = (\alpha, \beta)$ and let D be an open set in $\mathbb{R}^n$. Let f be a continuous function mapping $J \times D \to \mathbb{R}^n$, and let $(t_0, x_0) \in J \times D$. Then a continuous function $x: (\alpha_1, \beta_1) \to D$ where $t_0 \in (\alpha_1, \beta_1) \subset J$ is a solution of Eqs. (1) and (2) if and only if x satisfies the integral equation

$$x(t) = x_0 + \int_{t_0}^{t} f(s, x(s))ds \quad \text{for} \quad \alpha_1 < t < \beta_1. \quad (3)$$

As in Section 20, we shall actually prove the existence of a solution of the new problem (3).

Theorem A (Local Existence). Let $f: J \times D \to \mathbb{R}^n$ be continuous and locally Lipschitzian, and let $(t_0, x_0) \in J \times D$. Then there exists $\Delta > 0$ such that Eqs. (1) and (2) have a (unique) solution on $(t_0 - \Delta, t_0 + \Delta)$.

Proof. Choose numbers $a > 0$ and $b > 0$ (sufficiently small) such that $J_1 \equiv [t_0 - a, t_0 + a]$ is contained in J, $A \equiv \{\xi \in \mathbb{R}^n: ||\xi - x_0|| \leq b\}$ is a subset of D, and f is Lipschitzian on $J_1 \times A$. Since $J_1 \times A$ is closed and bounded,

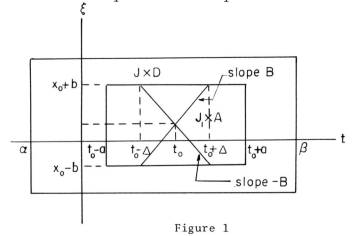

Figure 1

there exists a number $B > 0$ such that

$$||f(t,\xi)|| \leq B \quad \text{for all} \quad (t,\xi) \in J_1 \times A.$$

Let K be a Lipschitz constant for f there, so that

$$||f(t,\xi) - f(t,\tilde{\xi})|| \leq K||\xi - \tilde{\xi}|| \quad \text{for all} \quad (t,\xi),$$
$$(t,\tilde{\xi}) \in J_1 \times A.$$

Now choose

$$\Delta = \min \{a,b/B\}.$$

The situation is depicted in Figure 1 for the case $n = 1$.

As in the proof of Theorem 20-A, we hope to define successive approximations to a solution as follows. Let $x_{(0)}: [t_0-\Delta, t_0+\Delta] \to D$ be a rather arbitrary first guess. Then, for $t_0-\Delta \leq t \leq t_0+\Delta$, define

$$x_{(1)}(t) = x_0 + \int_{t_0}^t f(s, x_{(0)}(s))ds \tag{4}$$

$$x_{(2)}(t) = x_0 + \int_{t_0}^t f(s, x_{(1)}(s))ds \tag{5}$$

$$\vdots$$

$$x_{(\ell+1)}(t) = x_0 + \int_{t_0}^t f(s, x_{(\ell)}(s))ds \tag{6}$$

$$\vdots$$

Our first problem is to assure that each function $x_{(\ell)}$ so constructed will be continuous and take values in D. Actually, the values are going to remain in the smaller set A.

Let $x_{(0)}$ be any continuous function mapping $[t_0-\Delta, t_0+\Delta] \to A$. Then $(t, x_{(0)}(t)) \in J_1 \times A$ so that the integrand involved in Eq. (4) is a well-defined continuous function of s. Hence $x_{(1)}$ is well defined and continuous.

Moreover, for $t_0 - \Delta \leq t \leq t_0 + \Delta$

$$\|x_{(1)}(t) - x_0\| = \left\|\int_{t_0}^{t} f(s, x_{(0)}(s)) ds\right\|$$

$$\leq \left|\int_{t_0}^{t} B ds\right| \leq B\Delta \leq b.$$

This says that $x_{(1)}(t) \in A$ for $t_0 - \Delta \leq t \leq t_0 + \Delta$. In other words $x_{(1)}$ is another function of the same type as $x_{(0)}$.

Since Eq. (5) is entirely analogous to Eq. (4), it follows at once that $x_{(2)}$ will also be a well-defined continuous function mapping $[t_0 - \Delta, t_0 + \Delta] \to A$. Likewise, by mathematical induction, for $x_{(3)}, x_{(4)}, \ldots$

From this point onward the proof is much like that of Theorem 20-A.

Note that for $t_0 - \Delta \leq t \leq t_0 + \Delta$ and each $\ell = 0, 1, 2, \ldots,$

$$\|x_{(\ell+2)}(t) - x_{(\ell+1)}(t)\|$$

$$= \left\|\int_{t_0}^{t} [f(s, x_{(\ell+1)}(s)) - f(s, x_{(\ell)}(s))] ds\right\| \qquad (7)$$

$$\leq \left|\int_{t_0}^{t} K\|x_{(\ell+1)}(s) - x_{(\ell)}(s)\| ds\right|.$$

Note also that

$$\|x_{(1)}(t) - x_{(0)}(t)\| \leq 2b \quad \text{for} \quad t_0 - \Delta \leq t \leq t_0 + \Delta.$$

Then proceeding by induction one finds

$$\|x_{(\ell+1)}(t) - x_{(\ell)}(t)\| \leq 2b \, \frac{K^{\ell} |t - t_0|^{\ell}}{\ell!} \qquad (8)$$

$$\text{for} \quad t_0 - \Delta \leq t \leq t_0 + \Delta,$$

for $\ell = 0, 1, 2, \ldots$.

Again as in the proof of Theorem 20-A, inequality (8) establishes the uniform convergence on $[t_0 - \Delta, t_0 + \Delta]$ of the sequence of functions $\{x_{(\ell)k}\}$ for each $k = 1, \ldots, n$. Let

$$x(t) = \lim_{\ell \to \infty} x_{(\ell)}(t) \quad \text{for} \quad t_0 - \Delta \le t \le t_0 + \Delta.$$

Then indeed

$$||x(t) - x_{(\ell)}(t)|| \le 2b \sum_{p=\ell}^{\infty} \frac{(K\Delta)^p}{p!} \quad \text{for} \quad t_0 - \Delta \le t \le t_0 + \Delta. \quad (9)$$

It remains to show that x is a solution of (3). For $t_0 - \Delta \le t \le t_0 + \Delta$ and any $\ell = 0,1,2,\ldots$

$$||x(t) - x_0 - \int_{t_0}^{t} f(s, x(s))ds||$$

$$= ||x(t) - x_{(\ell+1)}(t) + \int_{t_0}^{t} [f(s, x_{(\ell)}(s)) - f(s, x(s))]ds||$$

$$\le ||x(t) - x_{(\ell+1)}(t)|| + |\int_{t_0}^{t} K||x_{(\ell)}(s) - x(s)||ds|.$$

Then, applying (9), we have for $\ell = 0,1,2,\ldots$

$$||x(t) - x_0 - \int_{t_0}^{t} f(s, x(s))ds||$$

$$\le 2b \sum_{p=\ell+1}^{\infty} \frac{(K\Delta)^p}{p!} + K\Delta \cdot 2b \sum_{p=\ell}^{\infty} \frac{(K\Delta)^p}{p!}. \quad (10)$$

Since the right hand side of (10) tends to zero as $\ell \to \infty$ it follows that the left hand side must be zero. Hence x satisfies (3) on $[t_0 - \Delta, t_0 + \Delta]$, which means that x is a solution of (1) and (2) on $(t_0 - \Delta, t_0 + \Delta)$.

The uniqueness of the solution on $[t_0, t_0 + \Delta)$ follows from Theorem 23-A. The reader should complete the proof by showing uniqueness on $(t_0 - \Delta, t_0]$. $\square$

Corollary B. Under the same hypotheses and using the same notation as in Theorem A and its proof except that $x_{(0)}(t) \equiv x_0$, we have for $\ell = 0,1,\ldots$

$$||x(t) - x_{(\ell)}(t)|| \le \frac{BK^{\ell}|t - t_0|^{\ell+1}}{(\ell+1)!} \quad \text{for} \quad t_0 - \Delta \le t \le t_0 + \Delta. \quad (11)$$

The proof is Problem 2.

Remarks. The construction of the successive approximations in the proof of Theorem A can be summarized as follows. We considered the set S of all continuous functions
mapping

$$[t_0-\Delta,t_0+\Delta] \rightarrow \{\xi \in \mathbb{R}^n: ||\xi - x_0|| \le b\}.$$

We chose an arbitrary function $x_{(0)}$ in S and described a
transformation (or mapping), T, which generates from $x_{(0)}$
a new function

$$x_{(1)} = Tx_{(0)}.$$

This abstract equation represents Eq. (4). Then we showed
that $Tx_{(0)}$ was also in S. It followed similarly that
$x_{(2)} \equiv Tx_{(1)}$ was in S, and so was $x_{(3)} \equiv Tx_{(2)}$, etc.

Existence proofs often follow this pattern. Thus we
try to define a suitable set of functions S and a transformation T such that whenever $\chi \in S$ then $T\chi \in S$. The
completion of the proof then involves constructing a sequence
$x_{(0)}, x_{(1)}, x_{(2)},\ldots$ of members of S where $x_{(\ell+1)} \equiv Tx_{(\ell)}$,
proving that it converges to some function $x \in S$, and fin-
ally showing that Tx = x. When Tx = x we say that x is
a fixed point for the mapping T.

Theorem A is of restricted usefulness because, in general, it asserts existence only on a small interval.

Of course, having proved existence on some small interval, say $[t_0,t_0+\Delta]$, one would naturally think of trying
to continue the solution further, say to $t_0+\Delta+\Delta_1$, by a similar argument. This process can be continued step by step, as

suggested in Problem 4. But the ultimate interval over which
we thus obtain a solution may be limited by our cleverness in
choosing the Δ's; or it could conceivably be limited by a defect in the method, for the particular equations under consideration.

How, then, can we ever tell when a solution has been
continued as far as possible?

Definitions. Let x on an interval J_1 and y on an interval J_2 be solutions of some system of differential equations with $J_1 \subset J_2$, $J_1 \neq J_2$, and $x(t) = y(t)$ for all t
in J_1. Then we say y is a continuation of x, or x can
be continued to J_2. A solution x of a system of differential equations is noncontinuable if it has no continuation.

The basic criterion for determining when a solution is
noncontinuable is contained in the following theorem.

Theorem C. Let $f: (\alpha,\beta) \times D \to \mathbb{R}^n$ be continuous and locally
Lipschitzian, and let $(t_0,x_0) \in (\alpha,\beta) \times D$. Then Eqs. (1)
and (2) have a unique noncontinuable solution x on $(\alpha_1,\beta_1) \subset$
(α,β); and if A is some closed bounded subset of D, then

$$x(t) \in A \quad \text{for} \quad t_0 \leq t < \beta_1 \quad \text{implies} \quad \beta_1 = \beta$$

and

$$x(t) \in A \quad \text{for} \quad \alpha_1 < t \leq t_0 \quad \text{implies} \quad \alpha_1 = \alpha.$$

[In other words, if $\beta_1 < \beta$ or $\alpha_1 > \alpha$, then for every
closed bounded set $A \subset D$ $x(t) \notin A$ for some $t \in (t_0,\beta_1)$
or for some $t \in (\alpha_1,t)$ respectively.]

The proof will use the following result from advanced
calculus.

<u>Lemma D</u>. Let x be a differentiable function mapping a bounded interval (a,b) into $\mathbb{R}^n$ with

$$||x'(t)|| \leq B \quad on \quad (a,b).$$

Then $\lim_{t \to a} x(t)$ and $\lim_{t \to b} x(t)$ exist.

<u>Proof of Lemma D</u>. We shall prove the existence of the limit at b. The proof at a is entirely analogous.

Consider any sequence $\{t_i\}$ of numbers in (a,b) with $\lim_{i \to \infty} t_i = b$. Let any $\varepsilon > 0$ be given and choose N such that $i \geq N$ implies

$$|t_i - b| < \varepsilon/2B.$$

Then it follows that for all $i, j \geq N$ and $k = 1,\ldots,n$,

$$|x_k(t_i) - x_k(t_j)| = |\int_{t_i}^{t_j} x_k'(s)ds|$$

$$\leq B|t_j - t_i| \leq B|t_j - b| + B|b - t_i| < \varepsilon.$$

Hence, by the Cauchy convergence criterion (Theorem A2-J), the sequences $\{x_k(t_i)\}_{i=1}^{\infty}$ converge for $k = 1,\ldots,n$. Let

$$\xi \equiv \lim_{i \to \infty} x(t_i).$$

We have not yet shown that $\lim_{t \to b} x(t) = \xi$. To establish this, again let any small $\varepsilon > 0$ be given. Consider any

$$t \in (b - \frac{\varepsilon}{2B}, b).$$

Choose t_i, from the sequence considered above, such that $t_i \in (b - \frac{\varepsilon}{2B}, b)$ and $|\xi_k - x_k(t_i)| < \frac{\varepsilon}{2}$ for each $k = 1,\ldots,n$. Then, for each k,

$$|\xi_k - x_k(t)| \le |\xi_k - x_k(t_i)| + |x_k(t_i) - x_k(t)|$$

$$< \frac{\epsilon}{2} + B \frac{\epsilon}{2B} = \epsilon.$$

This shows that $\lim_{t \to b} x(t) = \xi$. ☐

Proof of Theorem C.   The uniqueness of the solution on any interval of the form $[t_0, \beta_1)$ follows from Theorem 23-A.

Let

$$\beta_1 = \sup \{s \in \mathbb{R}: \text{ a solution exists on } [t_0, s)\}.$$

Then $\beta_1 > t_0$ and for every $s \in (t_0, \beta_1)$ a unique solution $y_{(s)}$ exists on $[t_0, s)$. We now define a function $x: [t_0, \beta_1) \to D$ as follows. For each t in $[t_0, \beta_1)$ let

$$x(t) = y_{(s)}(t) \quad \text{for some} \quad s \quad \text{in} \quad (t, \beta_1).$$

By uniqueness, this is an unambiguous definition of $x(t)$. Moreover, by construction, the resulting function x is a solution of Eqs. (1) and (2) on $[t_0, \beta_1)$, and it cannot be continued beyond β_1.

Now suppose (for contradiction) that $\beta_1 < \beta$ and yet, for some closed bounded set $A \subset D$, $x(t) \in A$ for $t_0 \le t < \beta_1$. Then, on the closed bounded set

$$[t_0, \beta_1] \times A \subset (\alpha, \beta) \times D, \quad ||f(t, \xi)|| \le B$$

for some constant B.   Hence

$$||x'(t)|| = ||f(t, x(t))|| \le B \quad \text{for} \quad t_0 < t < \beta_1.$$

So Lemma D assures the existence of

$$\lim_{t \to \beta_1} x(t) = \xi \in A \subset D.$$

Let us now extend the definition of x to a continuous function mapping $[t_0, \beta_1] \to D$ by setting $x(\beta_1) = \xi$. Then $f(t, x(t))$, being a composition of continuous functions, depends continuously on t in $[t_0, \beta_1]$. Thus Eq. (3), which holds on $[t_0, \beta_1)$ must also hold at $t = \beta_1$. Hence x satisfies Eq. (1) on $[t_0, \beta_1]$ (implying a left hand derivative at β_1 of course).

Apply Theorem A to assert the existence of a solution z on $(\beta_1 - \Delta, \beta_1 + \Delta)$ of Eq. (1) with $z(\beta_1) = x(\beta_1)$. Assume $\beta_1 - \Delta \geq t_0$. Then, by uniqueness, $z(t) = x(t)$ on $(\beta_1 - \Delta, \beta_1]$. So, if we define $x(t) \equiv z(t)$ on $(\beta_1, \beta_1 + \Delta)$ we have constructed a continuation of x to a solution on $[t_0, \beta_1 + \Delta)$. This contradicts the fact that x was noncontinuable. Hence we conclude that either $\beta_1 = \beta$ or else for each closed bounded set $A \subset D$, $x(t) \notin A$ for some $t \in (t_0, \beta_1)$.

The other assertions of the theorem, pertaining to the interval $(\alpha_1, t_0]$ are proved similarly. This is left to the reader. ☐

Example 1. The equation of motion for a simple pendulum, as in Figure 1, acting under the sole external influence of gravity can be taken to be

$$m\ell\theta''(t) = -mg \sin \theta(t) - b\ell\theta'(t). \tag{12}$$

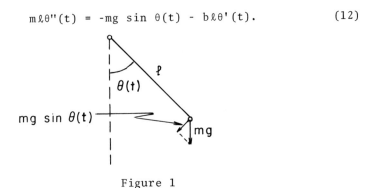

Figure 1

Here   m   is the mass of the bob, ℓ the length of the mass-less rod, g   the acceleration of gravity, and   b   a coefficient of friction -- all positive constants.   Let initial data

$$\theta(t_0) = \theta_0, \qquad \theta'(t_0) = \omega_0 \tag{13}$$

be specified.   To apply the theorems of this section, we define $x_1(t) = \theta(t)$ and $x_2(t) = \theta'(t)$ to transform Eq. (12) into the ordinary differential system

$$\begin{aligned} x_1'(t) &= x_2(t) \\ x_2'(t) &= -\frac{g}{\ell} \sin x_1(t) - \frac{b}{m} x_2(t), \end{aligned} \tag{14}$$

with

$$x_1(t_0) = \theta_0, \qquad x_2(t_0) = \omega_0. \tag{15}$$

We can take $(\alpha,\beta) = \mathbb{R}$ and $D = \mathbb{R}^2$.   Then the existence of a unique noncontinuable solution on some interval (α_1,β_1) follows from Theorem C.   We intend to show that $\beta_1 = \infty$.

Let us consider the function $v: (\alpha_1,\beta_1) \rightarrow \mathbb{R}$ defined by

$$v(t) = \frac{1}{2} x_2^2(t) + \frac{g}{\ell} [1 - \cos x_1(t)]. \tag{16}$$

Since x_1 and x_2 satisfy (14),

$$v'(t) = x_2(t)x_2'(t) + \frac{g}{\ell}[\sin x_1(t)]x_1'(t) = -\frac{b}{m} x_2^2(t) \leq 0.$$

Thus   v   is nonincreasing.   It follows that

$$|x_2(t)| \leq \sqrt{2v(t)} \leq \sqrt{2v(t_0)} \equiv M.$$

Using this and the first of Eqs. (14) we find

$$|x_1(t)| \leq |\theta_0| + M(t-t_0) \qquad \text{for} \quad t_0 \leq t < \beta_1.$$

Now suppose (for contradiction) that $\beta_1 < \infty$.   Then, on $[t_0, \beta_1)$,

$$|x_1(t)| \leq |\theta_0| + M(\beta_1 - t_0) \quad \text{and} \quad |x_2(t)| \leq M.$$

so that $x(t) = \text{col } (x_1(t), x_2(t))$ remains inside a closed

bounded set. By Theorem C, this contradicts the assumption

that $\beta_1 < \infty$. Hence $\beta_1 = \infty$.

The natural question regarding this example is, How

was the appropriate function v obtained? In the present

example, and in many others, the function v represents the

"energy" in the system.

More generally, for the second order equation of the

form

$$x''(t) + f_1(x(t))x'(t) + f_2(x(t)) = h(t), \tag{17}$$

one often uses, as in Example 1, the function

$$v(t) = \frac{1}{2}(x'(t))^2 + \int_0^{x(t)} f_2(\xi)d\xi. \tag{18}$$

Here $(x'(t))^2/2$ corresponds to the "kinetic energy" and

$\int_0^{x(t)} f_2(\xi)d\xi$, computed from the "restoring force" f_2, re-

presents "potential energy".

By modifying the proof in Example 1, one can also show

that $\alpha_1 = -\infty$ for the noncontinuable solution of Eqs. (12)

and (13). The procedure is outlined in Problem 8.

Example 2. In Theorem C, let $D = \mathbb{R}^n$ and assume that f

satisfies a (global) Lipschitz condition on all of $(\alpha, \beta) \times D$.

Then we shall show that the unique noncontinuable solution of

Eqs. (1) and (2) is defined on the entire interval (α, β).

If the Lipschitz constant is K, then at every point

in $(\alpha, \beta) \times \mathbb{R}^n$

$$||f(t,\xi)|| \leq ||f(t,0)|| + ||f(t,\xi) - f(t,0)||$$

$$\leq ||f(t,0)|| + K||\xi||.$$

Now suppose (for contradiction) that the unique noncontinuable solution x of Eqs. (1) and (2) is only defined on
(α_1,β_1) where $\beta_1 < \beta$. Let B be an upper bound for the
continuous function $||f(t,0)||$ for $t_0 \leq t \leq \beta_1$. Then,
from (3),

$$||x(t)|| \leq ||x_0|| + \int_{t_0}^{t} ||f(s,x(s))||ds$$

$$\leq ||x_0|| + \int_{t_0}^{t} (B + K||x(s)||)ds$$

$$\leq ||x_0|| + B(\beta_1 - t_0) + K\int_{t_0}^{t} ||x(s)||ds$$

on $[t_0,\beta_1)$. Hence, by Lemma 8-A,

$$||x(t)|| \leq [||x_0|| + B(\beta_1 - t_0)]e^{K(\beta_1 - t_0)}.$$

Thus $x(t)$ remains in a closed bounded subset of D -- contradicting the supposition that $\beta_1 < \beta$. Therefore $\beta_1 = \beta$.
A similar argument shows that $\alpha_1 = \alpha$.

The following corollary of Theorem C generalizes the
argument used in Example 2.

Corollary E. (Global Existence). Let $D = \mathbb{R}^n$. Assume the
hypotheses of Theorem C and assume further that

$$||f(t,\xi)|| \leq M(t) + N(t)||\xi|| \text{on} (\alpha,\beta) \times \mathbb{R}^n,$$

where M and N are continuous, positive functions on
(α,β). Then the unique noncontinuable solution of Eqs. (1)
and (2) exists on the entire interval (α,β).

Proof.   Let x on (α_1, β_1) be the unique noncontinuable solution of (1) and (2).   Suppose (for contradiction)
that $\beta_1 < \beta$.   Then $M(s) \leq M_1$ and $N(s) \leq N_1$ on $[t_0, \beta_1]$,
and so Eq. (3) yields

$$||x(t)|| \leq ||x_0|| + M_1(\beta_1 - t_0) + \int_{t_0}^{t} N_1||x(s)||ds$$

on $[t_0, \beta_1)$, which gives

$$||x(t)|| \leq [||x_0|| + M_1(\beta_1 - t_0)]e^{N_1(\beta_1 - t_0)}.$$

This shows that $x(t)$ remains in a closed bounded set, contradicting the supposition that $\beta_1 < \beta$.   Hence $\beta_1 = \beta$.
Similarly, one shows that $\alpha_1 = \alpha$.   □

Remark.   In particular, Corollary E applies to the
systems treated in Section 20 -- linear systems with continuous coefficients.   Problem 10.

Problems

1.   Obtain the first few successive approximations for the
     following problem and compare with the known solution ob-
     tained by another method:
     $$x'(t) = x^2(t)   \text{ with } x(0) = 1$$
     using (a) $x_{(0)}(t) \equiv 1$,   (b) $x_{(0)}(t) \equiv t$.

2.   Prove Corollary B.

3.   Find appropriate values of a, b, and Δ and then use
     Corollary B to estimate $|x(t) - x_{(4)}(t)|$ on $[-\Delta, \Delta]$ in
     Problem 1(a).

4. For the ordinary differential equation

$$x'(t) = x^2(t) \quad \text{with} \quad x(0) = 1$$

a unique solution exists on some interval $[-\Delta, \Delta]$ where $\Delta > 0$. Using Theorem A, how big can you choose Δ by making judicious choices for a and b? Solve the equation exactly on $[-\Delta, \Delta]$ and then seek a solution on $[\Delta, \Delta + \Delta_1]$ for $\Delta_1 > 0$. How big can you choose Δ_1? Discuss the possibility of continuing this process.

5. How would the proof of Theorem A be simplified if we chose Δ such that $0 < \Delta \leq \min \{a, b/B\}$ and $\Delta < 1/K$? Hint: Consider inequality (7) and the resulting simplifications.

*6. Let $f: (\alpha, \beta) \times \mathbb{R}^n \to \mathbb{R}^n$ be continuous and (globally) Lipschitzian. Prove that if $(t_0, x_0) \in (\alpha, \beta) \times \mathbb{R}^n$ then there exists a (unique) solution of Eqs. (1) and (2) on the entire interval (α, β), and this solution can be found by the method of successive approximations. Hint: Modify the proof of Theorem A as follows. Choose numbers α_1 and β_1 such that $\alpha < \alpha_1 < t_0 < \beta_1 < \beta$. Then note that the "successive approximations" can be defined on $[\alpha_1, \beta_1]$ instead of merely on $[t_0 - \Delta, t_0 + \Delta]$. Define

$$v_\ell = \sup_{\alpha_1 < t < \beta_1} [e^{-c|t - t_0|} ||x_{(\ell+1)}(t) - x_{(\ell)}(t)||]$$

where c is an unspecified positive constant. Show that $v_{\ell+1} \leq (K/c) v_\ell$. Now let $c > K$ and recall what happened in Problem 5. (This is an example of the use of a "weighted norm".)

7.   Apply the result of Problem 6 to

$$x'(t) = \sin x(t) - \sin t \quad \text{with} \quad x(1) = 1.$$

8.   Use the method of Example 1 to prove that the solution
     of Eqs. (12) and (13) can be continued back to $-\infty$.
     Hint: Define $v(t)$ as in (16) and show that

$$v'(t) = -\frac{b}{m} x_2^2(t) \geq -\frac{2b}{m} v(t).$$

From this inequality, show that

$$v(t) \leq v(t_0) e^{(2b/m)(t_0-t)} \quad \text{for} \quad t \leq t_0.$$

9.   Apply Corollary E (or the method of Example 2) to obtain
     another proof that the solution of Eqs. (12) and (13)
     can be continued to $(-\infty,\infty)$.

10. Show that Theorem 20-A follows from Corollary E.

11. The equation

$$x''(t) + \mu[x^2(t)-1]x'(t) + x(t) = 0, \quad \mu > 0$$

was studied by van der Pol as a model for the operation
of an electronic vacuum tube oscillator.   Prove that
every solution can be continued to $+\infty$.   Why does Corollary E not apply here?

12. Prove that every solution of

$$x''(t) - x'(t) + t^2 x(t) = 0$$

can be continued to $+\infty$.

*13.   Prove that the solution of $x''(t) - x^3(t) = 0$ with

$x(0) = x_0 > 0$ and $x'(0) = u_0 > 0$ cannot be continued

to $+\infty$.

*14.   Consider the scalar ordinary differential equation

$x'(t) = f(t,x(t))$.   Let   f, defined <u>only</u> on $\mathbb{R} \times (0,\infty)$

$\to \mathbb{R}$, be continuous and locally Lipschitzian.   If an

"initial condition" of the form $\lim_{t \to 0} x(t) = 0$ is

specified, we cannot directly apply our existence and

uniqueness theorems since $(0,0) \notin \mathbb{R} \times (0,\infty)$.   In fact

$(0,0)$ is on the boundary of $\mathbb{R} \times (0,\infty)$.   What can you

say about existence and uniqueness with this initial

condition for the following equations?

(a) $x' = 0$,   (b) $x' = 1/x$,   (c) $x' = -1/x$,

(d) $x' = t/x$,   (e) $x' = -t/x$,   (f) $x' = t^2/x^2$.

Now let $f: (0,\infty) \times \mathbb{R} \to \mathbb{R}$ and answer the same questions

for

(g) $x' = x/t$,   (h) $x' = x/t^2$,   (i) $x' = -x/t^2$,

(j) $x' = x^2/t^2$, (k) $x' = (x/t)^{1/3}$.

25.   <u>SYSTEMS</u> <u>WITH</u> <u>BOUNDED</u> <u>DELAYS</u>:   <u>NOTATION</u> <u>AND</u> <u>UNIQUENESS</u>

Many practical problems give rise to differential
equations having constant delays or delays which, if not
constant, are at least bounded.   Thus henceforth (unless an-
nounced otherwise), when we consider a delay differential
system such as

$$x'(t) = f(t,x(g_1(t)),\ldots,x(g_m(t))), \tag{1}$$

we shall assume that

$$t-r \leq g_j(t) \leq t \quad \text{for} \quad t \geq t_0, \quad j = 1,\ldots,m,$$

for some constant $r \geq 0$. Then the initial condition takes the form

$$x(t) = \theta(t) \quad \text{for} \quad t_0-r \leq t \leq t_0. \tag{2}$$

(Note that system (1) reduces to an ordinary differential system if $r = 0$.) We assume that f is defined on $[t_0,\beta) \times D^m \to \mathbb{R}^n$ for some $\beta > t_0$ and some open set $D \subset \mathbb{R}^n$.

In Section 23 we introduced a briefer notation for Eq. (1),

$$x'(t) = F(t,x_t). \tag{1'}$$

Recall that we did not define either F or x_t, but only the combination $F(t,x_t)$, which simply represented the right hand side of Eq. (1).

We shall now give meaning to F and x_t themselves for the special case of bounded delays; and the result will be that Eq. (1') is not only briefer but also potentially more general than Eq. (1). The following definition (introduced by Shimanov [1960]) is extensively used in the literature on delay differential equations.

Definition. If χ is a function defined at least on $[t-r,t] \to \mathbb{R}^n$, then we define a new function $\chi_t: [-r,0] \to \mathbb{R}^n$ by

$$\chi_t(\sigma) = \chi(t+\sigma) \quad \text{for} \quad -r \leq \sigma \leq 0.$$

Note that χ_t is obtained by considering only $\chi(s)$ for $t-r \leq s \leq t$ and then translating this segment of χ to the interval $[-r,0]$. (Here we have deliberately used the symbol χ instead of x in order to reserve x for those

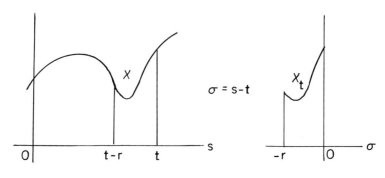

Figure 1

functions which satisfy the differential equation.) If χ is a continuous function, then χ_t is a continuous function on $[-r,0]$.

Notation. The set $C([-r,0],\mathbb{R}^n)$ of all continuous functions mapping $[-r,0] \to \mathbb{R}^n$ will be denoted by $\mathscr{C}$. And if A is any set in $\mathbb{R}^n$ we will let

$$\mathscr{C}_A = C([-r,0],A).$$

Thus if χ is continuous on $[t-r,t] \to A$ then $\chi_t \in \mathscr{C}_A$.

We shall sometimes consider Eq. (1) or (1') on a half-open interval $[t_0,\beta)$, and sometimes it will be desirable to be able to discuss such an equation on an open interval (α,β). So let us now use the symbol J to mean either $[t_0,\beta)$ or (α,β) as required.

If Eq. (1') is to represent Eq. (1) with f defined on $J \times D^m$, then we want $F(t,\chi_t)$ to make sense whenever $t \in J$ and $\chi_t \in \mathscr{C}_D$. Thus we should define

$$F: \ J \times \mathcal{L}_D \to \mathbb{R}^n.$$

In other words for each $(t,\psi) \in J \times \mathcal{L}_D$, $F(t,\psi)$ must be a well defined point in $\mathbb{R}^n$.

A mapping, such as F, defined on a set of functions is sometimes called a _functional_ instead of a function. Equation (1') is therefore often called a "functional differential equation". However we shall continue to refer to Eq. (1) or Eq. (1') as a delay differential equation to emphasize the fact that only present and past values of x are involved in determining $x'(t)$.

Example 1. If Eq. (1') is to be equivalent to Eq. (1), we must define

$$F(t,\psi) \ = \ f(t,\psi(g_1(t)-t),\ldots,\psi(g_m(t)-t)).$$

To see this, note that

$$x_t(g_j(t)-t) \ = \ x(t+g_j(t)-t) \ = \ x(g_j(t)),$$

and hence the right hand side of Eq. (1') is

$$F(t,x_t) \ = \ f(t,x_t(g_1(t)-t),\ldots,x_t(g_m(t)-t))$$

$$= \ f(t,x(g_1(t)),\ldots,x(g_m(t))).$$

But, as already suggested, Eq. (1') is potentially more general than the more cumbersome Eq. (1). For if the functional F is defined on $J \times \mathcal{L}_D$, then $F(t,x_t)$ can conceivably depend on any or all values of $x_t(\sigma) = x(t+\sigma)$, $-r \leq \sigma \leq 0$.

Example 2. If $F(t,\psi) \equiv \int_{-r}^{0} \psi(\sigma)d\sigma$, then Eq. (1') becomes a system with "infinitely many delays",

$$x'(t) = \int_{-r}^{0} x_t(\sigma) d\sigma = \int_{t-r}^{t} x(s) ds.$$

To gain further familiarity with the notation used in Eq. (1'), the reader should verify the assertions of the following two examples, referring to systems in Section 21. Actually, Examples 3 and 4 are both special cases of Example 1.

Example 3. Equation (1') becomes Eq. 21-(7) if we take $r = 1$, $J = \mathbb{R}$, and $D = \mathbb{R}$ and define $F: \mathbb{R} \times \mathscr{L} \to \mathbb{R}$ by

$$F(t,\psi) = -c\psi(-1)[1 + \psi(0)].$$

Example 4. Equation (1') becomes system 21-(11) if we take $D = \mathbb{R}^2$ and define $F: \mathbb{R} \times \mathscr{L} \to \mathbb{R}^2$ by

$$F(t,\psi) = \begin{pmatrix} a_1[1-\psi_1(0)/P]\psi_1(0) - b_1\psi_2(0)\psi_1(0) \\ -a_2\psi_2(0) + b_2\psi_1(-r)\psi_2(-r) \end{pmatrix}$$

When using the notation of Eq. (1'), we shall also re-write the initial condition (2). Equation (2) is equivalent to $x(t_0+\sigma) = \theta(t_0+\sigma)$ for $-r \leq \sigma \leq 0$, or simply $x_{t_0} = \theta_{t_0}$. Introducing $\phi = \theta_{t_0}$, this becomes

$$x_{t_0} = \phi. \qquad (2')$$

It is important to recognize that (2') means $x(t_0+\sigma) = \phi(\sigma)$ or, letting $t = t_0+\sigma$,

$$x(t) = \phi(t-t_0) \quad \text{for} \quad t_0-r \leq t \leq t_0. \qquad (2'')$$

In particular, note that

$$x(t_0) = \phi(0).$$

We shall assume $\phi \in \mathscr{L}_D$.

In previous cases, we have studied existence and uniqueness questions for a differential system with initial conditions by converting to an integral equation. This can also
be done for Eq. (1') provided the right hand side is continuous. The hypotheses, used previously for Eq. (1), of continuity of f and each g_j can now be replaced by the following
continuity condition on F in Eq. (1').

<u>Continuity</u> <u>Condition</u> <u>(C)</u> is satisfied if $F(t,\chi_t)$ is continu-
ous with respect to t in $[t_0,\beta)$ for each given continuous
function $\chi: [t_0-r,\beta) \to D$.

The significance of this condition is as follows. If
$F: [t_0,\beta) \times \mathscr{C}_D \to \mathbb{R}^n$ satisfies Continuity Condition (C), then
a continuous function x mapping $[t_0-r,\beta_1) \to D$, for some
$\beta_1 \in (t_0,\beta]$, is a solution of Eqs. (1') and (2') if and only
if

$$x(t) = \begin{cases} \phi(t-t_0) & \text{for } t_0-r \leq t \leq t_0 \\ \phi(0) + \int_{t_0}^{t} F(s,x_s)ds & \text{for } t_0 \leq t < \beta_1. \end{cases} \tag{3}$$

Note that the functional F of Example 2 satisfies
Condition (C). For if $\chi: \mathbb{R} \to \mathbb{R}$ is continuous, then

$$F(t,\chi_t) = \int_{-r}^{0} \chi_t(\sigma)d\sigma = \int_{t-r}^{t} \chi(s)ds$$

clearly depends continuously on t. (As a matter of fact,
it is even continuously differentiable.)

If F is defined as in Example 1 and if f and
$g_1,\ldots,g_m$ are all continuous, then, for any continuous function $\chi: [t_0-r,\beta) \to D$, $F(t,\chi_t)$ is a composition of continuous functions and hence is continuous for $t_0 \leq t < \beta$. Thus
this F also satisfies Condition (C).

The functionals F defined in Examples 3 and 4 can be viewed as special cases of Example 1 and thus are also seen to satisfy Condition (C).

In addition to Condition (C), we shall want a Lipschitz-type condition on F. This requires a means for measuring the magnitude of members of $\mathscr{L}_D$.

For a function $\psi \in \mathscr{L}_D$ it is convenient to define, as a measure of magnitude,

$$||\psi||_r = \sup_{-r \leq \sigma \leq 0} ||\psi(\sigma)||. \tag{4}$$

Lemma A. In the special case when $D = \mathbb{R}^n$, $\mathscr{L}_D = \mathscr{L}$ is a linear space and $||\cdot||_r$ is a norm on $\mathscr{L}$. This means that $||\cdot||_r$ satisfies the following conditions:

  (i) $||\psi||_r \geq 0$ for all $\psi \in \mathscr{L}$,

  (ii) $||\psi||_r = 0$ if and only if $\psi = 0$ (the zero function),

  (iii) $||c\psi||_r = |c| \cdot ||\psi||_r$ for all $\psi \in \mathscr{L}$ and all $c \in \mathbb{R}$, and

  (iv) $||\psi + \tilde{\psi}||_r \leq ||\psi||_r + ||\tilde{\psi}||_r$ for all $\psi, \tilde{\psi} \in \mathscr{L}$ -- the triangle inequality.

(We shall refer to $||\cdot||_r$ as the r-norm.)

Proof. The verification that $\mathscr{L}$ is a linear space and the proof of properties (i), (ii), and (iii) are left to the reader. To prove (iv) note that, for each $\sigma \in [-r, 0]$

$$||\psi(\sigma) + \tilde{\psi}(\sigma)|| \leq ||\psi(\sigma)|| + ||\tilde{\psi}(\sigma)|| \leq ||\psi||_r + ||\tilde{\psi}||_r.$$

Thus, letting σ vary over $[-r, 0]$,

$$\sup_{-r \leq \sigma \leq 0} ||\psi(\sigma) + \tilde{\psi}(\sigma)|| \leq ||\psi||_r + ||\tilde{\psi}||_r,$$

which is (iv). □

If $D \neq \mathbb{R}^n$, $\mathscr{L}_D$ may not be a linear space. Why? In that case $||\cdot||_r$ is not a norm on $\mathscr{L}_D$. Which property or properties in Lemma A fail? Problem 5. However, $||\cdot||_r$ can always be considered a norm on the space $\mathscr{L}$ which contains $\mathscr{L}_D$. So it is not wrong in any case to refer to $||\cdot||_r$ as a norm.

<u>Definition</u>. Let $F: J \times \mathscr{L}_D \rightarrow \mathbb{R}^n$ and let $\mathscr{L}$ be a subset of $J \times \mathscr{L}_D$. If for some $K \geq 0$

$$||F(t,\psi) - F(t,\tilde{\psi})|| \leq K||\psi-\tilde{\psi}||_r \tag{5}$$

whenever (t,ψ) and $(t,\tilde{\psi}) \in \mathscr{L}$ we say F satisfies a <u>Lip-schitz condition</u> (or F is <u>Lipschitzian</u>) on $\mathscr{L}$ with Lip-schitz constant K.

As Problem 6(a), the reader should show that if f satisfies the Lipschitz condition 23-(4) on $J \times D^m$, then the F defined in Example 1 is Lipschitzian on $J \times \mathscr{L}_D$. It should be remarked however, that the Lipschitz condition defined here for F does <u>not</u> imply 23-(4) for f. Problem 6(b).

<u>Example 5</u>. If F is defined, as in Example 2 (with $n = 1$), by $F(t,\psi) \equiv \int_{-r}^{0} \psi(\sigma)d\sigma$, then F is Lipschitzian on $R \times \mathscr{L}$. For, given any $t \in \mathbb{R}$ and any $\psi,\tilde{\psi} \in \mathscr{L}$,

$$||F(t,\psi) - F(t,\tilde{\psi})|| = |\int_{-r}^{0} [\psi(\sigma) - \tilde{\psi}(\sigma)]d\sigma|$$

$$\leq r||\psi-\tilde{\psi}||_r.$$

But, in general, a (global) Lipschitz condition on the entire set $J \times \mathscr{L}_D$ is too much to expect. It is not even

fulfilled for Examples 3 and 4.   Instead of this condition,
our uniqueness and existence theorems will assume the following much weaker "local Lipschitz condition".

Definition. The functional $F: J \times \mathscr{C}_D \to \mathbb{R}^n$ is said to be
locally Lipschitzian if for each given $(\bar{t}, \bar{\psi}) \in J \times \mathscr{C}_D$ there
exist numbers $a > 0$ and $b > 0$ such that

$$\mathscr{L} \equiv ([\bar{t}-a, \bar{t}+a] \cap J) \times \{\psi \in \mathscr{C}: ||\psi - \bar{\psi}||_r \leq b\}$$

is a subset of $J \times \mathscr{C}_D$ and F is Lipschitzian on $\mathscr{L}$. (The
Lipschitz constant for F depends, in general, on the parti-
cular set $\mathscr{L}$.)

    With respect to Example 1, we have already noted that
if f, g_1, ..., and g_m are all continuous, then F satisfies Condition (C).   Now we would like to show that if f is
locally Lipschitzian as defined in Section 23, then F is
locally Lipschitzian as defined above.   The proof of this, to
be given in Example 6, uses the following advanced calculus-
type result.

Lemma B.   Let $\bar{\psi} \in \mathscr{C}_D$ be given.   Then there exists $\delta > 0$
such that

$$\{\xi \in \mathbb{R}^n: ||\xi - \bar{\psi}(\sigma)|| \leq \delta \quad \text{for some} \quad \sigma \in [-r, 0]\}$$

is a subset of D, and hence in particular

$$\{\psi \in \mathscr{C}: ||\psi - \bar{\psi}||_r \leq \delta\} \subset \mathscr{C}_D.$$

    Proof.   Suppose (for contradiction) there is no such
$\delta > 0$.   Then for each $\ell = 1, 2, \ldots$ there must exist some
$\sigma_\ell \in [-r, 0]$ and some $\xi_\ell \in \mathbb{R}^n$-D (the complement of D) such

that

$$||\xi_\ell - \bar{\psi}(\sigma_\ell)|| \leq 1/\ell.$$

Why? But now, by the Bolzano-Weierstrass Theorem (A2-G) applied to $\{\sigma_\ell\}$, there exists a convergent subsequence $\{\sigma_{\ell_k}\}$. Say $\lim\limits_{k\to\infty} \sigma_{\ell_k} = \sigma_0$. Then $\lim\limits_{k\to\infty} \xi_{\ell_k} = \bar{\psi}(\sigma_0)$. Why? But $\bar{\psi}(\sigma_0)$ lies in D, an open set. Hence for all sufficiently large k, ξ_{ℓ_k} must also belong to D -- a contradiction. □

It might be worthwhile to consider Problem 8 before proceeding.

Example 6. Let $f: J \times D^m \to \mathbb{R}^n$ be locally Lipschitzian (as defined in Section 23). Then we will show that the functional F mapping $[t_0,\beta) \times \mathscr{C}_D \to \mathbb{R}^n$ defined in Example 1 is locally Lipschitzian.

Let any $(\bar{t},\bar{\psi}) \in J \times \mathscr{C}_D$ be given. Then, by Lemma B, there exists a number $\delta > 0$ such that the set

$$\{\xi \in \mathbb{R}^n: ||\xi - \bar{\psi}(\sigma)|| \leq \delta \text{ for some } \sigma \in [-r,0]\}$$

is contained in D. Now choose $a > 0$ and $b \in (0,\delta]$ sufficiently small so that f is Lipschitzian (with Lipschitz constant K) on

$$([\bar{t}-a,\bar{t}+a] \cap J) \times A_1 \times \cdots \times A_m,$$

where

$$A_j \equiv \{\xi \in \mathbb{R}^n: ||\xi - \bar{\psi}(g_j(\bar{t})-\bar{t})|| \leq b\} \text{ for } j = 1,\ldots,m.$$

Then

$$\mathscr{S} \equiv ([\bar{t}-a,\bar{t}+a] \cap J) \times \{\psi \in \mathscr{C}: ||\psi-\bar{\psi}||_r \leq b\}$$

is a subset of $J \times \mathscr{C}_D$, and if (t, ψ) and $(t, \tilde{\psi}) \in \mathscr{G}$ one finds

$$||F(t, \psi) - F(t, \tilde{\psi})||$$

$$= ||f(t, \ldots, \psi(g_j(t) - t), \ldots) - f(t, \ldots, \tilde{\psi}(g_j(t) - t), \ldots)||$$

$$\leq K \max_{j=1, \ldots, m} ||\psi(g_j(t) - t) - \tilde{\psi}(g_j(t) - t)||$$

$$\leq K||\psi - \tilde{\psi}||_r.$$

Combining Example 6 with Lemma 23-B, we see that for the F of Example 1 to be locally Lipschitzian it suffices that f be continuously differentiable.

The uniqueness theorem (Theorem D) to be proved below supercedes Theorem 23-A in the case of bounded delays. Its proof will use another result from advanced calculus stated and proved as follows.

<u>Lemma C</u>. Let $\chi: [t_0-r, \beta) \to \mathbb{R}^n$ be continuous. Then, given any $\bar{t} \in [t_0, \beta)$ and any $\varepsilon > 0$, there exists $\delta > 0$ such that

$$||\chi_t - \chi_{\bar{t}}||_r < \varepsilon \quad \text{whenever} \quad t \in [t_0, \beta) \quad \text{and} \quad |t - \bar{t}| < \delta.$$

<u>Proof</u>. Let $\bar{t} \in [t_0, \beta)$ and $\varepsilon > 0$ be given. Choose $\delta_1 > 0$ such that $\bar{t} + \delta_1 < \beta$. Since χ is <u>uniformly</u> continuous on the closed bounded interval $[t_0-r, \bar{t}+\delta_1]$ (Theorem A2-I), there exists $\delta \in (0, \delta_1]$ such that if s and $\tilde{s} \in [t_0-r, \bar{t}+\delta_1]$, then

$$||\chi(s) - \chi(\tilde{s})|| < \varepsilon/2 \quad \text{whenever} \quad |s - \tilde{s}| < \delta.$$

This in turn shows that whenever $t \in [t_0, \beta)$ and $|t - \bar{t}| < \delta$

$$||x_t - x_{\bar{t}}||_r = \sup_{-r \leq \sigma \leq 0} ||\chi(t+\sigma) - \chi(\bar{t}+\sigma)|| \leq \varepsilon/2 < \varepsilon. \qquad \square$$

Theorem D (Uniqueness). Let $F: [t_0,\beta) \times \mathscr{C}_D \to \mathbb{R}^n$ satisfy Condition (C) and let it be locally Lipschitzian. Then, given any $\phi \in \mathscr{C}_D$, Eqs. (1') and (2') have at most one solution on $[t_0-r,\beta_1)$ for any $\beta_1 \in (t_0,\beta]$.

Proof. Suppose (for contradiction) that for some $\beta_1 \in (t_0,\beta]$ there are two solutions x and $\tilde{x}$ mapping $[t_0-r,\beta_1) \to D$ with $x \neq \tilde{x}$. Let

$$t_1 = \inf \{t \in (t_0,\beta_1): x(t) \neq \tilde{x}(t)\}.$$

Then $t_0 \leq t_1 < \beta_1$ and

$$x(t) = \tilde{x}(t) \quad \text{for} \quad t_0-r \leq t \leq t_1.$$

Since $(t_1,x_{t_1}) \in [t_0,\beta_1) \times \mathscr{C}_D$, there exist numbers $a > 0$ and $b > 0$ such that the set

$$\mathscr{G} \equiv [t_1,t_1+a] \times \{\psi \in \mathscr{C}: ||\psi-x_{t_1}||_r \leq b\}$$

is contained in $[t_0,\beta) \times \mathscr{C}_D$ and F is Lipschitzian on $\mathscr{G}$ (with Lipschitz constant K).

By Lemma C, there exists $\delta \in (0,a]$ such that $(t,x_t) \in \mathscr{G}$ and $(t,\tilde{x}_t) \in \mathscr{G}$ for $t_1 \leq t < t_1+\delta$. Moreover, both x and $\tilde{x}$ satisfy Eq. (3) for $t_0-r \leq t < t_1+\delta$. Thus, for $t_1 \leq t < t_1+\delta$,

$$||x(t) - \tilde{x}(t)|| = ||\int_{t_0}^{t} [F(s,x_s) - F(s,\tilde{x}_s)]ds||$$

$$\leq \int_{t_1}^{t} K||x_s - \tilde{x}_s||_r ds.$$

Now, since the right hand side is an increasing function of t and since $||x(t)-\tilde{x}(t)|| = 0$ for $t_1-r \leq t \leq t_1$,

$$||x_t - \tilde{x}_t||_r \le \int_{t_1}^t K||x_s - \tilde{x}_s||_r ds \quad \text{for} \quad t_1 \le t < t_1 + \delta.$$

From this and Lemma 8-A, it follows that $x(t) = \tilde{x}(t)$ on $[t_1, t_1 + \delta)$, contradicting the definition of t_1. $\square$

Our next and final theorem of this section asserts, under the assumption of a global Lipschitz condition, that the solutions of Eq. (1') depend continuously on the initial function. This generalizes Theorem 23-D.

Theorem E. Let $F: [t_0, \beta) \times \mathscr{C}_D \to \mathbb{R}^n$ satisfy Continuity Condition (C) and let it be (globally) Lipschitzian (with Lipschitz constant K). Let ϕ and $\tilde{\phi} \in \mathscr{C}_D$ be given and let x and $\tilde{x}$ be unique solutions of Eq. (1') with $x_{t_0} = \phi$ and $\tilde{x}_{t_0} = \tilde{\phi}$. If x and $\tilde{x}$ are both valid on $[t_0 - r, \beta_1)$ then

$$||x(t) - \tilde{x}(t)|| \le ||\phi - \tilde{\phi}||_r e^{K(t - t_0)} \quad \text{for} \quad t_0 \le t < \beta_1. \quad (6)$$

Proof. On $[t_0 - r, \beta_1)$ x satisfies Eq. (3) and $\tilde{x}$ satisfies the same equation but with $\tilde{\phi}$ in place of ϕ. Subtraction then gives for $t_0 \le t < \beta_1$

$$||x(t) - \tilde{x}(t)|| = ||\phi(0) - \tilde{\phi}(0) + \int_{t_0}^t [F(s, x_s) - F(s, \tilde{x}_s)] ds||$$

$$\le ||\phi - \tilde{\phi}||_r + \int_{t_0}^t K||x_s - \tilde{x}_s||_r ds.$$

Hence

$$||x_t - \tilde{x}_t||_r \le ||\phi - \tilde{\phi}||_r + \int_{t_0}^t K||x_s - \tilde{x}_s||_r ds$$

for $t_0 \le t < \beta_1$. Why? Complete the proof by showing that this implies (6). $\square$

Remarks. Before leaving this section, it should be mentioned that Continuity Condition (C) is not the condition

usually found in the literature on delay differential equations.   It is usually assumed instead that   F   is <u>continuous</u>
on $J \times \mathscr{L}_D$.   This means that for every $(\bar{t}, \bar{\psi}) \in J \times \mathscr{L}_D$ and
every $\epsilon > 0$ there exists $\delta > 0$ such that

$$||F(t, \psi) - F(\bar{t}, \bar{\psi})|| < \epsilon$$

whenever $(t, \psi) \in J \times \mathscr{L}_D$ with

$$|t - \bar{t}| < \delta \quad \text{and} \quad ||\psi - \bar{\psi}||_r < \delta.$$

The fact is that if   F   is continuous on $J \times \mathscr{L}_D$ then
it does indeed satisfy Continuity Condition (C).   The interested
reader should write out a proof of this assertion.

We have used the weaker Continuity Condition (C) instead of ordinary continuity of   F   because the former
usually seems slightly easier to verify.

<u>Problems.</u>

1.   Find   F   so that Eq. (1') represents system 21-(12).

2.   Find   F   so that Eq. (1') represents a system equivalent
     to the second order equation 21-(15).

3.   Why is it not possible to represent Eq. 21-(16), $x'(t) =$
     $-x(t)x(t^{1/2})/2t$, as $x'(t) = F(t, x_t)$ on $[0, \infty)$ using
     the notation of this section?   Show that it would be
     possible on $[0, 100)$.

4.   Let $r = 1$ and let $\psi(\sigma) = \sigma^{1/3}$ on $[-1, 0]$.   Sketch ψ
     and sketch some other function $\tilde{\psi} \in \mathscr{L} = C([-1, 0], \mathbb{R})$ such
     that $0 < ||\tilde{\psi} - \psi||_r < 0.1$.

5. Show that if $D \neq \mathbb{R}^n$, $\mathscr{C}_D$ may not be a linear space and $||\cdot||_r$ may not satisfy all the properties of a norm on $\mathscr{C}_D$.

6. (a) Show that if f satisfies the Lipschitz condition 23-(4) on $J \times D^m$ then F defined as in Example 1 satisfies inequality (5) on $J \times \mathscr{C}_D$.

   (b) Show that inequality (5) does not imply 23-(4) by considering the following example. Define $f: \mathbb{R}^3 \to \mathbb{R}$ by $f(t,\xi_{(1)},\xi_{(2)}) = [t^2 + |\xi_{(1)} - \xi_{(2)}|]^{1/2}$ and assume that $g_1(t) = g_2(t)$ for $-10^{-6} \leq t \leq 10^{-6}$.

7. Find a Lipschitz constant for the functional F of Example 3 on the set $\mathbb{R} \times \mathscr{C}_A$ where $A = \{\xi \in \mathbb{R}: |\xi| \leq H\}$ for some $H > 0$.

8. Let $\bar{\psi} \in \mathscr{C}$ and let $\delta > 0$ be given. Define $A = \{\xi \in \mathbb{R}^n: ||\xi - \bar{\psi}(\sigma)|| \leq \delta$ for some $\sigma \in [-r,0]\}$ and define $\mathscr{B} = \{\psi \in \mathscr{C}: ||\psi - \bar{\psi}||_r \leq \delta\}$. (These two sets appear in the statement of Lemma B.) Which of the following is (or are) true: $\mathscr{C}_A \subset \mathscr{B}$, $\mathscr{C}_A = \mathscr{B}$, $\mathscr{B} \subset \mathscr{C}_A$, or none of these?

26. SYSTEMS WITH BOUNDED DELAYS: EXISTENCE

This section will generalize the existence theorems of Section 24 (for ordinary differential systems) to the case of delay differential systems with bounded delays. The notation is that of Section 25.

Let real numbers $r \geq 0$ and t_0 be given and let $t_0 < \beta \leq \infty$. Let D be an open set in $\mathbb{R}^n$, and let F be defined on $[t_0,\beta) \times \mathscr{C}_D \to \mathbb{R}^n$.   Then we shall consider the delay differential system

$$x'(t) = F(t,x_t). \tag{1}$$

Given any $\phi \in \mathscr{C}_D$, we seek a continuous function $x: [t_0-r,\beta_1) \to D$ for some $\beta_1 \in (t_0,\beta]$ such that Eq. (1) is satisfied on $[t_0,\beta_1)$ and such that

$$x_{t_0} = \phi. \tag{2}$$

As before we assume the following continuity condition on F.

<u>Continuity</u> <u>Condition</u> <u>(C)</u> is satisfied if $F(t,x_t)$ is continuous with respect to t in $[t_0,\beta)$ for each given continuous function $\chi: [t_0-r,\beta) \to D$.

If F satisfies Condition (C) then a continuous function $x: [t_0-r,\beta_1) \to D$ is a solution of Eqs. (1) and (2) if and only if

$$x(t) = \begin{cases} \phi(t-t_0) & \text{for } t_0-r \le t \le t_0 \\ \phi(0) + \int_{t_0}^t F(s,x_s)ds & \text{for } t_0 \le t < \beta_1. \end{cases} \tag{3}$$

The <u>simplest</u> existence theorem for Eqs. (1) and (2) is obtained by assuming that F satisfies a Lipschitz condition on all of $J \times \mathscr{C}_D$, where $J = [t_0,\beta)$.   But a much more reasonable assumption, which is still sufficient for a relatively elementary local existence proof, is a <u>local</u> Lipschitz condition.   This is defined as in Section 25:

<u>Definition.</u>   The functional $F: J \times \mathscr{C}_D \to \mathbb{R}^n$ is <u>locally</u> <u>Lipschitzian</u> if for each given $(\bar{t},\bar{\Psi}) \in J \times \mathscr{C}_D$ there exist

numbers   a > 0   and   b > 0   such that

$$\mathscr{B} \equiv ([\bar{t}-a, \bar{t}+a] \cap J) \times \{\psi \in \mathscr{C}: ||\psi - \bar{\psi}||_r \leq b\}$$

is a subset of $J \times \mathscr{C}_D$ and   F   is Lipschitzian on $\mathscr{B}$.   In
other words, for some number   K   (a Lipschitz constant depend-
ing on $\mathscr{B}$),

$$||F(t,\psi) - F(t,\tilde{\psi})|| \leq K||\psi - \tilde{\psi}||_r$$

whenever (t,ψ) and $(t,\tilde{\psi}) \in \mathscr{B}$.

The following local existence theorem uses exactly the
same hypotheses as Uniqueness Theorem 25-D.

<u>Theorem</u> <u>A</u> (<u>Local</u> <u>Existence</u>).   Let $F: [t_0, \beta) \times \mathscr{C}_D \to \mathbb{R}^n$ satis-
fy Condition (C) and let it be locally Lipschitzian.   Then,
for each $\phi \in \mathscr{C}_D$, Eqs. (1) and (2) have a unique solution on
$[t_0-r, t_0+\Delta)$ for some $\Delta > 0$.

<u>Remarks</u>.   This theorem essentially supercedes Theorem
24-A for ordinary differential systems.   And its proof is
quite similar to that of Theorem 24-A except for a small complication in establishing a bound for   F   on some appropriate
set.   Actually, in practical examples it is usually easy to
find explicit bounds for   F.   It is only because of the relatively weak hypotheses of our theorem that a few lines have
to be inserted in the beginning of the following proof which
were not needed for Theorem 24-A.

<u>Proof</u> <u>of</u> <u>Theorem</u> <u>A</u>.   Choose any   a > 0   and   b > 0
sufficiently small so that

$$\mathscr{B} \equiv [t_0, t_0+a] \times \{\psi \in \mathscr{C}: ||\psi - \phi||_r < b\}$$

is a subset of $[t_0,\beta) \times \mathscr{C}_D$ and F is Lipschitzian on $\mathscr{L}$ (say with Lipschitz constant K). Define a continuous function $\overline{\chi}$ on $[t_0-r,t_0+a] \to \mathbb{R}^n$ by

$$\overline{\chi}(t) = \begin{cases} \phi(t-t_0) & \text{for } t_0-r \le t \le t_0 \\ \phi(0) & \text{for } t_0 \le t \le t_0+a. \end{cases}$$

Then $F(t,\overline{\chi}_t)$ depends continuously on t, and hence $||F(t,\overline{\chi}_t)|| \le B_1$ on $[t_0,t_0+a]$ for some constant B_1.

Now define $B = Kb + B_1$. Choose $a_1 \in (0,a]$ such that

$$||\overline{\chi}_t - \phi||_r = ||\overline{\chi}_t - \overline{\chi}_{t_0}||_r \le b \quad \text{for } t_0 \le t \le t_0+a_1.$$

(The existence of such an $a_1 > 0$ is assured by Lemma 25-C.)

From this point onward, the proof will resemble that for an ordinary differential system (Theorem 24-A).

Choose $\Delta > 0$ such that

$$\Delta \le \min \{a_1, b/B\} \quad \text{and} \quad \Delta < 1/K.$$

The last restriction $(\Delta < 1/K)$ was not used in the proof of Theorem 24-A, and it is not necessary here. It only serves to simplify the proof slightly. Cf. Problem 24-5.

Let S be the set of all continuous functions $\chi: [t_0-r,t_0+\Delta] \to \mathbb{R}^n$ such that

$$\chi(t) = \phi(t-t_0) \quad \text{for } t_0-r \le t \le t_0, \quad \text{and}$$

$$||\chi(t) - \phi(0)|| \le b \quad \text{for } t_0 \le t \le t_0+\Delta.$$

Note that if $\chi \in S$ and $t \in [t_0,t_0+\Delta]$, then $||\chi_t - \overline{\chi}_t||_r \le b$; so that

$$||F(t,x_t)|| \leq ||F(t,x_t) - F(t,\overline{x}_t)|| + ||F(t,\overline{x}_t)||$$

$$\leq K||x_t - \overline{x}_t|| + B_1 \leq B.$$

For each $\chi \in S$ define a function $T\chi$ on $[t_0-r,t_0+\Delta]$ by

$$(T\chi)(t) = \begin{cases} \phi(t-t_0) & \text{for } t_0-r \leq t \leq t_0 \\ \phi(0) + \int_{t_0}^{t} F(s,\chi_s)ds & \text{for } t_0 \leq t \leq t_0+\Delta. \end{cases} \tag{4}$$

Then, since $||F(s,\chi_s)|| \leq B$,

$$||(T\chi)(t) - \phi(0)|| \leq B\Delta \leq b \quad \text{for } t_0 \leq t \leq t_0+\Delta.$$

Also $T\chi$ is continuous. Thus $T\chi \in S$ and we can say that T maps $S \to S$.

Let us now construct "successive approximations" in the usual manner -- choosing any $x_{(0)} \in S$ and then defining $x_{(1)} = Tx_{(0)}$, $x_{(2)} = Tx_{(1)}$, ... Bear in mind that each $x_{(\ell)}(t) = \phi(t-t_0)$ on $[t_0-r,t_0]$.

The proof that the sequence $\{x_{(\ell)}\}$ converges is now slightly simpler than it was for Theorem 24-A because of our additional assumption that $K\Delta < 1$. In fact, for each $\ell = 0,1,2,\ldots$, when $t_0 \leq t \leq t_0+\Delta$

$$||x_{(\ell+2)}(t) - x_{(\ell+1)}(t)||$$

$$= ||\int_{t_0}^{t} [F(s,x_{(\ell+1)s}) - F(s,x_{(\ell)s})]ds||$$

$$\leq K\Delta \sup_{t_0 \leq s \leq t_0+\Delta} ||x_{(\ell+1)s} - x_{(\ell)s}||.$$

From this and the fact that

$$||x_{(1)}(t) - x_{(0)}(t)|| \leq 2b$$

one finds, for $t_0 \leq t \leq t_0+\Delta$,

$$||x_{(2)}(t) - x_{(1)}(t)|| \leq 2bK\Delta,$$

$$||x_{(3)}(t) - x_{(2)}(t)|| \leq 2b(K\Delta)^2,$$

and, by induction,

$$||x_{(\ell+1)}(t) - x_{(\ell)}(t)|| \leq 2b(K\Delta)^{\ell} \quad \text{for} \quad \ell = 0,1,2,\dots$$

Now, since the series

$$\sum_{p=0}^{\infty} ||x_{(p+1)}(t) - x_{(p)}(t)|| \leq \sum_{p=0}^{\infty} 2b(K\Delta)^{\ell}$$

converges, the convergence of the sequence $\{x_{(\ell)}\}$ follows by application of the comparison test to each component of

$$x_{(\ell)}(t) = x_{(0)}(t) + \sum_{p=0}^{\ell-1} [x_{(p+1)}(t) - x_{(p)}(t)]$$

on $[t_0,t_0+\Delta]$. This and the proof that $x(t) \equiv \lim_{\ell\to\infty} x_{(\ell)}(t)$ satisfies Eq. (3), and hence (1) and (2), are just as in the proofs of Theorems 20-A and 24-A.

Uniqueness follows from Theorem 25-D. □

But Theorem A leaves unanswered the question, How far can the solution of Eqs. (1) and (2) be continued?

The following definitions are like their counterparts in Section 24, with notation adapted here to the case of delay differential systems with bounded delays.

Definitions. Let x on $[t_0-r,\beta_1)$ and y on $[t_0-r,\beta_2)$ both be solutions of Eqs. (1) and (2). If $\beta_2 > \beta_1$ we say y is a continuation of x, or x can be continued to $[t_0-r,\beta_2)$. A solution x of (1) and (2) is noncontinuable if it has no continuation.

In Theorem 24-C we proved the following for an ordinary

differential system. Under the hypotheses of continuity and a local Lipschitz condition, if a solution x cannot be continued beyond β_1, then either $\beta_1 = \beta$ or, for every closed bounded set $A \subset D$, $x(t) \notin A$ for some $t \in (t_0, \beta_1)$.

However, for Eq. (1) this is not quite true. The trouble is that a continuous functional F, unlike a continuous function f, is not necessarily bounded on "closed bounded" subsets of $J \times \mathcal{L}_D$. The reader should not expect to fully appreciate this statement unless he has studied some topology. For the reader who knows some topology, Problem 9 outlines an ingenious example due to Yorke [1969] showing that, for Eq. (1), the analog of Theorem 24-C is false.

The usual remedy for this situation is to simply add another hypothesis on F -- an hypothesis which is satisfied in the known practical examples:

<u>Definition</u>. The functional $F: [t_0, \beta) \times \mathcal{L}_D \to \mathbb{R}^n$ is said to be <u>quasi-bounded</u> if F is bounded on every set of the form $[t_0, \beta_1] \times \mathcal{L}_A$ where $t_0 < \beta_1 < \beta$ and A is a closed bounded subset of D.

Note that the functionals F defined in Examples 25-1 (with continuous f) and Examples 25-2, 3, and 4 are all quasi-bounded. Problem 2.

The proof of our forthcoming extended existence theorem (Theorem C) will use the following.

<u>Lemma B</u>. Let $\phi \in \mathcal{L} = C([-r, 0], \mathbb{R}^n)$. Then the set of values taken by ϕ,

$$A_\phi \equiv \{\phi(\sigma): -r \leq \sigma \leq 0\}$$

is a closed (and bounded) set in $\mathbb{R}^n$.

Proof. Let $\{\xi_{(i)}\}_{i=1}^\infty$ be a sequence of points (or vectors) in A_ϕ which converges to a point ξ in $\mathbb{R}^n$. (We must show that $\xi \in A_\phi$.) There exist numbers $\sigma_1, \sigma_2, \ldots$ in $[-r,0]$ such that $\xi_{(i)} = \phi(\sigma_i)$ for $i = 1,2,\ldots$ But, since $[-r,0]$ is a closed bounded set in $\mathbb{R}$, the Bolzano-Weierstrass Theorem assures the existence of a convergent subsequence $\{\sigma_{i_k}\}$ with limit, say σ_0, in $[-r,0]$. Thus it follows that

$$\xi = \lim_{i\to\infty} \phi(\sigma_i) = \lim_{k\to\infty} \phi(\sigma_{i_k}) = \phi(\sigma_0) \in A_\phi. \quad \square$$

Theorem C (Extended Existence). Let $F: [t_0,\beta) \times \mathscr{C}_D \to \mathbb{R}^n$ satisfy Condition (C), and let it be locally Lipschitzian and quasi-bounded. Then for each $\phi \in \mathscr{C}_D$, Eqs. (1) and (2) have a unique noncontinuable solution x on $[t_0-r,\beta_1)$; and if $\beta_1 < \beta$, then, for every closed bounded set $A \subset D$,

$$x(t) \notin A \quad \text{for some} \quad t \quad \text{in} \quad (t_0,\beta_1).$$

Proof. The uniqueness of the solution on any interval of the form $[t_0-r,\beta_1)$ follows at once from Theorem 25-D.

Define

$$\beta_1 = \sup \{s \in \mathbb{R}: \text{a solution exists on } [t_0-r,s)\}.$$

Then $\beta_1 > t_0$ and for every $s \in (t_0,\beta_1)$ a unique solution $y_{(s)}$ exists on $[t_0-r,s)$. Now define a function $x: [t_0-r,\beta_1) \to D$ as follows. For each t in $[t_0-r,\beta_1)$ let

$$x(t) = y_{(s)}(t) \quad \text{for some} \quad s \quad \text{in} \quad (t,\beta_1).$$

By uniqueness, this is an unambiguous definition of $x(t)$. Moreover, by construction, the resulting function x is a solution of Eqs. (1) and (2) on $[t_0-r,\beta_1)$, and it cannot be continued beyond β_1.

Now suppose (for contradiction) that $\beta_1 < \beta$ and yet, for some closed bounded set $A \subset D$, $x(t) \in A$ for $t_0 \le t < \beta_1$.

We can not assume that $\phi(\sigma) \in A$ for $-r \le \sigma \le 0$. So instead of A, let us consider the set $A \cup A_\phi$. By Lemma B, A_ϕ is closed; so it follows that $A \cup A_\phi$ is closed (and bounded). Problem 3.

Now $\dot{x}(t) \in A \cup A_\phi$ for $t_0-r \le t < \beta_1$, and, by the quasi-boundedness of F, there exists a constant B such that

$$||F(t,\psi)|| \le B \quad \text{for all} \quad (t,\psi) \in [t_0,\beta_1] \times \mathscr{C}_{A \cup A_\phi}.$$

Hence

$$||x'(t)|| = ||F(t,x_t)|| \le B \quad \text{for} \quad t_0 \le t < \beta_1,$$

and Lemma 24-D assures the existence of

$$\lim_{t \to \beta_1} x(t) = \xi \in A \subset D.$$

Let us now extend the definition of x to a continuous function on $[t_0-r,\beta_1] \to D$ by setting $x(\beta_1) = \xi$. Then it follows from Condition (C) that $F(t,x_t)$ is continuous for $t_0 \le t \le \beta_1$. (This is not quite obvious. Fill in the detail.) Thus Eq. (3) extends to include the point $t = \beta_1$, i.e.,

$$x(t) = \begin{cases} \phi(t-t_0) & \text{for } t_0-r \le t \le t_0 \\ \phi(0) + \int_{t_0}^{t} F(s,x_s)ds & \text{for } t_0 \le t \le \beta_1. \end{cases}$$

Now apply Theorem A to the new problem: $z'(t) = F(t,z_t)$ for $t \ge \beta_1$ with $z_{\beta_1} = x_{\beta_1}$. We conclude that this new problem has a solution z on $[\beta_1-r,\beta_1+\Delta)$ for some $\Delta > 0$.

Thus

$$z(t) = \begin{cases} x(t) & \text{for } \beta_1 - r \le t \le \beta_1 \\ x(\beta_1) + \int_{\beta_1}^{t} F(s,z_s)ds & \text{for } \beta_1 \le t < \beta_1 + \Delta. \end{cases}$$

Let $z(t) \equiv x(t)$ also on $[t_0 - r, \beta_1 - r]$, and we have

$$z(t) = \begin{cases} \phi(t - t_0) & \text{for } t_0 - r \le t \le t_0 \\ \phi(0) + \int_{t_0}^{t} F(s,z_s)ds & \text{for } t_0 \le t < \beta_1 \\ \phi(0) + \int_{t_0}^{\beta_1} F(s,z_s)ds + \int_{\beta_1}^{t} F(s,z_s)ds & \text{for } \beta_1 \le t < \beta_1 + \Delta \end{cases}$$

$$= \begin{cases} \phi(t - t_0) & \text{for } t_0 - r \le t \le t_0 \\ \phi(0) + \int_{t_0}^{t} F(s,z_s)ds & \text{for } t_0 \le t < \beta_1 + \Delta. \end{cases}$$

In other words z satisfies Eq. (3) and hence also Eqs. (1) and (2). The existence of this solution, continuing x to $\beta_1 + \Delta > \beta_1$, contradicts the definition of β_1. $\square$

Just as in the case of ordinary differential systems, we can now present a simple criterion for assuring the existence of a solution of (1) and (2) on the entire interval $[t_0 - r, \beta)$. Cf. Corollary 24-E.

Corollary D (Global Existence). Let $D = \mathbb{R}^n$. Let $F: [t_0, \beta) \times \mathscr{C} \to \mathbb{R}^n$ satisfy Condition (C) and let it be locally Lipschitzian. Assume further that

$$||F(t,\psi)|| \le M(t) + N(t)||\psi||_r \quad \text{on} \quad [t_0, \beta) \times \mathscr{C}, \tag{5}$$

where M and N are continuous positive functions on $[t_0, \beta)$. Then the unique noncontinuable solution of Eqs. (1) and (2) exists on the entire interval $[t_0 - r, \beta)$.

Proof. Condition (5) implies that F is quasi-
bounded. Why?

Let x on $[t_0-r,\beta_1)$ be the unique noncontinuable
solution of (1) and (2), and suppose (for contradiction) that
$\beta_1 < \beta$. Then $M(t) \leq M_1$ and $N(s) \leq N_1$ on $[t_0,\beta_1]$. So
Eq. (3) yields

$$||x(t)|| \leq ||\phi||_r + \int_{t_0}^t M_1 ds + \int_{t_0}^t N_1 ||x_s|| ds$$

for $t_0 \leq t < \beta_1$. From this we find

$$||x_t||_r \leq ||\phi||_r + M_1(\beta_1-t_0) + \int_{t_0}^t N_1 ||x_s|| ds$$

for $t_0 \leq t < \beta_1$. How? Thus

$$||x(t)|| \leq ||x_t||_r \leq [||\phi||_r + M_1(\beta_1-t_0)]e^{N_1(\beta_1-t_0)}$$

on $[t_0,\beta_1)$. This shows that $x(t)$ remains in a closed
bounded set, contradicting (by Theorem C) the supposition
that $\beta_1 < \beta$. Hence $\beta_1 = \beta$. ☐

Example 1. Consider the linear delay differential system

$$x'(t) = \sum_{j=1}^m A_j(t)x(g_j(t)) + h(t) \quad \text{on} \quad [t_0,\beta), \tag{6}$$

where each A_j is a continuous $n \times n$-matrix-valued function, h is a continuous n-vector-valued function, and each
g_j is a continuous real-valued function with $t-r \leq g_j(t) \leq t$.
Then, for each $\phi \in \mathscr{L}$, Eqs. (6) and (2) have a unique solution on $[t_0-r,\beta)$.

Example 2. If $D = \mathbb{R}^n$ and $F: [t_0,\beta) \times \mathscr{L} \to \mathbb{R}^n$ satisfies
Condition (C) and a global Lipschitz condition then, for each
$\phi \in \mathscr{L}$, Eqs. (1) and (2) have a unique solution on $[t_0-r,\beta)$.

This is a consequence of Corollary D since the global Lipschitz condition (with Lipschitz constant K) gives

$$||F(t,\psi)|| \leq ||F(t,0)|| + ||F(t,\psi) - F(t,0)||$$

$$\leq ||F(t,0)|| + K||\psi||_r,$$

which implies Condition (5).

Problems

1. What changes would be required in the proof of Theorem A if we put $\Delta = \min \{a_1, b/B\}$, i.e., if we did not assume $\Delta < 1/K$?

2. Show that the functionals F defined in Examples 25-1 (with continuous f) and Examples 25-2,3, and 4 are all quasi-bounded.

3. Prove that if A_1 and A_2 are two closed sets in $\mathbb{R}^n$, then $A_1 \cup A_2$ is closed. __Hint__. Let $\{\xi_{(i)}\}$ be a sequence in $A_1 \cup A_2$ such that $\lim_{i \to \infty} \xi_{(i)} = \xi$ exists. Then either A_1 or A_2 (or both) must contain infinitely many of the points $\{\xi_{(i)}\}$.

4. State and justify the best existence and uniqueness result you can for the scalar delay differential equation $x'(t) = \int_{t-r}^{t} x(s)ds$ of Example 25-2.

5. State and justify the best existence and uniqueness result you can for the scalar equation

$$x'(t) = \sum_{j=1}^{\infty} a_j x(t - \frac{1}{j}).$$

6. Convince yourself that Corollary D would be of no use in
    Problems 21-7 and 21-8.

7. Theorem. Let D be an open set in $\mathbb{R}^n$ and let
    $f: [t_0,\beta) \times D^n$ be continuous and locally Lipschitzian.
    Let each g_j be continuous on $[t_0,\beta)$ with $g_1(t) \equiv t$
    and $t-r \leq g_j(t) < t$ for $j = 2,\ldots,m$, and let $\phi \in \mathscr{C}_D$
    be given. Then there exists $\Delta > 0$ such that the delay
    differential equation

    $$x'(t) = f(t,x(g_1(t)),\ldots,x(g_m(t))) \quad \text{on} \quad [t_0,t_0+\Delta)$$

    with $x_{t_0} = \phi$ has a unique solution on $[t_0-r,t_0+\Delta)$.
    (Note that this is a special case of Theorem A, and that
    it covers most of the examples considered in Chapter V.)
        Give a proof of the above theorem using none of the re-
    sults of this section, but based on the method of steps
    together with Theorem 24-A.

*8. Give a direct proof of Corollary D, which is independent
    of Theorem C. Outline. If $x: [t_0-r,\beta_1) \to \mathbb{R}^n$ is any
    solution of Eqs. (1) and (2), where $t_0 < \beta_1 \leq \beta$, show
    that

    $$||x(t)|| \leq [||\phi||_r + \int_{t_0}^t M(s)ds]e^{\int_{t_0}^t N(s)ds} \equiv G(t)$$

    on $[t_0,\beta_1)$. Now let $\beta_1 \in (t_0,\beta)$ be arbitrary, and let
    S be the set of all continuous functions $\chi: [t_0-r,\beta_1] \to$
    $\mathbb{R}^n$ such that $\chi(t) = \phi(t-t_0)$ on $[t_0-r,t_0]$ and
    $||\chi(t)|| \leq G(t)$ on $[t_0,\beta_1]$. Show that T, defined as
    in Eq. (4) but with β_1 in place of $t_0+\Delta$, maps $S \to S$.
    Let K be a Lipschitz constant for F on $[t_0,\beta_1] \times \mathscr{C}_A$
    where $A = \{\xi \in \mathbb{R}^n: ||\xi|| \leq G(\beta_1)\}$. Define a sequence

$\{x_{(\ell)}\}$ of successive approximations in S in the usual way, and consider

$$v_\ell = \sup_{t_0 \leq t \leq \beta_1} [e^{-c(t-t_0)} ||x_{(\ell+1)}(t) - x_{(\ell)}(t)||]$$

where $c > K$. Show that $v_{\ell+1} \leq (K/c)v_\ell$ for $\ell = 0,1,\ldots$ Complete the proof.

*9.   (Assuming a little knowledge of topology.) Let real numbers $r > 0$ and $t_0 < 0$ be given. Let $h(t) \equiv \sin 1/t$ for $t < 0$ and let $D = \mathbb{R}$. Then define $F: [t_0,\infty) \times \mathscr{C} \to \mathbb{R}$ by

$$F(t,\psi) \equiv \begin{cases} h'(t)[1 + ||h_t-\psi||_r/t] & \text{whenever } ||h_t-\psi|| < -t \\ 0 & \text{otherwise} \end{cases}$$

Let $A = [-2,2]$ and let $\mathscr{B} = [t_0,0] \times \mathscr{C}_A$.

(a)   Show that $\mathscr{B}$ is a closed bounded subset of $[t_0,\infty) \times \mathscr{C}$.

(b)   Show that F is unbounded on $\mathscr{B}$ by considering

$$(t,\psi) = (t,h_t) \quad \text{for} \quad t_0 \leq t < 0.$$

(c)   Nevertheless, show that F is continuous on all of $\mathscr{B}$.

(d)   Show that F is locally Lipschitzian on $\mathscr{B}$.

(e)   Show that if $\phi \equiv h_{t_0}$ then the unique solution, x, through (t_0,ϕ) cannot be continuous past 0, even though $x(t) \in A$ for all $t \in [t_0,0)$.

This example is due to Yorke [1969]. The answers, if needed, can be found in Yorke's paper.

Chapter VII. <u>LINEAR</u> <u>DELAY</u> <u>DIFFERENTIAL</u> <u>SYSTEMS</u>

Chapter IV exploited the nice properties of linear
systems of ordinary differential equations. We shall now
show how at least some of these properties carry over to
linear systems of delay differential equations.

Our results will be restricted to a linear system with
a finite number of constant delays

$$x'(t) = \sum_{j=1}^{m} A_j(t)x(t-r_j) + h(t) \quad \text{on} \quad [t_0,\beta).$$

Each A_j is a continuous (if not constant) matrix-valued
function, and h is a continuous vector-valued function. Restriction to this special case will simplify the statement
and proof of results which actually are known to hold for
more general linear systems. Moreover, the special case described above does arise in practical applications.

But, even with the restriction to this special case we
will have to state without proof an important theorem in
Section 28 on the behavior of solutions in the case of constant coefficients.

Section 29 presents a theorem on variation of parameters. The variation-of-parameters results for linear systems, both with and without delays, will be of importance in
the study of "quasi-linear" systems in Chapter VIII.

313

27.   SUPERPOSITION

Consider a system of differential equations with constant delays $r_1,\ldots,r_m$ $(0 \le r_j \le r$ for each $j)$

$$x'(t) = \sum_{j=1}^{m} A_j(t)x(t-r_j) + h(t). \qquad (1)$$

Each A_j is a continuous $n \times n$-matrix-valued function and h is a continuous n-vector-valued function on $[t_0,\beta)$.

For brevity we can rewrite Eq. (1) in the form

$$x'(t) = L(t,x_t) + h(t), \qquad (1')$$

where we have defined

$$L(t,\psi) \equiv \sum_{j=1}^{m} A_j(t)\psi(-r_j)$$

for $t \in [t_0,\beta)$ and $\psi \in \mathscr{C} = C([-r,0],\mathbb{R}^n)$.

Note that Eq. (1) is <u>linear</u> (as defined in Section 22) since

$$L(t,c_1\psi+c_2\tilde{\psi}) = c_1 L(t,\psi) + c_2 L(t,\tilde{\psi})$$

for all $t \in [t_0,\beta)$, $\psi,\tilde{\psi} \in \mathscr{C}$, and $c_1,c_2 \in \mathbb{R}$.

As usual, Eq. (1) is to be solved subject to the initial condition

$$x_{t_0} = \phi, \qquad (2)$$

where $\phi \in \mathscr{C}$ is given.

Under these conditions, Example 26-1 asserts the existence of a unique solution of Eqs. (1) and (2) on the entire interval $[t_0-r,\beta)$.

In addition to system (1), we shall also consider the associated linear <u>homogeneous</u> system

$$y'(t) = L(t, y_t). \tag{3}$$

We shall sometimes talk about a <u>solution</u> of system (1) or of system (3) on $[t_0-r, \beta)$ without mentioning the initial function. Then we mean any continuous function mapping $[t_0-r, \beta) \to \mathbb{R}^n$ which satisfies (1) or (3) respectively on $[t_0, \beta)$.

In these terms, as in the case of ordinary differential systems, we get the following theorem on linear combinations of solutions.

<u>Theorem A</u> (<u>Superposition</u>). Let $h(t) = k_1 h_{(1)}(t) + \cdots + k_p h_{(p)}(t)$, where $k_1, \ldots, k_p$ are constants and $h_{(1)}, \ldots, h_{(p)}$ are given vector-valued functions on $[t_0, \beta)$. Let $x_{(q)}$ on $[t_0-r, \beta)$ be any (particular) solution of $x'(t) = L(t, x_t) + h_{(q)}(t)$ (for each $q = 1, \ldots, p$) and let $y_{(1)}, \ldots, y_{(\ell)}$ on $[t_0-r, \beta)$ be solutions of Eq. (3). Then

$$x = k_1 x_{(1)} + \cdots + k_q x_{(q)} + c_1 y_{(1)} + \cdots + c_\ell y_{(\ell)}$$

is a solution of Eq. (1) for every choice of the constants $c_1, \ldots, c_\ell$.

<u>Proof</u>. Using the linearity of the functional L, we compute

$$
\begin{aligned}
x'(t) &= \sum_{q=1}^{p} k_q x'_{(q)}(t) + \sum_{s=1}^{\ell} c_s y'_{(s)}(t) \\
&\quad + \sum_q k_q L(t, x_{(q)t}) + \sum_q k h_{(q)}(t) + \sum_s c_s L(t, y_{(s)t}) \\
&= L\left(t, \sum_q k_q x_{(q)t} + \sum_s c_s y_{(s)t}\right) + h(t) \\
&= L(t, x_t) + h(t). \qquad \square
\end{aligned}
$$

Example 1.   Consider the scalar equation

$$x'(t) = -2x(t) + x(t - \tfrac{\pi}{2}) + \sin t \quad \text{for} \quad t \geq 0 \tag{4}$$

with

$$x(t) = \phi(t) \quad \text{for} \quad -\pi/2 \leq t \leq 0, \tag{5}$$

where ϕ is a given continuous function.

Following the idea used for linear ordinary differential equations with constant coefficients, let us seek a particular solution of Eq. (4) of the form

$$\tilde{x}(t) = A \cos t + B \sin t.$$

Substitution into (4) gives

$$-A \sin t + B \cos t = -2A \cos t - 2B \sin t$$
$$+ A \sin t - B \cos t + \sin t.$$

This requires

$$-2A + 2B = 1 \quad \text{and} \quad 2A + 2B = 0,$$

or $A = -1/4$, $B = 1/4$. So a solution of (4) is defined by

$$\tilde{x}(t) = -\frac{1}{4} \cos t + \frac{1}{4} \sin t.$$

Letting x be the unique solution of (4) and (5) on $[-\pi/2, \infty)$, let us substract off $\tilde{x}$ and consider $y = x - \tilde{x}$. Then y satisfies the homogeneous equation

$$y'(t) = -2y(t) + y(t - \tfrac{\pi}{2}) \quad \text{for} \quad t \geq 0$$

with

$$y(t) = \phi(t) + \frac{1}{4} \cos t - \frac{1}{4} \sin t \quad \text{for} \quad -\pi/2 \leq t \leq 0,$$

and x is given by

$$x(t) = \tilde{x}(t) + y(t) \quad \text{for} \quad t \geq -\pi/2.$$

Now Problem 21-15(c) asserts that $y(t) \to 0$ as $t \to \infty$.
(If you did not work this starred problem, accept the assertion on faith for now. It will be proved in Example 32-4.)
Hence we might regard $\tilde{x}$ as the "steady state" part of the
solution of (4) and (5), and y as the "transient" part,
which dies out as $t \to \infty$. Note that only the transient part
depends on the initial function ϕ.

In applications one may encounter n'th order, linear,
scalar, delay differential equations such as

$$x^{(n)}(t) + \sum_{j=1}^{m} \sum_{i=0}^{n-1} a_{ij}(t)x^{(i)}(t-r_j) = h(t), \qquad (6)$$

where h and each a_{ij} are continuous real-valued functions
on $[t_0,\beta)$. Equation (6) is usually considered in conjunction
with an initial condition

$$x(t) = \phi(t-t_0) \qquad \text{on} \quad [t_0-r,t_0] \qquad (7)$$

where $\phi \in C^{n-1}([-r,0],\mathbb{R})$. The homogeneous equation associated with (6) is

$$y^{(n)}(t) + \sum_{j=1}^{m} \sum_{i=0}^{n-1} a_{ij}(t)y^{(i)}(t-r_j) = 0. \qquad (8)$$

Equations (6) and (8) can be transformed into equivalent first order systems and then discussed as special cases
of (1) and (3). The analog of Theorem A then is as follows.

Corollary B. Let $h(t) = k_1 h_1(t) + \cdots + k_p h_p(t)$, where
$k_1, \ldots, k_p$ are constants and $h_1, \ldots, h_p$ are given real-
valued functions on $[t_0,\beta)$. Let x_q on $[t_0-r,\beta)$ be any
solution of

$$x'(t) + \sum_{j=1}^{m} \sum_{i=0}^{n-1} a_{ij}(t)x^{(i)}(t-r_j) = h_q(t)$$

(for   q = 1,...,p) and let $y_1,...,y_\ell$ on $[t_0-r,\beta)$ be solutions of Eq. (8).   Then

$$x = k_1x_1 + \cdots + k_px_p + c_1y_1 + \cdots + c_\ell y_\ell$$

is a solution of Eq. (6) for every choice of the constants $c_1,...,c_\ell$.

The reader should find no difficulty in proving Corollary B by <u>direct</u> <u>substitution</u> of x into Eq. (6). (This is actually simpler than setting up the notation to convert Eqs. (6) and (8) into systems, and then invoking Theorem A.)

<u>Example</u> 2.   Let us seek a particular solution of

$$x''(t) + x'(t) + x'(t-1) + \pi^2x(t) = \cos \pi t. \tag{9}$$

It is natural to try $\tilde{x}(t) = A \cos \pi t + B \sin \pi t$. But, as the reader should verify, this $\tilde{x}$ satisfies the homogeneous equation associated with (9), and hence cannot satisfy Eq. (9) itself. A little reflection on similar cases in ordinary differential equations might lead one to try

$$\tilde{x}(t) = At \cos \pi t + Bt \sin \pi t.$$

Carry out the substitution, and you should find that this works with A = 0 and B = 1/3π.

<u>Problems</u>.

1.   Use your imagination (and analogy to ordinary differential equations) to find a particular solution, $\tilde{x}$ on $[-r,\infty)$, of each of the following equations.

   (a) $x'(t) + 2x(t) - x(t-\frac{\pi}{4}) = \sin t$,

(b) $x'(t) + x(t-\frac{\pi}{2}) = \sin t$,

(c) $x'(t) = -2x(t) + x(t-1) + t$,

(d) $x'(t) = -x(t) + x(t-1) + t$,

(e) $x'(t) = x(t) - x(t-1) + t$.

2.   Let $\tilde{x}(t) = -\frac{1}{4} \cos t + \frac{1}{4} \sin t$, the particular solution found for Eq. (4), and let λ be the unique number in $(-2,0)$ such that $\lambda+2 = e^{-\pi\lambda/2}$.

(a)   Show that $\hat{x}(t) = \tilde{x}(t) + 17e^{\lambda t}$ defines another particular solution of Eq. (4).

(b)   Suppose that somehow we had come upon $\hat{x}$ (instead of $\tilde{x}$) as our original particular solution of Eq. (4). Show that we would still have reached the same conclusion as in Example 1 about the steady state and transient parts of the solution of (4) and (5).   More specifically, we would still have concluded that the "steady state" part is $\tilde{x}$ (not $\hat{x}$).

28.   CONSTANT COEFFICIENTS

Most of the actual examples discussed thus far have been special cases of the linear differential system with finitely many constant coefficients and delays,

$$x'(t) = \sum_{j=1}^{m} A_j x(t-r_j) + h(t). \tag{1}$$

Here A_j is a (constant) $n \times n$ matrix and $0 \leq r_j \leq r$ for each $j = 1,\ldots,m$.   The vector-valued function h will be assumed continuous on $[t_0,\beta)$.

We usually consider Eq. (1) on $[t_0,\beta)$, together with an initial condition

$$x_{t_0} = \phi,$$
(2)

where $\phi \in \mathscr{C}$.

This section will emphasize the study of the homogeneous equation associated with (1),

$$y'(t) = \sum_{j=1}^{n} A_j y(t-r_j).$$
(3)

Recall that in the case of a system of n linear homogeneous ordinary differential equations with constant coefficients there are n linearly independent solutions. And we know that the general solution is expressible as an arbitrary linear combination of these n solutions.

But the situation is more complicated for Eq. (3) because, in general, (3) has infinitely many linearly independent solutions (valid actually on $\mathbb{R}$).

Let us approach Eq. (3) as we would a linear homogeneous ordinary differential system with constant coefficients. Observe first that one can consider complex-valued solutions of (3) just as in the ordinary-differential-systems case. Now if one seeks an "exponential" solution of Eq. (3) in the form $y(t) = e^{\lambda t}\xi$, where ξ is a constant vector, one is led to the equation

$$(\lambda I - \sum_{j=1}^{m} A_j e^{-\lambda r_j})\xi = 0.$$
(4)

Equation (4) has solutions $\xi \neq 0$ if and only if λ satisfies the <u>characteristic equation</u>

$$\det(\lambda I - \sum_{j=1}^{m} A_j e^{-\lambda r_j}) = 0.$$
(5)

Suppose $\lambda = \mu + i\omega$ (where μ and ω are real) is a solution of (5), and, using this value of λ, $\xi = \xi_{(1)} + i\xi_{(2)}$

is a solution of (4) where $\xi_{(1)}$ and $\xi_{(2)}$ are real n
vectors.   Then (as Problem 1) verify that both

$$y(t) = e^{\mu t}(\xi_{(1)}\cos \omega t - \xi_{(2)}\sin \omega t)$$
and   (6)
$$y(t) = e^{\mu t}(\xi_{(2)}\cos \omega t + \xi_{(1)}\sin \omega t)$$

are real-n-vector-valued solutions of (3).

Of course, by superposition, linear combinations of
the exponential solutions are also solutions of Eq. (3).   The
trouble is that Eq. (5), in general, has infinitely many
(complex) solutions, λ.   We will not prove this statement except in the case of the following optional example.   A complete understanding of this example requires knowledge of a
theorem of Picard in complex function theory.

*Example 1.   Consider the scalar equation

$$y'(t) = y(t-r),$$ (7)

where r is a positive constant.   Let us seek solutions of
the form $ce^{\lambda t}$.   Then we obtain the characteristic equation
for λ

$$\lambda = e^{-\lambda r}.$$ (8)

It is easily seen by graphing that Eq. (8) has exactly one
real solution.   To consider other possible roots, let us now
introduce $z = 1/\lambda$ and note that (8) is equivalent to
$w(z) = 0$, where
$$w(z) \equiv 1 - ze^{-r/z}.$$

But w has an isolated essential singularity at $z = 0$.
Thus, by Picard's Theorem, in every neighborhood of $z = 0$

$w(z)$ takes on every value (with one possible exception) infinitely often. It is easily seen that $w(z) \neq 1$ for all
$z \neq 0$, so the exceptional value in this case has to be 1.
Thus it follows that $w(z) = 0$ infinitely often; and hence
Eq. (8) must have infinitely many complex roots. If $\lambda =$
$\mu + i\omega$, where μ and ω are real, is a solution of Eq. (8),
the reader can verify that both

$$e^{\mu t} \cos \omega t \quad \text{and} \quad e^{\mu t} \sin \omega t \tag{9}$$

satisfy Eq. (7) for all t. Thus one obtains infinitely many
solutions of (7).

Remark. The superposition result (Theorem 27-A) can
be extended to cover the case of infinitely many solutions,
$y_{(1)}, y_{(2)}, \ldots$ of Eq. (3). Specifically, in this case, an
infinite series

$$y = c_1 y_{(1)} + c_2 y_{(2)} + \cdots$$

is also a solution of (3) provided the series converges and
admits term-by-term differentiation. (This remark would also
be valid for systems of ordinary differential equations. However it is of no interest in that case since such systems
have only a finite number of linearly independent solutions.)

The following is a basic property of the (complex)
roots of the characteristic equation (5).

Theorem A. Given any real number, ρ, Eq. (5) has no more
than a finite number of roots λ such that $\operatorname{Re} \lambda \geq \rho$.

*Proof. (assuming some knowledge of complex function
theory.)

Note that Eq. (5) has the form

$$\lambda^n + P_{n-1}(e^{-\lambda r_1}, \ldots, e^{-\lambda r_m})\lambda^{n-1} + \cdots$$

$$+ P_0(e^{-\lambda r_1}, \ldots, e^{-\lambda r_m}) = 0, \tag{5'}$$

where each $P_{n-1}, \ldots, P_0$ is a polynomial in $e^{-\lambda r_1}, \ldots, e^{-\lambda r_m}$.

Now let ρ be a given real number and assume $\text{Re } \lambda \geq \rho$. Then $|e^{-\lambda r_j}| = e^{-\text{Re } \lambda r_j} \leq e^{-\rho r_j}$, so that

$$|P_k(e^{-\lambda r_1}, \ldots, e^{-\lambda r_m})| \leq B_k$$

for some constant B_k (depending on ρ) for $k = 0, \ldots, n-1$. Next choose a positive number R so large that

$$B_{n-1}/R + \cdots + B_0/R^n < 1.$$

Equation (5') can have no roots with $\text{Re } \lambda \geq \rho$ and $|\lambda| \geq R$, since then one would have $|\lambda|^n > B_{n-1}|\lambda|^{n-1} + \cdots + B_0$, contradicting (5'). So all roots with $\text{Re } \lambda \geq \rho$ must have $|\lambda| < R$. But an analytic function which is not identically zero cannot have more than a finite number of zeros in any bounded set in the complex plane. Hence, Eq. (5') cannot have infinitely many roots with $\text{Re } \lambda \geq \rho$. ☐

Using Theorem A and other tools, various authors have made detailed studies of the roots of characteristic equations such as (8). Then they have investigated the possibility of finding solutions of equations such as (7) with a given initial function in terms of infinite series of the functions (9). See for example the papers of Pitt [1944], [1947], Wright [1948], [1949], and Zverkin [1965], and the books of Pinney [1958] and Bellman and Cooke [1963].

Such studies are complicated and will not be pursued here. But one related result for Eq. (3) is sufficiently important to be included here even though we shall omit the proof.

<u>Theorem B</u>. If $\text{Re }\lambda < \rho$ for every solution of the characteristic equation (5), then there exists a constant $M > 0$ such that, for each $\phi \in \mathcal{C}$ the solution of Eq. (3) with $y_{t_0} = \phi$ satisfies

$$||y(t;t_0,\phi)|| \leq M||\phi||_r e^{\rho(t-t_0)} \qquad \text{for all } t \geq t_0. \qquad (10)$$

The proof using functional analysis can be found in Stokes [1962], Hale [1971], Theorem 22.1, and essentially also in Krasovskiĭ [1959], Theorem 29.1. In fact, the proofs given there even cover systems with infinitely many delays.

<u>Remarks</u>. Theorem 16-A is a corollary of the above.

If $\text{Re }\lambda < \rho$ for each root of (5) then, for some $\gamma > 0$, each $\text{Re }\lambda < \rho-\gamma$ (since there are only a finite number of λ's say with $\text{Re }\lambda > \rho-1$). Thus the assertion of Theorem B could be rewritten as follows. There exist positive constants M and γ such that

$$||y(t;t_0,\phi)|| \leq M||\phi||_r e^{(\rho-\gamma)(t-t_0)} \qquad \text{for all } t \geq t_0. \quad (10')$$

Using (10') instead of (10), we see that the converse of Theorem B would also be true. For if $\text{Re }\lambda \geq \rho$ for some root of (5), then an appropriate initial function ϕ must lead to a solution of the form (6) which violates inequality (10').

Example 2. Let $\lambda = \mu + i\omega$ be a solution of Eq. (8), $\lambda = e^{-\lambda r}$. Then we must have

$$\mu = e^{-\mu r} \cos \omega r \le e^{-\mu r}$$

But, sketching the graphs of μ and $e^{-\mu r}$, one easily sees that this implies $\text{Re } \lambda = \mu \le \mu_0$ where μ_0 is the unique positive number such that $\mu_0 = e^{-\mu_0 r}$. Hence the solutions of Eq. (7) must grow more slowly than $e^{(\mu_0 + \varepsilon)t}$ as $t \to \infty$ for each $\varepsilon > 0$. Why?

Corollary C. If $\text{Re } \lambda < 0$ for every solution of Eq. (5), then there exist positive constants M and γ such that

$$||y(t)|| \le M ||\phi||_r e^{-\gamma(t-t_0)} \quad \text{for} \quad t \ge t_0$$

where y is the solution of (3) with $y_{t_0} = \phi \in \mathscr{L}$.

Instead of a system of n first-order linear equations, one may encounter an n'th order scalar equation

$$x^{(n)}(t) + \sum_{j=1}^{m} \sum_{i=0}^{n-1} a_{ij} x^{(i)}(t-r_j) = h(t)$$

or its associated homogeneous equation

$$y^{(n)}(t) + \sum_{j=1}^{m} \sum_{i=0}^{n-1} a_{ij} y^{(i)}(t-r_j) = 0, \tag{11}$$

where each $r_j \in [0,r]$ and each a_{ij} is a constant.

Theorem D. If Eq. (11) is transformed into a system of n first order equations (in a natural way), then the characteristic equation for that system is the same as that obtained by seeking exponential solutions, $e^{\lambda t}$, of (11), namely,

$$\lambda^n + \sum_{j=1}^{m} \sum_{i=0}^{n-1} a_{ij} \lambda^i e^{-\lambda r_j} = 0 \tag{12}$$

So, for example, if every root of (12) has negative real part, then every solution of Eq. (11) tends to zero as $t \to \infty$.

Proof. Without loss of generality, we can assume $r_1 = 0$. Then, choosing any nonzero constants $c_1, \ldots, c_n$ and defining $y_1 = c_1 y$, $y_2 = c_2 y'$, $\ldots$, $y_n = c_n y^{(n-1)}$, Eq. (11) is transformed into the system

$$y_1'(t) = (c_1/c_2)y_2(t),$$

$$y_2'(t) = (c_2/c_3)y_3(t),$$

$$\vdots$$

$$y_n'(t) = -\sum_{j=1}^{m}\sum_{k=1}^{n} a_{k-1,j}(c_n/c_k)y_k(t-r_j).$$

Regarding this as system (3), one can determine the matrices A_j and hence the matrix $\lambda I - \sum_{j=1}^{m} A_j e^{-\lambda r_j}$, which occurs in Eq. (5). To save some writing let us define

$$a_i = \sum_{j=1}^{m} a_{ij} e^{-\lambda r_j} \qquad (i = 0, \ldots, n-1).$$

Then $\lambda I - \sum_{j=1}^{m} A_j e^{-\lambda r_j}$ becomes

$$\begin{pmatrix} \lambda & -c_1/c_2 & \cdots & 0 & 0 \\ 0 & \lambda & \cdots & 0 & 0 \\ \cdot & \cdot & & \cdot & \cdot \\ \cdot & \cdot & & \cdot & \cdot \\ \cdot & \cdot & & \cdot & \cdot \\ 0 & 0 & \cdots & \lambda & -c_{n-1}/c_n \\ a_0 c_n/c_1 & a_1 c_n/c_2 & \cdots & a_{n-2}c_n/c_{n-1} & \lambda + a_{n-1} \end{pmatrix}$$

It is left as an exercise for the reader to show that when one sets the determinant of this matrix equal to zero, the result is exactly Eq. (12). In computing the required determinant you may find it helpful to rewrite the above matrix with a few more rows and columns written out explicitly. $\square$

Example 3.   Consider Eq. 21-(15) (with m = 1 for simplicity),

$$y''(t) + by'(t) + qy'(t-r) + ky(t) = 0, \tag{13}$$

where b, q, k, and r are non-negative numbers, and let us
assume that

$$b > q. \tag{14}$$

Then we shall show that every root of the characteristic equation,

$$\lambda^2 + b\lambda + q\lambda e^{-\lambda r} + k = 0, \tag{15}$$

must have negative real part.   Suppose for contradiction that
$\lambda = \mu + i\omega$ is a root of (15) with $\mu \geq 0$. We cannot have
$\omega = 0$, for then the left hand side of (15) would certainly be
positive.   So $\omega \neq 0$ and

$$\text{Im } (\lambda^2 + b\lambda + q\lambda e^{-\lambda r} + k)/\omega$$

$$= 2\mu + b + qe^{-\mu r}(\cos \omega r - \mu r \frac{\sin \omega r}{\omega r})$$

$$> b - qe^{-\mu r}(1 + \mu r) > b-q > 0.$$

This contradiction shows that every λ has negative real
part.   Hence every solution of Eq. (13) tends to zero exponentially as $t \to \infty$.

The condition $b > q$ is a sufficient condition, but
is certainly not necessary, for the solutions of Eq. (13) to
tend to zero.   Another sufficient condition (depending on the
smallness of r) is as follows.

Example 4.   Consider Eq. (13), where b, q, k, and r   are
non-negative numbers with

$$b+q > 0 \quad \text{and} \quad (b+q+k^{1/2})r < \pi/2. \tag{16}$$

Suppose (for contradiction) that $\lambda = \mu + i\omega$ is a root of Eq. (15) with $\mu \geq 0$. Then

$$0 = |\lambda^2 + b\lambda + q\lambda e^{-\lambda r} + k|$$

$$\geq |\lambda|^2 - (b+q)|\lambda| - k$$

which implies

$$|\lambda| \leq \frac{b + q + [(b+q)^2 + 4k]^{1/2}}{2}$$

$$\leq b + q + k^{1/2}.$$

Check this.   Hence $|\lambda r| < \pi/2$.

Now clearly Eq. (15) cannot be satisfied by a <u>real number</u> $\lambda \geq 0$. Thus we must have $0 < |\omega r| < \pi/2$. Using the above we can now compute

$$\text{Im}(\lambda^2 + b\lambda + q\lambda e^{-\lambda r} + k)/\omega$$

$$= 2\mu + b + qe^{-\mu r}(\cos \omega r - \frac{\mu}{\omega} \sin \omega r)$$

$$> 2\mu - qe^{-\mu r}\mu r \frac{\sin \omega r}{\omega r} \geq 2\mu - q\mu r \geq 0.$$

This contradiction shows that every root of Eq. (15) has negative real part.

Other methods, besides those illustrated in Examples 2, 3, and 4, for analyzing the roots of the characteristic equations can be found in the books of Bellman and Cooke [1963], El'sgol'c [1964], and Martynjuk [1971].

In closing this section we should introduce the following terminology which applies, in particular, to system (3).

Definition. A differential equation or system (with or with-out delays, linear or not) is said to be autonomous if it can be written in the form

$$x'(t) = F(x_t), \qquad (17)$$

where $F: \mathscr{C}_D \to \mathbb{R}^n$, i.e., if F does not depend explicitly on t.

Lemma E. Let x on $[t_0 - r, \beta)$ satisfy (17) on $[t_0, \beta)$. Choose any real number a and define $z(t) = x(t-a)$ for $t_0 + a - r \le t < \beta + a$. Then z satisfies (17) on $[t_0 + a, \beta + a)$. That is, any "time translation" of a solution of (17) is also a solution.

Proof. For $t_0 + a \le t < \beta + a$, $z'(t) = x'(t-a) = F(x_{t-a}) = F(z_t)$. $\square$

Thanks to Lemma E, there is no loss of generality in setting $t_0 = 0$ when studying an autonomous system.

Problems

1. Verify that Eqs. (6) define solutions of Eq. (3) valid on $\mathbb{R}$.

*2. Use Theorem A to show that Eq. (8) has a sequence of roots

$$\lambda_k = \mu_k + i\omega_k, \qquad k = 1, 2, \ldots$$

such that each $\mu_k \to -\infty$. Then show that the series

$$y(t) = \sum_{k=1}^{\infty} \frac{e^{\mu_k \tau}}{2^k (\mu_k^2 + \omega_k^2)^{1/2}} e^{\mu_k t} \cos \omega_k t \qquad \text{for} \quad t \ge -r$$

converges and can be differentiated term by term, and thus gives a solution of Eq. (7). (You may need to refer to a book on advanced calculus for this.)

You should now be able to construct other infinite-series solutions of Eq. (7).

3. Show that every solution of the following equations tends to zero as $t \to \infty$.

(a) $y'(t) = -by(t-r)$, where $b > 0$ and $0 \leq br < \pi/2$.

(b) $y''(t) + by'(t) + qy'(t-r_1) + ky(t) + py(t-r_2) = 0$, where b, q, k, p, r_1, and r_2 are non-negative constants with $b > q+pr_2$.

(c) The equation of part (b), where b, q, k, p, r_1, and r_2 are nonnegative constants with $b > pr_2$ and $[b + q + (k+p)^{1/2}]r_1 < \pi/2$.

4. Show that the assertion of Problem 3(a), that every solution tends to zero, is false if $br = \pi/2$.

5. Determine the nature of the solution (for large t) of

$$x'(t) = -\frac{3}{2} x(t-\frac{\pi}{4}) + \sin 2t \qquad t \geq 0$$

with $x_0 = \phi \in \mathscr{C}$ given.

6. (a) Show that system (3) is autonomous but system (1) is not (in general).

(b) Show that Lemma E is false for a nonautonomous system, $x'(t) = F(t,x_t)$.

29. VARIATION OF PARAMETERS

In the theory of linear systems of ordinary differential equations, one of the most important tools is "variation of parameters", Theorem 17-A. We shall now generalize this theorem to cover the linear delay differential system

$$x'(t) = L(t,x_t) + h(t) = \sum_{j=1}^{m} A_j(t)x(t-r_j) + h(t), \qquad (1)$$

where h is a continuous n-vector-valued function on $J = [t_0,\beta)$, each A_j is a continuous $n \times n$-matrix-valued function on J, and each $r_j \in [0,r]$ is constant.

Given any $\phi \in \mathscr{C}$, Example 26-1 asserted the existence of a unique function $x: [t_0-r,\beta) \to \mathbb{R}^n$ such that

$$x_{t_0} = \phi \qquad (2)$$

and Eq. (1) is satisfied on $[t_0,\beta)$. To emphasize the dependence of x on t_0 and ϕ, we will sometimes denote $x(t)$ by $x(t;t_0,\phi)$.

In the variation-of-parameters theorem we wish to express $x(t;t_0,\phi)$ in terms of solutions of the homogeneous equation

$$y'(t) = L(t,y_t) = \sum_{j=1}^{m} A_j(t)y(t-r_j). \qquad (3)$$

But now we must consider Eq. (3) in conjunction with a certain type of <u>discontinuous</u> initial function. Define the "unit step function" $u: [-r,0] \to \mathbb{R}$ by

$$u(\sigma) = \begin{cases} 0 & \text{for} \quad -r \le \sigma < 0 \\ 1 & \text{for} \quad \sigma = 0. \end{cases}$$

Then, given $s \in [t_0,\beta)$ and given $\xi \in \mathbb{R}^n$, we seek a function $y: [s-r,\beta) \to \mathbb{R}^n$ such that

$$y_s = \xi u, \tag{4}$$

y is continuous on $[s,\beta)$, and Eq. (3) is satisfied on $[s,\beta)$ except at those points t where $t-r_j = s$ for some $j = 1,\ldots,$ or m.

Example 1.   Consider the scalar equation

$$y'(t) = ay(t) + by(t-r) \quad \text{for} \quad t \geq 0$$

with

$$y(t) = \begin{cases} 0 & \text{for} \quad -r \leq t < 0 \\ 1 & \text{for} \quad t = 0. \end{cases}$$

This problem is easily seen to be solvable by the method of steps.   Thus one finds

$$y(t) = e^{at} \quad \text{for} \quad 0 \leq t \leq r.$$

Then one solves

$$y'(t) = ay(t) + be^{a(t-r)} \quad \text{for} \quad r \leq t \leq 2r$$
$$\text{with} \quad y(r) = e^{ar},$$

etc.

Theorem A.   Let $s \in [t_0,\beta)$ and $\xi \in \mathbb{R}^n$ be given.   Then Eqs. (3) and (4) have a unique solution y on $[s-r,\beta)$. Moreover

$$\|y(t)\| \leq \|\xi\| e^{\int_s^t K(\sigma)d\sigma} \quad \text{for} \quad s \leq t < \beta, \tag{5}$$

where $K(t) = \sum_{j=1}^m \|A_j(t)\|.$

Proof.   Without loss of generality we can assume that

$$0 = r_1 < r_2 < \cdots < r_m = r.$$

If this is not the case, it can always be achieved by rede-
fining the A_j's and r_j's.

Case 1. Assume that $\beta > s+r$. On the interval
$[s,s+r_2]$ Eqs. (3) and (4) reduce to the ordinary differential system

$$y'(t) = A_1(t)y(t)$$

with initial condition

$$y(s) = \xi.$$

Why? Consider this problem on the larger interval $[s,\beta)$.
Apply Example 26-1 to see that a unique solution exists on
$[s,\beta)$ and so, in particular, on $[s,s+r_2]$. Denote this
function by $y_{(2)}$.

Next consider the interval $[s+r_2, s+r_3]$. There Eqs.
(3) and (4) are equivalent to the delay differential system
with maximum delay r_2

$$y'(t) = A_1(t)y(t) + A_2(t)y(t-r_2),$$

and initial condition

$$y(t) = y_{(2)}(t) \quad \text{on} \quad [s,s+r_2].$$

Now Example 26-1 assures the existence of a unique solution,
say $y_{(3)}$, on $[s,s+r_3]$. Here we have $y_{(3)}(t) = y_{(2)}(t)$ on
$[s,s+r_2]$.

On $[s+r_3, s+r_4]$ one considers

$$y'(t) = A_1(t)y(t) + A_2(t)y(t-r_2) + A_3(t)y(t-r_3)$$

with $y(t) = y_{(3)}(t)$ on $[s,s+r_3]$. This yields a unique continuation of the solution to a function $y_{(4)}$ on $[s,s+r_4]$.

Continuing this procedure, one constructs a solution y of Eqs. (3) and (4) on $[s-r,s+r]$. Then, since y is continuous on $[s,s+r]$ another application of Example 26-1 gives the desired continuation to $[s-r,\beta)$.

Case 2. If $\beta \leq s+r$, one proceeds exactly as in the first stages of Case 1. The argument ends as soon as one reaches β.

It remains to obtain inequality (5). Note that

$$||y'(t)|| \leq K(t)||y_t||_r \quad \text{for} \quad s < t < \beta,$$

except at $t = s+r_1, s+r_2, \ldots, s+r_m$. Thus

$$||y(t)|| \leq ||\xi|| + \int_s^t K(\sigma)||y_\sigma||_r d\sigma,$$

and hence

$$||y_t||_r \leq ||\xi|| + \int_s^t K(\sigma)||y_\sigma||_r d\sigma \quad \text{for} \quad s \leq t < \beta.$$

From this and Reid's Lemma (8-A), one obtains (5). $\square$

If y is the unique solution of Eqs. (3) and (4) on $[s-r,\beta)$ we denote $y(t)$ by $y(t;s,\xi u)$.

Theorem B (Variation of Parameters). Let $A_1,\ldots,A_m$ and h be continuous on $[t_0,\beta)$, and let $0 \leq r_j \leq r$ for $j = 1,\ldots,m$. Then for each $\phi \in \mathscr{L}$ the unique solution of Eqs. (1) and (2) is given by

$$x(t) = y(t;t_0,\phi) + \int_{t_0}^t y(t;s,h(s)u)ds \quad \text{for} \quad t_0-r \leq t < \beta. \quad (6)$$

Remark. Before reading the proof, try to "guess" Eq. (6) using the argument presented following Eq. 17-(7)

together with the superposition concept (Theorem 27-A).

Proof of Theorem B. Let us introduce

$$z(t) \equiv \int_{t_0}^{t} y(t;s,h(s)u)ds \quad \text{for} \quad t_0-r \le t < \beta. \tag{7}$$

Note that $z(t) = 0$ on $[t_0-r,t_0]$. Why? It will suffice to show that z satisfies Eq. (1) on $[t_0,\beta)$. For then, by superposition, (6) defines another solution of Eq. (1), and, from (6),

$$x(t) = \phi(t-t_0) \quad \text{for} \quad t_0-r \le t \le t_0.$$

Hence (6) must define the solution of Eqs. (1) and (2).

We will actually show that z satisfies the integral equation

$$z(t) = \int_{t_0}^{t} [L(v,z_v) + h(v)]dv \quad \text{for} \quad t_0 \le t < \beta, \tag{8}$$

from which $z'(t) = L(t,z_t) + h(t)$ follows.

Since $y(t;s,h(s)) = 0$ for $s-r \le t < s$, we find for $t_0 \le t < \beta$

$$z(t-r_j) = \int_{t_0}^{t-r_j} y(t-r_j;s,h(s)u)ds$$

$$= \int_{t_0}^{t} y(t-r_j;s,h(s)u)ds.$$

Check this. Thus for $t_0 \le t < \beta$

$$\int_{t_0}^{t} [L(v,z_v) + h(v)]dv$$

$$= \sum_{j=1}^{m} \int_{t_0}^{t} A_j(v)z(v-r_j)dv + \int_{t_0}^{t} h(v)dv$$

$$= \sum_{j=1}^{m} \int_{t_0}^{t} \int_{t_0}^{v} A_j(v)y(v-r_j;s,h(s)u)dsdv + \int_{t_0}^{t} h(v)dv.$$

Now, using Corollary A2-N to interchange the order of integration, this becomes

$$\sum_{j=1}^{m} \int_{t_0}^{t} \int_{s}^{t} A_j(v) y(v-r_j;s,h(s)u) dvds + \int_{t_0}^{t} h(v) dv$$

$$= \int_{t_0}^{t} \int_{s}^{t} y'(v;s,h(s)u) dvds + \int_{t_0}^{t} h(v) dv$$

$$= \int_{t_0}^{t} [y(t;s,h(s)u) - h(s)] ds + \int_{t_0}^{t} h(v) dv$$

$$= z(t). \quad \square$$

In principle Eq. (6) provides a very general method
for solving system (1) provided the solutions of (3) are
known.   The method does not depend upon h being of a
special form as it was in Problem 27-1.

However, in practice the expressions for $y(t;t_0,\phi)$
and $y(t;s,\xi u)$ are not available in any simple form.   So one
does not use the variation-of-parameters formula to calculate
explicit solutions as we did for linear ordinary differential
systems.

But, on several occasions, we have seen that one can
have useful information about the solutions of a differential
system without actually knowing the solutions.   In this
spirit, the following corollary is a useful consequence of
Theorem B.   It says, in particular, that if system (1) has
constant coefficients, if the roots of the characteristic
equation all have negative real parts, and if h is bounded,
then every solution, x, of (1) is bounded.

Corollary C.   Assume there exist constants $M \geq 1$ and $\gamma > 0$
such that, for each $\phi \in \mathscr{C}$ and each $s \geq t_0$,

$$||y(t;s,\phi)|| \le M||\phi||_r e^{-\gamma(t-s)} \qquad \text{for} \quad t \ge s; \qquad (9)$$

and assume there exist constants   K   and   c, such that

$$\sum_{j=1}^{m} ||A_j(t)|| \le K \quad \text{and} \quad ||h(t)|| \le c \quad \text{for} \quad t \ge t_0. \quad (10)$$

Then the solution, x, of Eqs. (1) and (2) is bounded on $[t_0, \infty)$.

Proof.   First observe that the solution of Eq. (3) with $y_s = \xi u$ is continuous and, by (5),

$$||y(t;s,\xi u)|| \le ||\xi|| e^{Kr} \qquad \text{for} \quad s \le t \le s+r.$$

Thus

$$||y(t;s,\xi u)|| \le M||\xi|| e^{Kr} e^{-\gamma(t-s-r)} \qquad \text{for} \quad t \ge s+r;$$

and this holds also for $s \le t \le s+r$.

We now apply the above estimates to Eq. (6) and find, for $t \ge t_0$,

$$||x(t;t_0,\phi)|| \le M||\phi||_r e^{-\gamma(t-t_0)}$$
$$+ \int_{t_0}^{t} Mce^{(K+\gamma)r} e^{-\gamma(t-s)} ds \qquad (11)$$
$$\le M||\phi||_r + Mce^{(K+\gamma)r}/\gamma.$$

This is the asserted bound.   □

Example 2.   Consider the equation

$$z''(t) + bz'(t) + qz'(t-r) + kz(t) = h(t) \quad \text{for} \quad t \ge 0 \quad (12)$$

where   b, q, k, and   r   are non-negative constants with $b > q$, and $h(t)$ is continuous and bounded for $t \ge 0$.

Convert Eq. (12) into a system by defining $x_1(t) \equiv z(t)$, $x_2(t) \equiv z'(t)$.   Then

$$x_1'(t) = x_2(t)$$

$$x_2'(t) = -kx_1(t) - bx_2(t) - qx_2(t-r) + h(t).$$

$$(13)$$

The associated homogeneous system is

$$y'(t) = A_1 y(t) + A_2 y(t-r),$$

$$(14)$$

where

$$A_1 = \begin{pmatrix} 0 & 1 \\ -k & -b \end{pmatrix} \quad \text{and} \quad A_2 = \begin{pmatrix} 0 & 0 \\ 0 & -q \end{pmatrix}.$$

The characteristic equation for system (14) was examined in Example 28-3, and it was shown that all roots have negative real parts. Thus, by Corollary 28-C, inequality (9) holds. Inequalities (10) also hold since A_1 and A_2 are constant matrices and col $(0,h(t))$ is bounded. So it follows from Corollary C that every solution of system (13) is bounded, and hence every solution of (12) is bounded.

Actually, the variation-of-parameters theorem is known to hold for every general linear systems with infinitely many bounded delays. But the proof for that case, due to Banks [1969], assumes a rather sophisticated knowledge of integration theory.

Problems

1. Referring to Eq. (14), show that

$$||A_1|| + ||A_2|| = \max \{k, 1+b\} + q.$$

2. The definitions of x_1 and x_2 used in Example 2 are "dimensionally inconsistent". What does this mean? It would have been better to define $x_1(t) = x(t)$ and

$x_2(t) = k^{-1/2}x'(t)$. If x_1 and x_2 are defined this way,

(a) find the resulting system

$$x'(t) = A_1^* x(t) + A_2^* x(t-r) + \text{col }(0,h(t))$$

which replaces (13),

(b) show that $||A_1^*|| + ||A_2^*|| = k^{1/2} + b + q$,

(c) show that $k^{1/2}$, b, and q all have the same "dimen-sions", while k, 1, b, and q (which occur in Problem 1) do not.

3. Assuming, in each case, that h is continuous and bounded, which of the following has only bounded solu-tions and which can have unbounded solutions? Justify your answers.

(a) $x'(t) = x(t-r) + h(t)$ (scalar),

(b) $x'(t) = -x(t-r) + h(t)$ where $0 \le r < \pi/2$ (scalar),

(c) $x'(t) = -x(t-\pi/2) + h(t)$ (scalar),

(d) Eq. (12) with $b+q > 0$ and $(b + q + k^{1/2})r < \pi/2$.

Chapter VIII.   STABILITY

Throughout this chapter we shall consider differential systems with bounded delays,

$$x'(t) = F(t, x_t)$$

where $F: (\alpha, \infty) \times \mathscr{C}_D \to \mathbb{R}^n$ with D an open set in $\mathbb{R}^n$. The special case of an ordinary differential system is always obtained by setting $r = 0$.

We wish to examine the effect which a small change in the initial conditions has on a solution.

Results along this line have already been encountered in Theorems 8-B, 8-C, 18-E, 23-D, and 25-E. Those theorems assert that (under appropriate conditions on F), if some positive number T is prescribed in advance, then a sufficiently small change in the initial conditions over the interval $[t_0 - r, t_0]$ causes only a small change in the solution on the interval $[t_0, t_0 + T]$. Such results are generally said to show that the solution "depends continuously on the initial data".

But now we will ask more. If a certain solution $\bar{x}$ is valid on the semi-infinite interval $[t_0 - r, \infty)$, does a small change in the values of $\bar{x}$ on $[t_0 - r, t_0]$ lead to a new solution still valid on the entire interval $[t_0 - r, \infty)$ and always remaining close to $\bar{x}$? If so we shall say the solution $\bar{x}$ is "stable".

This and other concepts of stability will be made precise and illustrated with examples and theorems.

340

30. DEFINITIONS AND EXAMPLES

Consider the delay differential system

$$x'(t) = F(t, x_t), \tag{1}$$

where $F: (\alpha, \infty) \times \mathscr{L}_D \to \mathbb{R}^n$ with $\mathscr{L}_D = C([-r, 0], D)$, D being an open set in $\mathbb{R}^n$. Given any $t_0 > \alpha$ and any $\phi \in \mathscr{L}_D$, we shall study Eq. (1) in conjunction with the initial condition

$$x_{t_0} = \phi. \tag{2}$$

Henceforth, let us always understand that $\phi \in \mathscr{L}_D$, without necessarily repeating it each time.

Let us assume that, for each $t_0 > \alpha$, F on $[t_0, \infty) \times \mathscr{L}_D$ satisfies Continuity Condition (C), and is locally Lipschitzian and quasi-bounded (as defined in Section 26). These hypotheses assure the existence and uniqueness of a noncontinuable solution of Eqs. (1) and (2). (See Theorem 26-C).

We now ask the following question. If a certain solution $\bar{x} = x(\cdot; t_0, \bar{\phi})$ is valid on the interval $[t_0-r, \infty)$, does a small change in the initial function $\bar{\phi}$ give a new solution still valid on the entire interval $[t_0-r, \infty)$ and remaining close to $\bar{x}$?

An assertion that an initial function ϕ is "close" to $\bar{\phi}$ or that ϕ differs from $\bar{\phi}$ by a "small amount" will imply smallness of $||\phi - \bar{\phi}||_r$.

Definitions. Let $\bar{x}: (\alpha-r, \infty) \to D$ satisfy Eq. (1) on (α, ∞). We say $\bar{x}$ is stable at $t_0 > \alpha$, (in the sense of Lyapunov) if for each $\varepsilon > 0$ there exists $\delta = \delta(\varepsilon, t_0) > 0$ such that

whenever $||\phi - \bar{x}_{t_0}|| < \delta$ it follows that $x(\cdot\,;t_0,\phi)$ exists
on $[t_0-r,\infty)$ and

$$||x(t;t_0,\phi) - \bar{x}(t)|| < \varepsilon \quad \text{for all} \quad t \geq t_0-r.$$

(One must always have $0 < \delta \leq \varepsilon$. Why?)

Otherwise the solution $\bar{x}$ is said to be <u>unstable at</u>
t_0 (in the sense of Lyapunov).

Note. In the special case of an ordinary differential
system (the case $r = 0$) ϕ is replaced by x_0, and x_t is
replaced by $x(t)$ in all definitions and theorems.

Example 1. We shall demonstrate the stability (at every t_0)
of every solution of the ordinary differential system

$$x'(t) = \begin{pmatrix} 1 & 2 \\ -1 & -1 \end{pmatrix} x(t).$$

Since the system is autonomous, we might as well take $t_0 = 0$
for simplicity. (See Lemma 28-E.) This system was encountered in Problem 15-5(b). Its solution is

$$x(t;0,x_0) = \begin{pmatrix} x_{01}(\cos t + \sin t) + 2x_{02}\sin t \\ -x_{01}\sin t + x_{02}(\cos t - \sin t) \end{pmatrix}$$

for all t. Similarly for a solution $\bar{x} = x(\cdot\,;0,\bar{x}_0)$. Subtracting, one finds

$$||x(t;0,x_0) - x(t,0,\bar{x}_0)||$$

$$= |(x_{01}-\bar{x}_{01})(\cos t + \sin t) + 2(x_{02}-\bar{x}_{02})\sin t|$$

$$+ |(-x_{01}+\bar{x}_{01})\sin t + (x_{02}-\bar{x}_{02})(\cos t - \sin t)|$$

$$\leq 3|x_{01} - \bar{x}_{01}| + 4|x_{02} - \bar{x}_{02}| \leq 4||x_0 - \bar{x}_0||.$$

Thus, the definition of stability at $t_0 = 0$ is satisfied if for each $\varepsilon > 0$, we choose $\delta = \varepsilon/4$.

Example 2. Every solution of the scalar equation

$$x'(t) = x(t)$$

is unstable at every t_0. This can be seen by examining the general solution $x(t) = ce^t$.

The discussion of stability is simplified if we consider only the stability of a constant solution, and more specifically the zero solution. But perhaps Eq. (1) does not have a zero solution, or perhaps some other solution is of interest. Then one can proceed as follows.

Let $\bar{x}$ be a solution of (1) whose stability is to be considered. If x is any other solution, define $y(t) = x(t) - \bar{x}(t)$ and find

$$y'(t) = F(t, \bar{x}_t + y_t) - F(t, \bar{x}_t),$$

or, introducing

$$G(t, \psi) \equiv F(t, \bar{x}_t + \psi) - F(t, \bar{x}_t),$$

we have

$$y'(t) = G(t, y_t). \tag{3}$$

Now $G(t, 0) \equiv 0$ so that $y = 0$ is a solution of Eq. (3); and the question of stability of the solution $\bar{x}$ of (1) becomes a question of the stability of the zero solution (the trivial solution) of (3).

Thus we shall now assume, usually, without loss of generality, that

$$0 \in D \quad \text{and} \quad F(t, 0) = 0 \quad \text{for all} \quad t > \alpha, \tag{4}$$

i.e., $F(t,\psi) = 0$ if $\psi(\sigma) = 0$ for $-r \leq \sigma \leq 0$. Then we con-

sider the stability of the trivial solution of (1).

We can also simplify matters, without serious loss of

generality, by assuming for D the special form

$$D \equiv \{\xi \in \mathbb{R}^n: ||\xi|| < H\} \qquad (5)$$

for some $H > 0$. This special form of D will be used through

throughout the chapter. In case $H = \infty$, $D = \mathbb{R}^n$.

For the special case of the trivial solution let us

now restate the stability definition, and add another to it.

<u>Definitions</u>. The trivial solution of (1) is said to be

<u>stable at</u> $t_0 > \alpha$ if for each $\varepsilon > 0$ there exists

$\delta = \delta(\varepsilon,t_0) > 0$ such that whenever $||\phi||_r < \delta$, the solu-

tion $x(t;t_0,\phi)$ exists and

$$||x(t;t_0,\phi)|| < \varepsilon \qquad \text{for all}\quad t \geq t_0-r.$$

Otherwise the trivial solution is said to be <u>unstable</u>

<u>at</u> t_0.

The trivial solution of (1) is said to be <u>uniformly</u>

<u>stable</u> on (α,∞) if it is stable at each $t_0 > \alpha$ and the

number δ is independent of t_0, i.e., $\delta = \delta(\varepsilon)$ depends

only on ε.

One can easily see that in an autonomous ordinary dif-

ferential system, such as that of Example 1, stability at

some t_0 necessarily implies uniform stability. Thus every

solution in Example 1 is uniformly stable. More generally,

we can make the analogous statement for an autonomous delay

differential system.

<u>Theorem A</u>.   If Eq. (1) is autonomous then stability of the trivial solution at some $t_0 \in \mathbb{R}$ implies uniform stability.

The proof (Problem 1) follows easily from Lemma 28-E and the stability definitions.

However, it is easy to produce a nonautonomous example in which the trivial solution is stable at every $t_0 > \alpha$ but not uniformly stable.

<u>Example 3</u>.   The trivial solution of the scalar ordinary differential equation

$$x' = t^{-2}x \quad \text{with} \quad f(t,\xi) \equiv t^{-2}\xi \quad \text{on} \quad (0,\infty) \times \mathbb{R}$$

will be shown to be stable at every $t_0 > 0$, but not uniformly stable.   If $t_0 > 0$ and $x(t_0) = x_0$, then the solution on $[t_0,\infty)$ is found (by the method of Section 2) to be

$$x(t) = x(t;t_0,x_0) = x_0 e^{t_0^{-1} - t^{-1}}$$

so that

$$|x(t)| \leq |x_0| e^{1/t_0} \quad \text{for all} \quad t \geq t_0.$$

Given any $\varepsilon > 0$, choose $\delta = \varepsilon e^{-1/t_0}$, and observe that no larger value of δ will satisfy the requirements of the definition of stability.   Since $\delta \to 0$ as $t_0 \to 0$, it follows that the stability cannot be uniform.

On the other hand, if we consider the same differential equation, but with f defined on $(1,\infty) \times \mathbb{R}$ (i.e., $\alpha = 1$ instead of 0), then

$$|x(t)| \leq |x_0| e \quad \text{for all} \quad t \geq t_0 > \alpha = 1,$$

and we do have uniform stability.   Thus <u>a change</u> <u>in</u> <u>the</u> <u>domain</u>

of f can affect stability.

In Examples 1, 2, and 3 the stability questions were
resolved by explicitly solving the given differential equations. It is more interesting and more useful to consider
methods for proving stability without actually solving the
differential equations. In fact we want methods which will
apply to cases when we cannot (or would rather not) solve
precisely.

We illustrate with a simple and familiar example.

Example 4. Consider the system of ordinary differential
equations

$$x_1'(t) = (\tfrac{k}{m})^{1/2} x_2(t)$$

$$x_2'(t) = -(\tfrac{k}{m})^{1/2} x_1(t) - \tfrac{b}{m} x_2(t)$$

(6)

arising from $mz''(t) + bz'(t) + kz(t) = 0$, the equation for
damped motion of a mass on a spring, by setting $x_1 = z$ and
$x_2 = (m/k)^{1/2}z'$. We take m, b, and k to be positive constants, and we can let $\alpha = -\infty$ and $D = \mathbb{R}^2$.

Given any t_0 in $\mathbb{R}$ and x_0 in $\mathbb{R}^2$, let x be the
unique solution of system (6) with $x(t_0) = x_0$. (Since (6)
is linear we know the solution exists on all of $\mathbb{R}$.) Define

$$v(t) \equiv \tfrac{1}{2}kx_1^2(t) + \tfrac{1}{2}kx_2^2(t).$$

(This represents the total "potential and kinetic energy",
$\tfrac{1}{2}kz^2(t) + \tfrac{1}{2}mz'^2(t)$, in the system at time t.) We can now
compute

$$v'(t) = kx_1(t)x_1'(t) + kx_2(t)x_2'(t)$$

$$= - (kb/m)x_2^2(t) \le 0,$$

and learn that v(t) is nonincreasing as t increases.
Furthermore, as the reader should verify

$$\frac{1}{4} k||x(t)||^2 \leq v(t) \leq \frac{1}{2} k||x(t)||^2, \qquad (7)$$

where $||x(t)|| = |x_1(t)| + |x_2(t)|$ as usual. Problem 2.
The proof of the first inequality in (7) uses the fact that
$2|ab| \leq a^2 + b^2$ for any real numbers a and b.

It follows that for $t \geq t_0$,

$$\frac{1}{4} k||x(t)||^2 \leq v(t_0) \leq \frac{1}{2} k||x_0||^2$$

and so

$$||x(t)|| \leq \sqrt{2} \, ||x_0|| \qquad \text{for} \quad t \geq t_0.$$

Thus if $\varepsilon > 0$ is given we can fulfill the requirements for
uniform stability of the trivial solution by choosing
$\delta = \varepsilon/\sqrt{2}.$

Even though we know how to find the exact solution of
system (6), a proof of the above result by direct examination
of the known solution would be more difficult. Try it. Problem 3.

Example 5   (Krasovskiĭ [1959]). Consider the scalar delay
differential equation

$$x'(t) = ax(t) + bx(t-r), \qquad (8)$$

where a, b, and r are constants with $a < 0$, $|b| \leq |a|$,
and $r \geq 0$. Let $\alpha = -\infty$ and let $H = +\infty$.

Let $(t_0, \phi) \in \mathbb{R} \times \mathscr{C}$, and let $x: [t_0-r, \infty) \to \mathbb{R}$ be the
unique noncontinuable solution of Eq. (8) with $x_{t_0} = \phi$. De-
fine a new function

$$v(t) \equiv x^2(t) + |a| \int_{t-r}^{t} x^2(s)ds \quad \text{for} \quad t \geq t_0,$$

as suggested in Problem 21-15. Then it is not difficult to verify that for $t \geq t_0$

$$v'(t) = 2x(t)x'(t) + |a|x^2(t) - |a|x^2(t-r)$$

$$= -|a|x^2(t) + 2bx(t)x(t-r) - |a|x^2(t-r)$$

$$\leq (-|a| + |b|)[x^2(t) + x^2(t-r)] \leq 0.$$

Thus $v(t)$ is nonincreasing. Furthermore, as the reader can verify,

$$|x(t)|^2 \leq v(t) \leq (1+|a|r)||x_t||_r^2. \tag{9}$$

Since $x_{t_0} = \phi$, it follows from the above that for $t \geq t_0$,

$$|x(t)| \leq [v(t_0)]^{1/2} \leq (1+|a|r)^{1/2}||\phi||_r.$$

Consequently for each $\varepsilon > 0$, if we take $||\phi||_r < \varepsilon(1+|a|r)^{-1/2}$ it follows that $|x(t;t_0,\phi)| < \varepsilon$ for all $t \geq t_0$-r. Thus the trivial solution is uniformly stable.

Our definition of (ordinary) stability of the trivial solution of (1) talks about stability at a particular value of $t_0 > \alpha$. The following example shows that, for a delay differential equation, the trivial solution may be stable at one $t_0 > \alpha$ but unstable at another.

Example 6 (Zverkin [1959]). Let $\alpha = -\infty$, $D = \mathbb{R}$, and $r = 3\pi/2$. Consider the equation

$$x'(t) = b(t)x(t-3\pi/2) \tag{10}$$

where b is the continuous function defined by

$$b(t) = \begin{cases} 0 & \text{for } t \le 3\pi/2 \\ -\cos t & \text{for } 3\pi/2 < t \le 3\pi \\ 1 & \text{for } t > 3\pi. \end{cases}$$

Then, taking $t_0 = 0$ and any $\phi \in \mathscr{C} = C([-3\pi/2,0],\mathbb{R})$, one can verify that

$$x(t;0,\phi) = \begin{cases} \phi(0) & \text{for } 0 \le t < 3\pi/2 \\ -\phi(0)\sin t & \text{for } t \ge 3\pi/2. \end{cases}$$

Thus $|x(t;0,\phi)| \le |\phi(0)|$ for $t \ge 0$ and hence the small-ness of $||\phi||_r$ certainly assures the smallness of $|x(t;0,\phi)|$ for all $t \ge -3\pi/2$. However, if we consider any $\tilde{t}_0 \ge 3\pi$ and if we define ϕ by $\phi(s) = \delta e^{\lambda s}$ for $s \in [-3\pi/2,0]$ where λ is the positive root of the equation $\lambda = e^{-3\pi\lambda/2}$ we have

$$x(t;\tilde{t}_0,\phi) = \delta e^{\lambda(t-\tilde{t}_0)} \qquad \text{for all } t \ge \tilde{t}_0 - r.$$

Thus, even though $||\phi||_r = \delta > 0$ be made arbitrarily small, the solution can be unbounded.

These arguments show that the trivial solution of (10) is stable at $t_0 = 0$, but unstable at every $\tilde{t}_0 \ge 3\pi$.

Of course, this sort of thing cannot occur for a system with constant coefficients such as Eq. (8) or, more generally, for an autonomous system. (Theorem A).

It also cannot occur for ordinary differential systems, and it can occur only "in one direction" for delay differential systems. These remarks are made precise in Theorem C, the proof of which uses the following lemma. Lemma B is interesting in its own right since it is actually a "continuous dependence theorem" in which the hypothesis of a global

Lipschitz condition (cf. Theorem 25-E) is reduced to a local
Lipschitz condition. Consistently with our simplified approach to stability, Lemma B only considers solutions near
the trivial solution.

Since Lemma B and Theorem C will not be used in our
examples, their proofs can be considered optional.

Lemma B (Continuous Dependence). Let $t_0 > \alpha$, $T > t_0$ and
$\varepsilon > 0$ be given. Then there exists $\delta > 0$ such that if
$||\phi||_r < \delta$, $x(t;t_0,\phi)$ exists and satisfies

$$||x(t;t_0,\phi)|| < \varepsilon \quad \text{for} \quad t_0 - r \leq t \leq T.$$

(for the special case of ordinary differential systems, the
analogous result holds also for $T < t \leq t_0$.)

*Proof. Without loss of generality let $\varepsilon < H$. For
each $t \in [t_0,T]$ choose $a_t > 0$ and $b_t \in (0,H)$ sufficiently small so that $t - a_t > \alpha$ and F is Lipschitzian with
Lipschitz constant K_t on $[t-a_t,t+a_t] \times \{\psi \in \mathscr{C}: ||\psi||_r \leq b_t\}$.
By the Heine-Borel Theorem (A2-H), there exists a finite set
of points $\{t_1,\ldots,t_\ell\} \subset [t_0,T]$ such that

$$[t_0,T] \subset \bigcup_{k=1}^{\ell} (t_k - a_{t_k}, t_k + a_{t_k}).$$

Let $K = \min\{K_{t_1},\ldots,K_{t_\ell}\}$, let $b = \min\{b_{t_1},\ldots,b_{t_\ell}\}$, and
let $D_1 = \{\xi \in \mathbb{R}^n: ||\xi|| < b\}$. Then F is (globally) Lip-
schitzian with Lipschitz constant K on $[t_0,T] \times \mathscr{C}_D$.
Choose $\delta > 0$ such that $\delta e^{K(T-t_0)} \leq \varepsilon$. Let $||\phi||_r < \delta$ and
let $x(\cdot;t_0,\phi)$ on $[t_0-r,\beta_1)$ be the noncontinuable solution
of Eqs. (1) and (2). Then, by Theorem 25-E, $||x(t;t_0,\phi)|| \leq$
$||\phi||_r e^{K(T-t_0)} < \varepsilon$ for $t_0 - r \leq t < \min\{\beta_1,T\}$. Finally we
apply Theorem 26-C to conclude that $\min\{\beta_1,T\} < \beta_1$ and so

min $\{\beta_1,T\} = T$ and

$$||x(t;t_0,\phi)|| \leq ||\phi||_r e^{K(T-t_0)} < \varepsilon \quad \text{for} \quad t_0-r \leq t \leq T. \quad \square$$

__Theorem C.__ Let the trivial solution of Eq. (1) be stable at some $t_0 > \alpha$. Then it is also stable at all $\tilde{t}_0 \in (\alpha,t_0)$. (In the special case of ordinary differential systems, stability at some $t_0 > \alpha$ implies stability at all $\tilde{t}_0 > \alpha$.)

  __Remarks.__ Note that in Example 6, $\tilde{t}_0 > t_0$.

  For ordinary differential systems, one can simply say the trivial solution is "stable" rather than "stable at t_0".

  *__Proof of Theorem C.__ For each $\varepsilon > 0$ there exists $\delta = \delta(\varepsilon,t_0) > 0$ such that if $||\phi||_r < \delta$, then $||x(t;t_0,\phi)|| < \varepsilon$ for all $t \geq t_0$. Given any $\tilde{t}_0 \in (\alpha,t_0)$, choose $\tilde{\delta} > 0$ such that if $||\tilde{\phi}||_r < \tilde{\delta}$ then $||x(t;\tilde{t}_0,\tilde{\phi})|| < \delta$ for $\tilde{t}_0-r \leq t \leq t_0$. It follows that $||x(t;\tilde{t}_0,\tilde{\phi})|| < \varepsilon$ for all $t \geq \tilde{t}_0-r$. Why? $\square$

Problems

1. Prove Theorem A.

2. Verify inequalities (7) in Example 4.

3. Why would there be any difficulty proving stability of the trivial solution in Example 4 directly from the general solution of (6)?

4. Prove that the trivial solution of Eq. (8) is unstable (at every t_0) if $a + b > 0$. __Hint:__ Seek a solution of the form $ce^{\lambda t}$ where λ is a positive constant.

5. Apply the result of Example 5 to prove that if $a < 0$,

$|b| \leq |a|$ and $r \geq 0$, then every solution λ of the

transcendental equation $\lambda = a + be^{-\lambda r}$ must have non-

positive real part.

31. LYAPUNOV METHOD FOR UNIFORM STABILITY

In several examples of the previous section stability
of the trivial solution of

$$x'(t) = F(t,x_t) \tag{1}$$

was proved by considering the behavior of an auxiliary function v on $[t_0,\infty)$, where v was defined in terms of the unknown solution x. This procedure is formalized in "Lyapunov's direct method". This method for ordinary differential
systems, named after A. M. Lyapunov [1892], has been extended
to delay differential systems by Krasovskiǐ [1959] and
others.

Let

$$D = \{\xi \in \mathbb{R}^n : ||\xi|| < H\},$$

as before, let $F: (\alpha,\infty) \times \mathscr{C}_D \to \mathbb{R}^n$ satisfy Continuity Condition (C), and let it be locally Lipschitzian and quasi-
bounded. Assuming $F(t,0) = 0$, we try to determine the stability of the trivial solution of (1).

In the Lyapunov-Krasovskiǐ method one seeks to find a
functional V mapping $(\alpha,\infty) \times \mathscr{C}_D \to [0,\infty)$ such that $V(t,\psi)$
in some sense measures the size of ψ. Then one considers

$$v(t) \equiv V(t,x_t) \quad \text{for} \quad t \geq t_0,$$

where $x = x(\cdot\,; t_0, \phi)$ for some $t_0 > \alpha$. In order to prove
stability at t_0 of the trivial solution we wish to show
that if $v(t_0) = V(t_0, \phi)$ is sufficiently small then $v(t) = V(t, x_t)$ remains small for $t \geq t_0$.

A functional V meeting these requirements is called
a "Lyapunov functional". For the case of an ordinary differential system $(r = 0)$, V becomes a "Lyapunov function"
defined on $(\alpha, \infty) \times D$.

For instance, in Example 30-4 we would define V on
$\mathbb{R} \times \mathbb{R}^2$ by

$$V(t, \xi) = \frac{1}{2} k \xi_1^2 + \frac{1}{2} k \xi_2^2.$$

In Example 30-5 we used $v(t) = x^2(t) + |a| \int_{t-r}^{t} x^2(s)\,ds$,
which can be re-written as

$$v(t) = x_t^2(0) + |a| \int_{-r}^{0} x_t^2(\sigma)\,d\sigma.$$

Thus a suitable functional V on $\mathbb{R} \times \mathscr{C}$ is defined by

$$V(t, \psi) = \psi^2(0) + |a| \int_{-r}^{0} \psi^2(s)\,ds.$$

In physically motivated examples, Lyapunov functions
or functionals often represent the "energy" in the corresponding physical system, or something similar to energy. The
expression $v(t) = V(t, x_t)$, being related to the unknown solutions, will generally not be specifically evaluated. But
hopefully its study will be easier than that of the solution
of the differential system itself.

The simplest Lyapunov theorem, guaranteeing uniform
stability of the trivial solution of (1), is merely an abstrac-
tion of the argument used in Examples 30-4 and 30-5.

Theorem A.   Let   w   and   W   be continuous nondecreasing functions on   [0,H]   which are zero at   0   and positive on   (0,H).
If there exists a nonnegative functional   V   on $(\alpha,\infty) \times \mathscr{C}_D$
such that

  (a) $V(t,\psi) \geq w(||\psi(0)||)$,

  (b) $V(t,\psi) \leq W(||\psi||_r)$,   and

  (c)   whenever $x = x(\cdot;t_0,\phi)$ on $[t_0-r,\beta_1)$ is the

    noncontinuable solution of Eq. (1) through some

    $(t_0,\phi) \in (\alpha,\infty) \times \mathscr{C}_D$, $V(t,x_t)$ defines a nonin-

    creasing function of   t   on $[t_0,\beta_1)$,

then the trivial solution of (1) is uniformly stable.

  Remarks.   When   V   satisfies condition (a),   V   is said
to be positive definite.   When   V   satisfies condition (b),
V   is said to have an infinitesimal upper bound.

  By considering a constant function ψ, one can easily
see that we must have $w(s) \leq W(s)$ for all   s   in   [0,H].

  Condition (c) is usually verified by showing that
$V(t,x_t)$ is continuous for $t \geq t_0$ and that $\frac{d}{dt} V(t,x_t) \leq 0$
for $t > t_0$.   Then the nonincreasing property for $V(t,x_t)$
follows from the mean value theorem.   Problem 1.

  In the special case of an ordinary differential system
$x'(t) = f(t,x(t))$, if the function   V   has continuous partial
derivatives, then, by Theorem A2-C,

$$\frac{d}{dt} V(t,x(t)) = D_1 V(t,x(t)) + \sum_{i=1}^{n} D_{i+1} V(t,x(t)) f_i(t,x(t)).$$

  Proof of Theorem A.   Let $\varepsilon > 0$ be given.   Without
loss of generality we shall assume $0 < \varepsilon < H$.   Then
$w(\varepsilon) > 0$, and by the continuity of   W, we can choose
$\delta = \delta(\varepsilon) \in (0,\varepsilon)$ such that

$$W(\delta) < w(\varepsilon). \tag{2}$$

Now consider any $(t_0,\phi) \in (\alpha,\infty) \times \mathcal{L}_D$ with $||\phi||_r < \delta$. Equation (1) has a unique noncontinuable solution $x = x(\cdot\,;t_0,\phi)$ through (t_0,ϕ) on $[t_0-r,\beta_1)$ for some $\beta_1 > t_0$. Thus, using hypotheses (a), (b), and (c) together with condition (2), we find for $t_0 \le t < \beta_1$

$$w(||x(t)||) \le V(t,x_t) \le V(t_0,\phi) \le W(||\phi||_r) \le W(\delta) < w(\varepsilon).$$

Now, since w is a nondecreasing function, this can hold only if

$$||x(t)|| < \varepsilon \quad \text{for} \quad t_0 \le t < \beta_1$$

Thus, from 26-C, it follows that $\beta_1 = \infty$ and the assertion of the theorem is proved. □

The examples given thus far which illustrate the Lyapunov method (Examples 30-4 and 30-5) involved linear systems. So it was appropriate and natural to take $H = +\infty$. On the other hand, for a nonlinear equation or system, one often has a finite value for H. Moreover it is always permissible and often helpful to reduce H to some smaller value than that originally given. A reduction of H does not restrict the generality of the conclusions because, when considering stability of the trivial solution, we need only consider "small" solutions. Thus we can and will always assume the ε of the stability definitions to be less than H, and we will feel free to reduce the size of H whenever necessary or convenient.

For brevity, let us agree that, henceforth, "solution" means noncontinuable solution.

<u>Example 1</u>.   Consider the ordinary differential equation

$$x'(t) = (t^{-2} - 1)x(t) + x^3(t) \tag{3}$$

first with $\alpha = 1$ and then with $\alpha = 0$. We can take $D = \mathbb{R}$, or $D = (-H,H)$ for some finite $H > 0$ if that is more convenient.

Let us define
$$V(t,\xi) = e^{2/t}\xi^2 \quad \text{on} \quad (\alpha,\infty) \times D.$$

Then if $\alpha = 1$

(a) $V(t,\xi) \geq \xi^2$,

(b) $V(t,\xi) \leq e^2\xi^2$, and

(c) if $x = x(\cdot;t_0,x_0)$ on (α_1,β_1) is the solution of (3) through (t_0,x_0) we have for $t_0 \leq t < \beta_1$

$$\frac{d}{dt} V(t,x(t)) = \frac{d}{dt} [e^{2/t}x^2(t)]$$

$$= -2t^{-2}e^{2/t}x^2(t) + 2e^{2/t}x(t)x'(t)$$

$$= -2e^{2/t}x^2(t)[1 - x^2(t)].$$

(At $t = t_0$ we understand $\frac{d}{dt}V(t,x(t))$ to mean the right hand derivative.) The above derivative is nonpositive provided $x^2(t) \leq 1$. Let us therefore take $H = 1$, i.e., $D = (-1,1)$. Then

$$\frac{d}{dt} V(t,x(t)) \leq 0 \quad \text{on} \quad [t_0,\beta_1).$$

Thus, if $\alpha = 1$, uniform stability of the trivial solution of (3) follows from Theorem A.

If $\alpha = 0$ and $H = 1$, conditions (a) and (c) are unchanged, but instead of (b) we can only say

$$V(t,0) = 0 \quad \text{for each} \quad t > 0.$$

Thus we cannot apply Theorem A directly. However, if $t_0 > 0$ and $\varepsilon \in (0,H)$ are specified, then (invoking the continuity of V) we can choose $\delta = \delta(\varepsilon,t_0) > 0$ such that

$$V(t_0,x_0) < \varepsilon^2 \quad \text{whenever} \quad |x_0| < \delta.$$

It follows that if $t_0 > 0$ and $|x_0| < \delta$ we will have

$$x^2(t) \leq V(t,x(t)) \leq V(t_0,x_0) < \varepsilon^2 \quad \text{for} \quad t_0 \leq t < \beta_1.$$

Thus $|x(t)| < \varepsilon$ for all $t \geq t_0$. This proves the stability at each t_0 of the trivial solution of (3) when $\alpha = 0$, but not uniform stability.

Note that we have not shown that the stability is not uniform when $\alpha = 0$. We have only failed to prove that it is. See Problem 3.

The reader might well ask how the function V was invented in this example. The answer illustrates a method which is often used in similar situations: Since we hope to show that certain solutions of (3) remain "small", let us assume that we can ignore the term x^3 in (3) in comparison with $(t^{-2}-1)x$. Then we consider the equation

$$y' = (t^{-2}-1)y. \tag{4}$$

But (4) is a first order linear equation, which can be solved exactly as in Section 2. We find

$$e^{1/t}e^t y(t) = \text{constant}.$$

Thus non-negative, nonincreasing (with respect to t) functions which "measure the size" of the solution of (4) are

$$e^{2/t}e^{2t}y^2(t), \quad \text{or} \quad e^{2/t}y^2(t),$$

or similar expressions. In Example 1 we adopted the defini-
tion $V(t,\xi) = e^{2/t}\xi^2$ because $V(t,\zeta) = e^{2/t}e^{2t}\xi^2$ would not
satisfy condition (c) for Eq. (3).

The reader will soon discover that the biggest challenge in applying the Lyapunov method is often the selection
of an appropriate Lyapunov function or functional. You
might say this is the only challenge.

The next example introduces a trick which is sometimes
useful in studying stability for delay differential systems.

Example 2. Consider again the delay differential equation of
Example 30-5,

$$x'(t) = ax(t) + bx(t-r), \tag{5}$$

with $\alpha = -\infty$ and $D = \mathbb{R}$. This time we shall show that the
trivial solution of (5) is uniformly stable if $a \leq 0$, $r > 0$,
and $-1/r \leq b \leq 0$.

The key trick which makes the argument work is this.
Following Krasovskiĭ [1956] let us note that if $x = x(\cdot\,;t_0,\phi)$ is the solution of (5) with $x_{t_0} = \phi$, then for
$t \geq t_0+r$

$$x(t-r) = x(t) - \int_{t-r}^{t} x'(s)ds$$

$$= x(t) - \int_{t-r}^{t} [ax(s) + bx(s-r)]ds.$$

(We know, of course, from the method of steps that x exists
on $[t_0-r,\infty)$.) Thus for $t \geq t_0+r$

$$x'(t) = (a+b)x(t) - b\int_{t-r}^{t} [ax(s) + bx(s-r)]ds. \tag{6}$$

Equation (6) can be considered as a delay differential equation with infinitely many distributed delays.   The maximum
delay in (6) is not   r   but   2r.   Why?

Allowing for the doubling of the delay in (6), let us
define a functional V on $\mathbb{R} \times C([-2\tau,0],\mathbb{R})$ by

$$V(t,\psi) \equiv \psi^2(0) + \frac{|a|}{r}\int_{-r}^{0}\int_{\sigma}^{0}\psi^2(s)\,ds\,d\sigma + \frac{|b|}{r}\int_{-r}^{0}\int_{\sigma-r}^{0}\psi^2(s)\,ds\,d\sigma.$$

Then it is a straightforward exercise to verify that

$$|\psi(0)|^2 \le V(t,\psi) \le (1 + \tfrac{1}{2}|a|r + \tfrac{3}{2}|b|r)||\psi||_{2r}^2, \tag{7}$$

where

$$||\psi||_{2r} \equiv \sup_{-2r \le s \le 0} |\psi(\sigma)|.$$

Inequalities (7) correspond to (a) and (b) of Theorem A.

Now we shall not exactly obtain condition (c) of
Theorem A.   Instead let us at first consider the solution
$x = x(\cdot;t_0,\phi)$ of Eq. (5) only for $t \ge t_0+r$.   Then, understanding $x_t(\sigma) \equiv x(t+\sigma)$ for $-2r \le \sigma \le 0$,

$$V(t,x_t) = x^2(t) + \frac{|a|}{r}\int_{-r}^{0}\int_{t+\sigma}^{t}x^2(u)\,du\,d\sigma + \frac{|b|}{r}\int_{-r}^{0}\int_{t+\sigma-r}^{t}x^2(u)\,du\,d\sigma.$$

So for $t \ge t_0+r$, invoking Eq. (6), we find

$$\frac{d}{dt}V(t,x_t) = 2(a+b)x^2(t) - 2abx(t)\int_{t-r}^{t}x(s)\,ds$$

$$-2b^2x(t)\int_{t-r}^{t}x(s-r)\,ds + \frac{|a|}{r}\int_{-r}^{0}[x^2(t)-x^2(t+\sigma)]\,d\sigma$$

$$+ \frac{|b|}{r}\int_{-r}^{0}[x^2(t) - x^2(t+\sigma-r)]\,d\sigma$$

$$= \frac{1}{r}\int_{-r}^{0}\{(a+b)x^2(t) - 2abrx(t)x(t+\sigma)$$

$$- 2b^2rx(t)x(t+\sigma-r) + ax^2(t+\sigma) + bx^2(t+\sigma-r)\}\,d\sigma$$

$$\leq \frac{1}{r}\int_{-r}^{0} \{a(1+br)[x^2(t) + x^2(t+\sigma)]$$

$$+ b(1+br)[x^2(t) + x^2(t+\sigma-r)]\}d\sigma$$

$$\leq 0.$$

Since we have not literally verified condition (c), we cannot directly invoke Theorem A.   However it should be clear from (7), by the same argument as in the proof of Theorem A, that if

$$\sup_{t_0-r\leq t\leq t_0+r} |x(t;t_0,\phi)| < \epsilon(1 + \tfrac{1}{2}|a|r + \tfrac{3}{2}|b|r)^{-1/2} \tag{8}$$

for some $\epsilon > 0$, then

$$|x(t;t_0,\phi)| < \epsilon \quad \text{for all} \quad t \geq t_0-r.$$

Finally we note that, by continuous dependence Theorem 25-E, we can achieve condition (8) by simply requiring

$$||\phi||_r < \delta = \epsilon(1 + \tfrac{1}{2}|a|r + \tfrac{3}{2}|b|r)^{-1/2}e^{-Kr},$$

where K is a Lipschitz constant for Eq. (5), say $K = |a| +$ $|b|$.   It follows that if $||\phi||_r < \delta$, then $|x(t;t_0,\phi)| < \epsilon$ for all $t \geq t_0-r$.

For reference, we display in Figure 1 the set of all values of a and b (for a given $r > 0$) for which the trivial solution of Eq. (5) is (uniformly) stable.   This region (shaded) can be determined by analyzing the roots of the characteristic equation, $\lambda = a + be^{-\lambda r}$, discussed in Section 28.   See for example Èl'sgol'c   [1955] or [1964]. The doubly shaded part of the figure is that which we have obtained combining the result of Example 2 above with that of

Example 30-5.

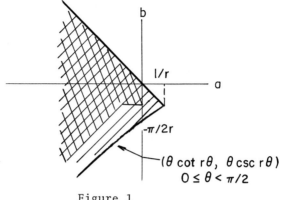

Figure 1

<u>Problems.</u>

1. Show that if $v: [t_0, \beta) \to \mathbb{R}$ is continuous and $v'(t) \leq 0$
   for $t_0 < t < \beta$ then v is nonincreasing on $[t_0, \beta)$, as
   noted in the remark concerning condition (c) of Theorem A.

2. Find the functions w and W in Theorem A when using
   the Lyapunov function(a1) appropriate for
   (a) Example 30-4.   (b) Example 30-5.

*3. Show that the stability of the trivial solution in
   Example 1 is not uniform when $\alpha = 0$. <u>Hint:</u> If $x(t) \geq 0$
   then $x'(t) \geq (t^{-2} - 1)x(t)$.

4. Use Theorem A to prove the uniform stability of the trivial solution of the following differential systems (with
   $\alpha = -\infty$ in each case):
   (a) $x' = -3x + (\sin t)x^2$

(b) $x_1' = x_2$

$x_2' = -\frac{g}{\ell} \sin x_1 - \frac{b}{m} x_2$

where   m, ℓ, g, and   b   are positive constants and

$\alpha = -\infty$.   <u>Hint</u>:   Take $H = \pi$ and $V(\xi) = \frac{1}{2} \xi_2^2 +$

$\frac{g}{\ell} (1 - \cos \xi_1)$.

(c) $x_1'(t) = x_2(t)$

$x_2'(t) = -kx_1(t) - bx_2(t) - qx_2(t-r)$

(previously discussed in Example 28-3).   Here   k, b,

q, and   r   are positive constants with $q \leq b$.

<u>Hint</u>:   Define

$$V(t,\psi) = \frac{1}{2} k\psi_1^2(0) + \frac{1}{2} \psi_2^2(0) + \frac{1}{2} b\int_{-r}^{0} \psi_2^2(s)ds.$$

Krasovskiĭ [1959].

(d)   System 16-(17).

32.   ASYMPTOTIC STABILITY

We now define a stricter condition than stability.   In
this we shall require that a small change or "error" in the
initial values cause only a small error in the solution of
our differential system, <u>and</u> that the error in the solution
actually tend to zero as $t \to \infty$.   When this concept is made
precise we shall call it "asymptotic stability".

The system to be considered is again

$$x'(t) = F(t,x_t) \tag{1}$$

where $F: (\alpha,\infty) \times \mathscr{C}_D \to \mathbb{R}^n$ is as described in the previous
two sections with $D = \{\xi \in \mathbb{R}^n : ||\xi|| < H\}$ and $F(t,0) = 0$.
Thus we continue to assume that   F   is locally Lipschitzian

and quasi-bounded, and that it satisfies Continuity Condition (C).

Definition. The trivial solution of (1) is said to be asymptotically stable at $t_0 > \alpha$ if it is stable at t_0 and there exists $\delta_1 = \delta_1(t_0) > 0$ such that whenever $||\phi||_r < \delta_1$,

$$\lim_{t \to \infty} x(t;t_0,\phi) = 0.$$

Example 1. The trivial solution of the linear system

$$x'(t) = \sum_{j=1}^{m} A_j x(t-r_j) \tag{2}$$

with constant coefficient matrices and constant delays $r_j \in [0,r]$ is asymptotically stable at each $t_0 \in \mathbb{R}$ if all roots of the characteristic equation

$$\det \left(\lambda I - \sum_{j=1}^{m} A_j e^{-\lambda r_j} \right) = 0$$

have negative real parts. This follows at once since Corollary 28-C asserts the existence of positive constants M and γ such that

$$||x(t;t_0,\phi)|| \le M ||\phi||_r e^{-\gamma(t-t_0)} \quad \text{for} \quad t \ge t_0. \tag{3}$$

In practice, of course, it may be difficult to determine whether or not all roots of Eq. (3) have negative real parts. For this reason and because of the need for examples, we shall continue to discuss, by other methods, systems which are actually special cases of (2).

The strongest type of stability which we will use is as follows.

Definition. The trivial solution of (1) is said to be

uniformly asymptotically stable if it is uniformly stable
and, furthermore, there exists $\delta_1 > 0$ (independent of t_0)
such that whenever $t_0 > \alpha$ and $||\phi||_r < \delta_1$, $x(t;t_0,\phi) \to 0$
as $t \to \infty$ in the following explicit (and uniform) manner.
For each $\eta > 0$, there exists $T = T(\eta) > 0$ (independent of
t_0) such that

$$||x(t;t_0,\phi)|| < \eta \quad \text{for all} \quad t \geq t_0+T.$$

Example 2. The trivial solution in Example 1 is uniformly
asymptotically stable on $\mathbb{R} \times \mathscr{C}$. In fact one can use any
value for $\delta_1 > 0$. Then if $t_0 \in \mathbb{R}$ and $||\phi||_r < \delta_1$, Cor-
ollary 28-C gives

$$||x(t;t_0,\phi)|| \leq M\delta_1 e^{-\gamma(t-t_0)} \quad \text{for} \quad t \geq t_0.$$

So now if $\eta > 0$ is given we will have

$$||x(t;t_0,x_0)|| < \eta \quad \text{whenever} \quad t-t_0 > \gamma^{-1}\ln(M\delta_1/\eta).$$

Thus one can use any $T(\eta) > \gamma^{-1}\ln(M\delta_1/\eta)$.

Actually Example 2 is merely an illustration of the
following property of autonomous systems.

Theorem A. If Eq. (1) is autonomous (with $\alpha = -\infty$), then
asymptotic stability of the trivial solution at some $t_0 \in \mathbb{R}$
implies uniform asymptotic stability.

The proof of this analog of Theorem 30-A is left to
the reader. Problem 2.

Another observation is the analog of Theorem 30-C.
Again the proof is left to the interested reader.

Theorem B. If the trivial solution of (1) is asymptotically
stable at some $t_0 > \alpha$. Then the trivial solution is also
asymptotically stable at each $\tilde{t}_0 \in (\alpha, t_0)$.

The next example illustrates, in a simple case, the
use of a Lyapunov function in proving asymptotic stability.

Example 3. Consider the ordinary differential equation of
Example 31-1,

$$x'(t) = (t^{-2} - 1)x(t) + x^3(t). \tag{4}$$

Let $\alpha = 0$ and let us now take $H = 1/2$ (instead of $H = 1$).
As in Example 31-1 we define

$$V(t,\xi) = e^{2/t}\xi^2,$$

but now the domain is $(0,\infty) \times (-1/2, 1/2)$. Let t_0 be
positive and let $|x_0| < \delta_1 \equiv \delta(1/2, t_0)$ where $\delta(\epsilon, t_0)$ is
as described in the stability proof in Example 31-1. Then it
follows that the solution $x = x(\cdot; t_0, x_0)$ exists and satis-
fies $|x(t)| < 1/2$ for all $t \geq t_0$. Thus, using calculation
(c) of Example 31-1,

$$\frac{d}{dt}V(t,x(t)) = -2e^{2/t}x^2(t)[1-x^2(t)] \leq -\frac{3}{2}V(t,x(t)). \tag{5}$$

Applying the usual technique to solve the resulting differential inequality we get

$$\frac{d}{dt}V(t,x(t)) + \frac{3}{2}V(t,x(t)) \leq 0$$

or

$$\frac{d}{dt}[V(t,x(t))e^{3t/2}] \leq 0,$$

so that

$$V(t,x(t)) \leq V(t_0,x_0)e^{-3(t-t_0)/2} \quad \text{for}\quad t \geq t_0.$$

Thus

$$|x(t)| \leq [V(t,x(t))]^{1/2} \to 0 \quad \text{as} \quad t \to \infty,$$

which, together with the stability already proved, establishes the asymptotic stability at each $t_0 > 0$ of the trivial solution of Eq. (4).

If one takes $\alpha = 1$ (instead of $\alpha = 0$) the trivial solution of Eq. (4) is uniformly asymptotically stable. Problem 4(a).

Unfortunately, for general differential and delay differential systems, one seldom gets an inequality as nice as (5). Thus the proof of the following theorem, which assumes a condition weaker than (5), is more difficult than the analysis of Example 3. This theorem is probably the simplest Lyapunov theorem on asymptotic stability which is widely applicable to delay differential systems.

Theorem C. Let w, W, and w_1 be continuous nondecreasing functions on $[0,H]$ which are zero at 0 and positive on $(0,H)$. Let $||F(t,\psi)|| \leq B$ for some constant $B > 0$ for all $(t,\psi) \in (\alpha,\infty) \times \mathcal{L}_D$. If there exists a functional V on $(\alpha,\infty) \times \mathcal{L}_D$ such that

    (a) $V(t,\psi) \geq w(||\psi(0)||)$,

    (b) $V(t,\psi) \leq W(||\psi||_r)$, and

    (c) whenever $(t_0,\phi) \in (\alpha,\infty) \times \mathcal{L}_D$ and $x = x(\cdot;t_0,\phi)$

        on $[t_0-r,\beta_1)$ one has

$$\frac{d}{dt} V(t,x_t) \leq -w_1(||x(t)||) \quad \text{for} \quad t_0 \leq t < \beta_1,$$

then the trivial solution of (1) is uniformly asymptotically stable.

Remarks. As usual, at $t = t_0$ the derivative in condition (c) should be regarded as the right hand derivative.

The theorem would be easier to prove if, instead of (c)
we assumed $(d/dt)V(t,x_t) \leq -w_1(||x_t||_r)$. But that stronger
hypothesis would mean a much less useful theorem for equations with delays.

Proof of Theorem C (adapted from proofs of Massera
[1949] for ordinary differential systems and Krasovskii [1959]
for systems with delays).

The required uniform stability follows from Theorem
31-A.

Select and fix some $H_1 \in (0,H)$. Then choose $\delta_1 > 0$,
independent of t_0, such that

$$W(\delta_1) < w(H_1).$$

(It follows that $\delta_1 < H_1$. Why?)

Let arbitrary $\eta \in (0,\delta_1]$ be given and let $\gamma > 0$
satisfy $W(\gamma) < w(\eta)$. Then $0 < \gamma < \eta \leq \delta_1$. Choose a positive integer

$$K > \frac{W(\delta_1)}{w_1(\gamma/2)\gamma} \, 2B,$$

and define $T(\eta) = Kr_1$ where $r_1 = \max\{r,\gamma/B\}$.

Now let $x = x(\cdot\,;t_0,\phi)$ be the solution of Eq. (1)
through any $(t_0,\phi) \in (\alpha,\infty) \times \mathscr{C}_D$ with $||\phi||_r < \delta_1$. Then
it follows from the uniform stability proof and the choice
of δ_1 that x exists and

$$||x(t)|| \leq H_1 < H \quad \text{for all} \quad t \geq t_0\text{-r}.$$

We are next going to show that, for some $t_1 \in$

$[t_0, t_0 + T(\eta)]$, $||x_{t_1}||_r < \gamma$. Suppose (for contradiction) that $||x_t||_r \geq \gamma$ for all $t \in [t_0, t_0 + T(\eta)]$. Then in each inter-val $[t_0 + kr_1, t_0 + (k+1)r_1]$ for $k = 0, 1, \ldots, K-1$ there must be a subinterval of length at least $\gamma/2B$ on which $||x(t)|| \geq \gamma/2$. Why? Problem 3. On these intervals,

$$\frac{d}{dt} V(t, x_t) \leq -w_1(\gamma/2).$$

Therefore for $t = t_0 + T(\eta) = t_0 + Kr_1$,

$$V(t, x_t) \leq V(t_0, \phi) - Kw_1(\gamma/2)\gamma/2B$$

$$< W(\delta_1) - W(\delta_1) = 0,$$

contradicting the nonnegativity of $V(t, \psi)$.

From the fact that $||x_{t_1}||_r < \gamma$ for some $t_1 \in$ $[t_0, t_0 + T(\eta)]$, it follows that for all $t \geq t_1$

$$V(t, x_t) \leq V(t_1, x_{t_1}) \leq W(\gamma) < w(\eta).$$

Thus for all $t \geq t_1$, and in particular for all $t \geq t_0 + T(\eta)$,

$$||x(t)|| < \eta,$$

as required for uniform asymptotic stability. ☐

Example 4. We shall prove the uniform asymptotic stability of the trivial solution of

$$x'(t) = ax(t) + bx(t-r)$$

when $a < 0$, $|b| < |a|$, and $r \geq 0$. Taking $\alpha = -\infty$ and $D = (-H, H)$ for some $H > 0$, it follows that $|F(t, \psi)| \leq$ $(|a| + |b|)H$ for all $(t, \psi) \in \mathbb{R} \times \mathscr{C}_D$. Moreover, the func-tional V suggested by Example 30-5,

$$V(t,\psi) = \psi^2(0) + |a| \int_{-r}^0 \psi^2(\sigma)d\sigma \quad \text{for} \quad (t,\psi) \in \mathbb{R} \times \mathscr{L}_D,$$

satisfies conditions (a), (b), and (c) of Theorem C. To verify condition (c), let $x = x(\cdot\,;t_0,\phi)$ for arbitrary $(t_0,\phi) \in \mathbb{R} \times \mathscr{L}_D$. Then for all $t \geq t_0$

$$V(t,x_t) = x^2(t) + |a| \int_{t-r}^t x^2(s)ds$$

and, using the calculation of Example 30-5,

$$\frac{d}{dt}V(t,x_t) \leq (-|a| + |b|)[x^2(t) + x^2(t-r)]$$

$$\leq (-|a| + |b|)x^2(t).$$

This is condition (c) with $w_1(s) = (|a| - |b|)s^2$ so uniform asymptotic stability is proved.

<u>Example 5</u>. Consider the ordinary differential system

$$x_1' = \left(\frac{k}{m}\right)^{1/2}x_2 \tag{7}$$

$$x_2' = -\left(\frac{k}{m}\right)^{1/2}x_1 - \frac{b}{m}x_2$$

obtained from the second-order equation $mz'' + bz' + kz = 0$. Assume that m, b, and k are given positive constants and let $\alpha = -\infty$.

The uniform asymptotic stability of the trivial solution actually follows as a special case of Example 2. But, suppose we wanted to prove this with the use of Theorem C. It would be natural to try the Lyapunov function of Example 30-4,

$$V(t,\xi) \equiv \frac{1}{2}k\xi_1^2 + \frac{1}{2}k\xi_2^2 \quad \text{for} \quad (t,\xi) \in \mathbb{R}^3. \tag{8}$$

Then conditions (a) and (b) of Theorem C are fulfilled. However when we try to obtain condition (c) we find, as in

Example 30-4,

$$\frac{d}{dt}V(t,x(t)) = -(kb/m)x_2^2(t).$$ (9)

This is not good enough since it is possible to have

$x_2(t) = 0$ while $||x(t)|| \neq 0$.

It _is_ possible to apply Theorem C to Eq. (7) by using

a different Lyapunov function.   For example let us try to

choose a constant $c > 0$ such that the conditions of Theorem

C are fulfilled when

$$V(t,\xi) \equiv \tfrac{1}{2}k\xi_1^2 + c\xi_1\xi_2 + \tfrac{1}{2}k\xi_2^2.$$ (10)

In order to satisfy conditions (a) and (b) of the theorem it

suffices to require

$$c < k.$$ (11)

Why?   Then, in trying to verify condition (c), we compute

$$\frac{d}{dt}V(t,x(t)) = -[c(\tfrac{k}{m})^{1/2}x_1^2 + \tfrac{cb}{m}x_1x_2 + (\tfrac{kb}{m} - c(\tfrac{k}{m})^{1/2})x_2^2].$$

Now the quadratic expression in brackets will be positive

definite (i.e., positive whenever $x_1^2 + x_2^2 \neq 0$) if all its

coefficients are positive and its discriminant is negative,

i.e., if

$$\tfrac{kb}{m} - c(\tfrac{k}{m})^{1/2} > 0 \quad \text{and} \quad (\tfrac{cb}{m})^2 - 4[\tfrac{ckb}{m}(\tfrac{k}{m})^{1/2} - \tfrac{c^2k}{m}] < 0.$$

These two conditions and (11) hold provided

$$0 < c < \min \{k, \frac{4kb}{4k+b^2/m} (\tfrac{k}{m})^{1/2}\}.$$ (12)

Then   V   defined by (10) satisfies the conditions of Theorem

C for Eq. (7).   Problem 5(a).

Motivated by the difficulty we encountered in applying

Theorem C to system (7), we are now going to present another theorem for uniform asymptotic stability. This theorem, in which condition (c) is relaxed, is applicable to autonomous systems. Unfortunately we cannot give a proof here, since that would require some results not covered in this text.

This important result (actually in more general form) was proved by E. A. Barbašin and Krasovskiĭ in 1952 for ordinary differential systems. It was generalized to delay differential systems by Krasovskiĭ [1956], [1959].

Theorem D (Barbašin-Krasovskiĭ). Let Eq. (1) be autonomous so that $\alpha = -\infty$ and $F: \mathscr{C}_D \to \mathbb{R}^n$ with $F(\psi) = 0$ when $\psi = 0$. Let w and W be as in Theorem C. If there exists a functional $V: \mathscr{C}_D \to [0,\infty)$ with the properties

(a) $V(\psi) \geq w(||\psi(0)||)$,

(b) $V(\psi) \leq W(||\psi||_r)$, and

(c') whenever $(t_0,\phi) \in \mathbb{R} \times \mathscr{C}_D$ and $x = x(\cdot;t_0,\phi)$ on $[t_0-r,\beta)$ one has, for $t_0 \leq t < \beta$,

$$\frac{d}{dt}V(x_t) \leq 0 \quad \text{with} \equiv 0 \quad \text{only if} \quad \phi = 0,$$

then the trivial solution of (1) is uniformly asymptotically stable.

Example 6. The uniform asymptotic stability of the trivial solution of system (7) can be proved using the Lyapunov function defined in (8). For let $(t_0,x_0) \in \mathbb{R}^3$ and let $x = x(\cdot;t_0,x_0)$. Then from (9) it is clear that

$$\frac{d}{dt} V(x(t)) \leq 0 \quad \text{for} \quad t \geq t_0$$

and if $(d/dt)V(x(t)) \equiv 0$ then $x_2(t) \equiv 0$. But this, in

system (7), implies $x_1(t) \equiv 0$ which means $x_0 = 0$. Thus
condition (c') of Theorem D is satisfied.

Example 7 (Krasovskiĭ [1959]). Consider the system with
delay

$$x_1'(t) = x_2(t) \tag{13}$$

$$x_2'(t) = -\frac{k}{m}x_1(t) - \frac{b}{m}x_2(t) - \frac{q}{m}x_2(t-r),$$

obtained from Eq. 21-(15),

$$mz''(t) + bz'(t) + qz'(t-r) + kz(t) = 0.$$

Let   m, b, q, k, and   r   be positive constants and let
$\alpha = -\infty$ and $H = +\infty$.   Define

$$V(\psi) = \frac{1}{2}k\psi_1^2(0) + \frac{1}{2}m\psi_2^2(0) + \frac{1}{2}b\int_{-r}^{0} \psi_2^2(s)ds.$$

The reader should verify that   V   satisfies conditions (a)
and (b) of Theorem D and that, if $(t_0,\phi) \in \mathbb{R} \times \mathscr{C}$ and   x =
$x(\cdot;t_0,\phi)$ then

$$\frac{d}{dt} V(x_t) = -\frac{1}{2}bx_2^2(t) - qx_2(t)x_2(t-r) - \frac{1}{2}bx_2^2(t-r)$$

$$\text{for   } t \geq t_0.$$

When $q < b$ this gives condition (c') of Theorem D and the
uniform asymptotic stability of the trivial solution of (13)
follows.   Problem 5(b).

Further results and examples in the Lyapunov method for
delay differential equations can be found in Krasovskiĭ
[1959], Driver [1962], Halanay [1963], Lakshmikantham and
Leela [1969], and Hale [1971].

Problems

1. The ordinary differential system

$$x'(t) = \begin{pmatrix} 0 & a \\ -a & -2b \end{pmatrix} x(t)$$

is a special case of Eq. (2). Assuming $0 < b < a$, verify
inequality (3) by actually solving the system with
$x(0) = x_0$. You should find

$$||x(t;0,x_0)|| \le 2a(a^2-b^2)^{-1/2}||x_0||e^{-bt}$$

for $t \ge 0$. (We shall use this specific estimate in some
problems in Sections 33 and 34.)

2. Prove Theorem A.

3. Why, in the proof of Theorem C, does it follow that
    $||x(t)|| \ge \gamma/2$ for all t in some subinterval of length
    $\gamma/2B$ of each interval $[t_0+kr_1, t_0+(k+1)r_1]$?

4. Prove the uniform asymptotic stability of the trivial
    solution of each of the following systems.
    (a) $x'(t) = (t^{-2}-1)x(t) + x^3(t)$ with $\alpha = 1$. Cf.
         Example 3.
    (b) $x'(t) = -3x(t) + (\sin t)x^2(t)$ with $\alpha = -\infty$.
         Cf. Problem 31-4(a).
    (c) $x'(t) = ax(t) + bx(t-r)$ with $\alpha = -\infty$, where $a < 0$
         and $-1/r < b \le 0$. Cf. Example 31-2.

5. Prove the assertions of (a) Example 5, (b) Example 7.

6. Prove the uniform asymptotic stability of the trivial
    solution of (a) Problem 31-4(b), (b) System 16-(17).

(Perhaps you can use the same   V's you used in Problem
31-4.)

7.   Use results of this section to show that each solution
     of the following transcendental equations (with constant
     real coefficients) must have negative real part.

     (a) $\lambda = a + be^{-\lambda r}$, where $a < 0$, $r > 0$, and $|b| < |a|$.
           See Example 4.

     (b) $\lambda = a + be^{-\lambda r}$, where $a < 0$, $r > 0$, and $-1/r < b \leq 0$.
           See Problem 4(c).

     (c) $m\lambda^2 + b\lambda + q\lambda e^{-\lambda r} + k = 0$, where   m, b, q, k, and   r
           are positive with $q < b$.   See Example 7.

33.   LINEAR AND QUASI-LINEAR ORDINARY DIFFERENTIAL SYSTEMS

     Suppose the trivial solution of some linear ordinary
differential system is known to be uniformly asymptotically
stable.   Then we shall find that if suitably "small" nonlinear terms are added to the right hand side of the differential system the trivial solution of the new nonlinear system will also be uniformly asymptotically stable.

Example 1.   Let us try to determine the stability (or in-
stability) of the trivial solution of the scalar equation

$$x' = -3x + \sin t \sin x^2.$$

The trick which will work here and in similar cases is as
follows.   Regard the equation as linear $(x' = -3x)$ with a
"perturbation" $(\sin t \sin x^2)$ which is small in case $|x|$
is small.   If the perturbation term were a given function of
t   alone we would proceed as follows to find $x(t) =$

$x(t;t_0,x_0)$, with the aid of the integrating factor e^{3t}. From

$$x' + 3x = \sin t \sin x^2$$

we find

$$\frac{d}{dt}[e^{3t}x] = e^{3t} \sin t \sin x^2,$$

and hence

$$e^{3t}x(t) - e^{3t_0}x_0 = \int_{t_0}^{t} e^{3s} \sin s \sin x^2(s)ds.$$

Now if $|x(s)| \le H$ (some constant) for $t_0 \le s \le t$, the above gives

$$e^{3(t-t_0)}|x(t)| \le |x_0| + \int_{t_0}^{t} He^{3(s-t_0)}|x(s)|ds.$$

Why? Applying, Lemma 8-A to $v(t) \equiv e^{3(t-t_0)}|x(t)|$, we find

$$e^{3(t-t_0)}|x(t)| \le |x_0|e^{H(t-t_0)} \qquad \text{for} \quad t \ge t_0$$

as long as $|x(t)|$ remains $\le H$. Thus, for $t \ge t_0$, we have

$$|x(t)| \le |x_0|e^{-(3-H)(t-t_0)} \quad \text{as long as} \quad |x(t)| \text{ remains} \le H.$$

So, if $0 < H < 3$ and if $|x_0| < H$, it follows that

$$|x(t)| \le |x_0|e^{-(3-H)(t-t_0)} \qquad \text{for all} \quad t \ge t_0.$$

From this the uniform stability and uniform asymptotic stability of the trivial solution follow easily.

The general linear homogeneous ordinary differential
system has the form

$$y' = A(t)y. \tag{1}$$

Assume that the matrix-valued function A is continuous on
(α,∞). Then consider (1) on $(\alpha,\infty) \times \mathbb{R}^n$.

We shall also consider perturbations of system (1) of

the form

$$x' = A(t)x + h(t,x) \quad \text{on} \quad (\alpha,\infty) \times D, \tag{2}$$

where h is continuous and locally Lipschitzian with $h(t,0) = 0$ and $D = \{\xi \in \mathbb{R}^n: ||\xi|| < H\}$ for some $H > 0$. The goal of this section is to show that if the trivial solution of (1) is uniformly asymptotically stable, then the trivial solution of the perturbed system (2) is also asymptotically stable provided the perturbation h is "small" in some sense.

We remark that an ordinary differential system $x' = f(t,x)$ can be written in the form (2) if $f(t,0) = 0$ and $f(t,\xi)$ has continuous partial derivatives. In this case $||h(t,\xi)||/||\xi|| \to 0$ as $||\xi|| \to 0$.

We now proceed to develop the theorems which will generalize the argument in Example 1 to system (2).

<u>Lemma A.</u> The trivial solution of Eq. (1) is uniformly stable if and only if there exists a constant M_1 such that

$$||y(t;t_0,x_0)|| \leq M_1 ||x_0|| \quad \text{for all} \quad t \geq t_0 > \alpha$$
$$\text{and} \quad x_0 \in \mathbb{R}^n. \tag{3}$$

Note that M_1 must be ≥ 1.

<u>Proof.</u> From condition (3), the uniform stability of the trivial solution follows easily. (Given $\varepsilon > 0$, choose $\delta = \varepsilon/M_1$.)

For the converse, assume that the trivial solution of (1) is uniformly stable. Taking $\varepsilon = 1$, let $\delta = \delta(1) > 0$ so that $||y(t;t_0,x_0)|| < 1$ whenever $t \geq t_0 > \alpha$ and $||x_0|| < \delta$. Define $M_1 = 1/\delta$. Clearly (3) holds if $x_0 = 0$.

So consider an arbitrary $x_0 \neq 0$. Then for each $\delta^* \in (0,\delta)$,
$||(\delta^* x_0/||x_0||)|| = \delta^* < \delta$. Thus $||y(t;t_0,\delta^* x_0/||x_0||)||$
< 1, or by linearity,

$$||y(t;t_0,x_0)|| < ||x_0||/\delta^* \quad \text{for all} \quad t \geq t_0 > \alpha.$$

Now, for each fixed $t_0 > \alpha$ and $t \geq t_0$, let $\delta^* \to \delta$ to find

$$||y(t;t_0,x_0)|| \leq ||x_0||/\delta = M_1||x_0||. \quad \square$$

Theorem B. The trivial solution of Eq. (1) is uniformly asymptotically stable if and only if there exist constants $M \geq 1$ and $\gamma > 0$ such that for all $(t_0,x_0) \in (\alpha,\infty) \times \mathbb{R}^n$

$$||y(t;t_0,x_0)|| \leq M||x_0||e^{-\gamma(t-t_0)} \quad \text{for all} \quad t \geq t_0. \quad (4)$$

Remark. It is rather surprising that the exponential decay estimate (4) obtained here, under the assumption of uniform asymptotic stability, is just like the one found in the special case of constant coefficients. Example 32-1.

Proof of Theorem B. Condition (4) certainly guarantees the uniform stability of the trivial solution via Lemma A. To show that the other requirement of the definition of uniform asymptotic stability is also fulfilled, take $\delta_1 = 1$. Then if $(t_0,x_0) \in (\alpha,\infty) \times \mathbb{R}^n$ with $||x_0|| \leq 1 = \delta_1$,

$$||y(t;t_0,x_0)|| \leq Me^{-\gamma(t-t_0)} \quad \text{for} \quad t \geq t_0.$$

Now if $\eta > 0$ is given, we choose $T = T(\eta) > 0$ sufficiently large so that

$$Me^{-\gamma T} < \eta.$$

Then for $t \geq t_0 + T$,

$$||y(t;t_0,x_0)|| \leq Me^{-\gamma T} < \eta.$$

For the converse, assume the trivial solution of (1)
to be uniformly asymptotically stable. Then, by Lemma A,
there exists a constant $M_1 \geq 1$ such that

$$||y(t;t_0,x_0)|| \leq M_1||x_0|| \quad \text{for all} \quad t \geq t_0 > \alpha$$

$$\text{and} \quad x_0 \in \mathbb{R}^n. \qquad (3')$$

Now let $\delta_1 > 0$ be as in the definition of uniform
asymptotic stability. Letting $\eta = \delta_1/2$, there exists
$T = T(\delta_1/2) > 0$ such that if $t_0 > \alpha$ and $||x_0|| < \delta_1$, then

$$||y(t;t_0,x_0)|| < \delta_1/2 \quad \text{for all} \quad t \geq t_0+T.$$

Consider any $x_0 \neq 0$. Then for each $\delta_1^* \in (0,\delta_1)$,
$||(\delta_1^*/||x_0||)x_0|| = \delta_1^* < \delta_1$. Thus

$$||y(t;t_0,\delta_1^*x_0/||x_0||)|| < \delta_1/2$$

or, by linearity,

$$||y(t;t_0,x_0)|| < \frac{\delta_1}{2\delta_1^*}||x_0|| \quad \text{for all} \quad t \geq t_0+T.$$

Now for each fixed $t_0 > \alpha$ and $t \geq t_0+T$, let $\delta_1^* \to \delta_1$ to
find

$$||y(t;t_0,x_0)|| \leq \tfrac{1}{2}||x_0|| \quad \text{for} \quad t \geq t_0+T. \qquad (5)$$

Letting t_0+T play the role of t_0, (5) now gives for
$t \geq t_0+2T$

$$||y(t;t_0,x_0)|| = ||y(t;t_0+T,y(t_0+T;t_0,x_0))||$$

$$\leq 2^{-1}||y(t_0+T;t_0,x_0)||$$

$$\leq 2^{-2}||x_0||.$$

And by induction one finds

$$||y(t;t_0,x_0)|| \leq 2^{-k}||x_0|| \quad \text{for} \quad t \geq t_0+kT$$

for k = 1,2,... . This can be rewritten as

$$||y(t;t_0,x_0)|| \leq ||x_0||e^{-(\ln 2)k} \quad \text{for} \quad t_0+kT \leq t < t_0+(k+1)T,$$

and it follows that

$$||y(t;t_0,x_0)|| \leq ||x_0||e^{-(\ln 2)(t-t_0-T)/T}$$

$$= 2||x_0||e^{-(\ln 2)(t-t_0)/T} \quad \text{for} \quad t \geq t_0+T. \tag{6}$$

This would be assertion (4) except for the fact that it has not been proved for $t_0 \leq t < t_0+T$. But, combining (3') with (6) we find

$$||y(t;t_0,x_0)|| \leq 2M_1||x_0||e^{-(\ln 2)(t-t_0)/T} \quad \text{for} \quad t \geq t_0 > \alpha.$$

Check this carefully for $t_0 \leq t < t_0+T$. We have proved (4) with $M = 2M_1$ and $\gamma = (\ln 2)/T$. □

Now we are ready to generalize Example 1 to give a simple criterion for stability for a system which represents only a small perturbation from a linear system.

Theorem C (Perron). Let the trivial solution of system (1) be uniformly asymptotically stable. Then there exists a con- stant N > 0 such that, if

$$||h(t,\xi)|| \leq N||\xi|| \quad \text{for all} \quad t > \alpha \quad \text{and} \quad \xi \in D \tag{7}$$

the trivial solution of system (2) is also uniformly asymp- totically stable. (Any $N \in (0,\gamma/M)$ will suffice, where M and γ are as in Theorem B.)

Remark. Condition (7) is often satisfied by reducing the size of H in the definition of D.

Proof.   Let   h   satisfy (7) for some $N \in (0,\gamma/M)$.
Let $(t_0,x_0) \in (\alpha,\infty) \times D$ be given, and let $x = x(\cdot;t_0,x_0)$
on $[t_0,\beta_1)$ be the solution of (2) continued as far as possible to the right.   We are now going to treat system (2) as
though the perturbation $h(t,x(t))$ were a known function of
t.   Then "variation of parameters" (Theorem 17-A), gives for
$t_0 \leq t < \beta_1$

$$x(t) = y(t;t_0,x_0) + \int_{t_0}^{t} y(t;s,h(s,x(s)))ds$$

But, because of the assumed uniform asymptotic stability of
the trivial solution of (1), we have condition (4).   Thus,
for $t_0 \leq t < \beta_1$,

$$||x(t)|| \leq Me^{-\gamma(t-t_0)}||x_0|| + \int_{t_0}^{t} Me^{-\gamma(t-s)}N||x(s)||ds,$$

or

$$e^{\gamma(t-t_0)}||x(t)|| \leq M||x_0|| + \int_{t_0}^{t} MNe^{\gamma(s-t_0)}||x(s)||ds$$

And Lemma 8-A then gives

$$e^{\gamma(t-t_0)}||x(t)|| \leq M||x_0||e^{MN(t-t_0)},$$

or

$$||x(t)|| \leq M||x_0||e^{-(\gamma-MN)(t-t_0)}. \tag{8}$$

Since $\gamma-MN > 0$, it follows that if we only make sure that
$||x_0|| < H/M$ then $x(t)$ will exist for all $t \geq t_0$.

The uniform asymptotic stability of the trivial solution of (2) now follows from (8) by an argument quite like
that used in the first part of the proof of Theorem B.   One
takes $\delta_1 \leq H/M$ and chooses $T = T(\eta) > 0$ such that
$M\delta_1 e^{-(\gamma-MN)T} < \eta$.   □

Example 2.   The familiar system

$$x_1' = x_2$$

$$x_2' = -\frac{g}{\ell} \sin x_1 - \frac{b}{m} x_2, \tag{9}$$

where m, ℓ, g, and b are positive constants, describes the motion of a pendulum with friction. We can take $\alpha = -\infty$. Existence of all solutions to $+\infty$ was established in Example 24-1. Considering that example or invoking our "physical intuition" about friction, it seems that solutions <u>ought</u> to tend to the rest position as $t \rightarrow \infty$. But how can we prove that? More specifically, how can we prove that the trivial solution of (9) is asymptotically stable? Or, is it? The following is an alternative argument to that suggested in Problem 32-6(a).

If only small solutions are considered, then it seems reasonable to approximate $\sin x_1$ by x_1 in (9). This would lead to a linear system (1) with constant coefficient matrix

$$A = \begin{pmatrix} 0 & 1 \\ -g/\ell & -b/m \end{pmatrix},$$

the eigenvalues of which are the solutions λ_1 and λ_2 of

$$\det(A - \lambda I) = \lambda^2 + \frac{b}{m}\lambda + \frac{g}{\ell} = 0.$$

Since m, ℓ, g, and b are all positive, it follows that λ_1 and λ_2 both have negative real parts. Then it follows from Corollary 28-C that, for any positive $\gamma < \min \{|\mathrm{Re}\, \lambda_1|, |\mathrm{Re}\, \lambda_2|\}$, the solutions of system (1) satisfy (4) for some constant $M \geq 1$.

But what does all this have to do with the original system (9)?

We can rewrite (9) as

$$x' = Ax + h(t,x) \quad \text{where} \quad h(t,x) = \frac{g}{\ell}\begin{pmatrix} 0 \\ x_1 - \sin x_1 \end{pmatrix}.$$

This is an example of Eq. (2), a perturbation of a linear system. Our problem is to show that h satisfies condition
(8) with some appropriate choice of H.

By Taylor's theorem, for any real number ξ_1,

$$\sin \xi_1 = \xi_1 - (\cos \theta) \frac{1}{3!} \xi_1^3$$

for some θ between 0 and ξ_1. So for any $(t,\xi) \in \mathbb{R}^3$

$$\|h(t,\xi)\| = \frac{g}{\ell} |\xi_1 - \sin \xi_1|$$

$$\leq \frac{g}{6\ell} |\xi_1|^3 \leq \frac{g}{6\ell} \|\xi\|^3.$$

Thus condition (7) is satisfied whenever H is so small that

$$\frac{gH^2}{6\ell} = N < \frac{\gamma}{M}.$$

The idea used above can be generalized as follows.

Corollary D.   The assertion of Theorem C remains true if, instead of condition (7),

$$\|h(t,\xi)\| \leq w(\|\xi\|)\|\xi\| \quad \text{for all} \quad t > \alpha \quad \text{and} \quad \xi \in D, \quad (10)$$

where $w: [0,\infty) \to [0,\infty)$ satisfies $\lim_{\rho \to 0+} w(\rho) = 0$.

Proof.   Let M and γ be as in Theorem B. Then reduce H, if necessary, so that

$$\sup_{0 < \rho < H} w(\rho) < \gamma/M.$$

Now (10) gives condition (7). $\square$

Problems

1. Let $T(t,t_0)$ be the transition matrix for system (1) so
   that $y(t;t_0,x_0) = T(t,t_0)x_0$. (See Section 17.) Then
   show that (4) holds for all $t_0 > \alpha$ and $x_0 \in \mathbb{R}^n$ if and
   only if

   $$||T(t,t_0)|| \le Me^{-\gamma(t-t_0)} \quad \text{for all} \quad t \ge t_0 > \alpha.$$

   Hint: You will need Lemma 18-C.

2. System (9) came from Eq. 24-(12), $m\ell\theta'' + b\ell\theta' +$
   $mg \sin \theta = 0$, where m, ℓ, g, and b are positive. Now
   let $x_1 = \theta$ and $x_2 = (\ell/g)^{1/2}\theta'$, and show that

   (a) $x' = \begin{pmatrix} 0 & (g/\ell)^{1/2} \\ -(g/\ell)^{1/2} & -b/m \end{pmatrix} x + (g/\ell)^{1/2} \begin{pmatrix} 0 \\ x_1 - \sin x_1 \end{pmatrix}.$

   (b) If $b^2/m^2 < 4g/\ell$ then the solutions of the associated linear system (1) satisfy (4) with
   $M = 4(4 - b^2\ell/m^2g)^{-1/2}$ and $\gamma = b/2m$. You may use
   the assertion of Problem 32-1.

3. Prove that the trivial solution of each of the following
   is uniformly asymptotically stable (taking $\alpha = -\infty$).
   Furthermore, in each case find an explicit value for
   δ_1 $(= H/M)$ in the definition of uniform asymptotic stability

   (a) $x' = -2x + x^2 + 5x^3 \sin t.$

   (b) $x_1' = -2x_1 + 2x_2 + 4x_1 x_2$
   $x_2' = 2x_1 - 5x_2 - 6x_2^2.$

   You may use the results of Example 15-1.

(c) $x_1' = -x_1 - x_2 + 2x_1x_2$

$x_2' = x_1 - 3x_2 - 3x_2^2.$

You may use the result of Example 19-4.

4.   (a)   Prove that the solution $N(t) = P$ of the equation

$$N' = kN - \frac{k}{P}N^2,$$

where k and P are positive constants, is uniformly asymptotically stable.

(b)   Prove that the solution $N(t) = 0$ of the same equation is unstable.   You may refer to the solution of
Eq. 21-(5).

34.   LINEAR AND QUASI-LINEAR DELAY DIFFERENTIAL SYSTEMS

The results to be developed here will generalize, and
hence supercede, those of Section 33.   However, it seemed
worthwhile to cover ordinary differential equations first in
order to present the basic ideas of the perturbation method
without the additional complication of delays.

Just as the variation of parameters theorem for ordinary differential systems was a key tool used in Section 33,
we will now depend heavily on the variation of parameters
theorem for delay differential systems.   Since that result
was only proved (in Section 29) for a linear system with
finitely many constant delays, such will be the nature of our
basic linear system here too.

Thus our (unperturbed) linear system will be

$$y'(t) = \sum_{j=1}^{m} A_j(t)y(t-r_j), \tag{1}$$

where each A_j is a continuous $n \times n$-matrix-valued function on (α,∞) and each r_j is a constant in $[0,r]$.

The perturbed system to be considered will be denoted by

$$x'(t) = F(t,x_t),   \tag{2}$$

where $F: (\alpha,\infty) \times \mathscr{S}_D \to \mathbb{R}^n$ with $D = \{\xi \in \mathbb{R}^n: ||\xi|| < H\}$. We assume that F satisfies Continuity Condition (C), that it is locally Lipschitzian and quasi-bounded, and that $F(t,0) = 0$.

Note that we do not require the retarded arguments in (2) to agree with those in (1). There may not even be the same number of delays in the two equations, and those in Eq. (2) may not be constant.

To make the notation more analogous to that in Section 33 we introduce a functional $h: (\alpha,\infty) \times \mathscr{S}_D \to \mathbb{R}^n$ defined by

$$h(t,\psi) = F(t,\psi) - \sum_{j=1}^{m} A_j(t)\psi(-r_j),$$

so that Eq. (2) becomes

$$x'(t) = \sum_{j=1}^{m} A_j(t)x(t-r_j) + h(t,x_t).   \tag{2'}$$

Our first results generalize Lemma 33-A and Theorem 33-B to the linear system (1) with delays

Lemma A.   The trivial solution of Eq. (1) is uniformly stable if and only if there exists a constant M_1 such that

$$||y(t;t_0,\phi)|| \le M_1||\phi||_r \quad \text{for all} \quad t \ge t_0 > \alpha \quad \text{and} \quad \phi \in \mathscr{S}.   \tag{3}$$

It follows that $M_1 \ge 1$, and so (3) actually holds for all $t \ge t_0 - r$.

Proof.   The proof that condition (3) implies uniform

stability is easy.

Conversely, if the trivial solution of (1) is uniformly
stable, let $\delta = \delta(1)$. Then $||y(t;t_0,\phi)|| < 1$ whenever
$||\phi||_r < \delta$ and $t \geq t_0 > \alpha$. Define $M_1 = 1/\delta$. Clearly (3)
holds if $\phi = 0$. Consider an arbitrary $\phi \neq 0$. Then for each
$\delta^* \in (0,\delta)$, $||y(t;t_0,\delta^*\phi/||\phi||_r)|| < 1$, or by linearity

$$||y(t;t_0,\phi)|| < ||\phi||_r/\delta^* \quad \text{for all} \quad t \geq t_0.$$

Now for each fixed $t_0 > \alpha$ and $t \geq t_0$, let $\delta^* \to \delta$ to find

$$||y(t;t_0,\phi)|| \leq ||\phi||_r/\delta = M_1||\phi||_r.$$

Note that this implies $M_1 \geq 1$ so that (3) holds also for
$t_0 - r \leq t \leq t_0$. □

Theorem B. The trivial solution of Eq. (1) is uniformly
asymptotically stable if and only if there exist constants
$M \geq 1$ and $\gamma > 0$ such that for every $(t_0,\phi) \in (\alpha,\infty) \times \mathscr{C}$

$$||y(t;t_0,\phi)|| \leq M||\phi||_r e^{-\gamma(t-t_0)} \quad \text{for all} \quad t \geq t_0, \qquad (4)$$

or equivalently,

$$||y_t(t;t_0,\phi)||_r \leq Me^{\gamma r}||\phi||_r e^{-\gamma(t-t_0)} \quad \text{for all} \quad t \geq t_0. \quad (5)$$

Proof. The proofs that (4) and (5) are equivalent and
that either one assures the uniform asymptotic stability of
the trivial solution of (1) are straightforward exercises.
Problem 1.

For the converse, let us assume that the trivial solution of (1) is uniformly asymptotically stable. Then, by
Lemma A, there exists a constant $M_1 \geq 1$ such that

$$||y(t;t_0,\phi)|| \leq M_1||\phi||_r \quad \text{for all} \quad t > t_0 > \alpha$$
$$\text{and} \quad \phi \in \mathcal{L}. \tag{3'}$$

Now let $\delta_1 > 0$ be as in the definition of uniform asymptotic stability. Then if we take $\eta = \delta_1/2$, there exists $T > 0$ such that for each $t_0 > \alpha$ and $\phi \in \mathcal{L}$ with $||\phi||_r < \delta_1$

$$||y(t;t_0,\phi)|| < \delta_1/2 \quad \text{for all} \quad t \geq t_0+T.$$

Consider any $\phi \neq 0$. Then for each $\delta_1^* \in (0,\delta_1)$,
$$||y(t;t_0,\delta_1^*\phi/||\phi||_r)|| < \delta_1/2 \quad \text{or, by linearity,}$$

$$||y(t;t_0,\phi)|| < \frac{\delta_1}{2\delta_1^*}||\phi||_r \quad \text{for all} \quad t \geq t_0+T.$$

Now, for fixed $t_0 > \alpha$ and $t \geq t_0+T$, let $\delta_1^* \to \delta_1$ to find

$$||y(t;t_0,\phi)|| \leq \tfrac{1}{2}||\phi||_r \quad \text{for} \quad t \geq t_0+T.$$

This implies

$$||y_t(\cdot;t_0,\phi)||_r \leq \tfrac{1}{2}||\phi||_r \quad \text{for all} \quad t \geq t_0+T+r. \tag{6}$$

If we now follow the solution into the future, repeated application of (6) gives for $t_0+k(T+r) \leq t < t_0+(k+1)(T+r)$ $(k = 1,2,\ldots)$

$$||y_t(\cdot;t_0,\phi)||_r \leq 2^{-k}||\phi||_r = ||\phi||_r e^{-(\ell n\ 2)k}.$$

It follows that

$$||y_t(\cdot;t_0,\phi)||_r \leq ||\phi||_r e^{-(\ell n\ 2)(t-t_0-T-r)/(T+r)}$$

$$= 2||\phi||_r e^{-(\ell n\ 2)(t-t_0)/(T+r)} \quad \text{for} \quad t \geq t_0+T+r.$$

This combined with (3') gives for all $t \geq t_0$,

$$||y_t(\cdot;t_0,\phi)||_r \leq 2M_1||\phi||_r e^{-(\ell n\ 2)(t-t_0)/(T+r)}.$$

Why? This is (5) $\gamma = (\ell n\ 2)/(T+r)$ and $M = 2M_1 e^{-\gamma r}$. $\square$

Our objective now is to obtain results, generalizing Theorem 33-C, which state that if the trivial solution of (1) is uniformly asymptotically stable then so too will be the trivial solution of (2) provided $h(t,\psi)$ is small in some sense.

In doing this we will want to apply inequality (4), not only when $y_{t_0} = \phi$ is continuous, but also when $y_{t_0} = \xi u$ where $\xi \in \mathbb{R}^n$ and u is the unit step function defined, as in Section 29, by

$$u(\sigma) \equiv \begin{cases} 0 & \text{for} \quad -r \leq \sigma < 0 \\ 1 & \text{for} \quad \sigma = 0. \end{cases}$$

Corollary C. If the trivial solution of Eq. (1) is uniformly asymptotically stable, then for each $t_0 > \alpha$ and each $\xi \in \mathbb{R}^n$,

$$||y(t;t_0,\xi u)|| \leq M||\xi||e^{-\gamma(t-t_0)} \quad \text{for all} \quad t \geq t_0 \tag{7}$$

where the constants $M \geq 1$ and $\gamma > 0$ are the same as in (4). (Note that $||\xi u||_r = ||\xi||$.)

Proof. We shall proceed by "approximating" the function u by continuous functions $u_{(\ell)}$ ($\ell = 1, 2, \ldots$), where

$$u_{(\ell)}(\sigma) \equiv \begin{cases} 0 & \text{for} \quad -r \leq \sigma < -r/\ell \\ 1 + \ell\sigma/r & \text{for} \quad -r/\ell \leq \sigma \leq 0. \end{cases}$$

Sketch a graph of $u_{(\ell)}$ and compare with u. Now let $y(t) = y(t;t_0,\xi u)$ and $y_{(\ell)}(t) = y(t;t_0,\xi u_{(\ell)})$, and consider for

$t \geq t_0$

$$y(t) - y_{(\ell)}(t) = \int_{t_0}^{t} \sum_{j=1}^{m} A_j(s)[y(s-r_j) - y_{(\ell)}(s-r_j)]ds.$$

Letting $K_j(t) = \max_{t_0 \leq s \leq t} ||A_j(s)||$ for each j and $K(t) = K_1(t) + \cdots + K_m(t)$, we find for $t \geq t_0$

$$||y(t) - y_{(\ell)}(t)|| = ||\sum_{j=1}^{m} \int_{t_0-r_j}^{t-r_j} A_j(s+r_j)[y(s) - y_{(\ell)}(s)]ds||$$

$$\leq \sum_{j=1}^{m} \int_{t_0-r/\ell}^{t} K_j(t)||y(s) - y_{(\ell)}(s)||ds$$

$$\leq K(t)||\xi||r/\ell + \int_{t_0}^{t} K(t)||y(s)-y_{(\ell)}(s)||ds.$$

Thus, by Lemma 8-A,

$$||y(t) - y_{(\ell)}(t)|| \leq \frac{K(t)r}{\ell}||\xi||e^{K(t)(t-t_0)} \quad \text{for } t \geq t_0.$$

But, since $\xi u_{(\ell)} \in \mathcal{C}$,

$$||y(t)|| \leq ||y(t) - y_{(\ell)}(t)|| + ||y_{(\ell)}(t)||$$

$$\leq \frac{K(t)r}{\ell}||\xi||e^{K(t)(t-t_0)} + M||\xi||e^{-\gamma(t-t_0)} \quad \text{for } t \geq t_0.$$

Since this holds for each ℓ, we can simply let $\ell \to \infty$ to obtain (7).   □

In the course of the stability analysis for (2) we will encounter a scalar "delay differential inequality" of the form

$$v'(t) \leq -\gamma v(t) + p||v_t||_r \quad \text{for } t \geq t_0, \tag{8}$$

where γ and p are given constants with $0 < p < \gamma$ and $v(t) \geq 0$.

Lemma D (Halanay [1966]). Let γ and p be constants with

$0 < p < \gamma$.   Let v be a continuous nonnegative function on $[t_0 - r, \beta)$ satisfying inequality (8) for $t_0 \le t < \beta$.   Then

$$v(t) \le ||v_{t_0}||_r e^{-\lambda(t-t_0)} \quad \text{for} \quad t_0 \le t < \beta, \tag{9}$$

where λ is the (unique) positive solution of

$$\lambda = \gamma - pe^{\lambda r}. \tag{10}$$

Proof.   To show that Eq. (10) has a unique positive solution, we consider the function defined by

$$\Delta(\mu) = \mu - \gamma + pe^{\mu r}.$$

Since $\Delta(0) < 0$, $\Delta(\gamma) > 0$, and $\Delta'(\mu) > 0$, it follows that there is a unique $\lambda \in \mathbb{R}$ for which $\Delta(\lambda) = 0$.   Moreover $0 < \lambda < \gamma$.

Define $w(t) = ||v_{t_0}||_r e^{-\lambda(t-t_0)}$ for $t_0 - r \le t < \beta$ and let $k > 1$ be arbitrary.   Then $v(t) < kw(t)$ for $t_0 - r \le t \le t_0$.   Check this.   Now suppose (for contradiction) that $v(t) = kw(t)$ for some $t \in (t_0, \beta)$.   Then, since v and w are continuous functions, there must exist some $t_1 \in (t_0, \beta)$ such that

$$v(t) < kw(t) \quad \text{for} \quad t_0 - r \le t < t_1 \quad \text{and} \quad v(t_1) = kw(t_1).$$

This could not occur if $v'(t_1)$ were $< kw'(t_1)$.   But, on the other hand, we find

$$v'(t_1) \le -\gamma v(t_1) + p||v_{t_1}||_r$$

$$< -\gamma kw(t_1) + pkw(t_1 - r) = kw'(t_1)$$

-- a contradiction.   Thus we conclude that

$$v(t) < kw(t) \quad \text{for} \quad t_0 \le t < \beta.$$

Finally let $k \to 1$ to find $v(t) \le w(t)$, which is (9). $\square$

__Theorem E.__   Let the trivial solution of system (1) be uniformly asymptotically stable and let M and γ be the constants in (4).   If for some constant $N \in (0,\gamma/M)$

$$||h(t,\psi)|| \le N||\psi||_r \quad \text{for all} \quad (t,\psi) \in (\alpha,\infty) \times \mathscr{C}_D, \tag{11}$$

then the trivial solution of system (2) is also uniformly
asymptotically stable.

__Proof.__   Let $0 < N < \gamma/M$ and let h satisfy condi-
tion (11).   Let $(t_0,\phi) \in (\alpha,\infty) \times \mathscr{C}_D$ be given, and let
$x = x(\cdot\,;t_0,\phi)$ on $[t_0\text{-}r,\beta_1)$ be the unique noncontinuable
solution of (2) with $x_{t_0} = \phi$.
     Regarding (2') a system (1) with an added "forcing
term", $h(t,x_t)$, variation of parameters (Theorem 29-B) gives

$$x(t) = y(t;t_0,\phi) + \int_{t_0}^{t} y(t;s,h(s,x_s)u)ds$$

for $t_0\text{-}r \le t < \beta_1$.   Applying estimates (4) and (7) and condition (11) to this, we find

$$||x(t)|| \le M||\phi||_r e^{-\gamma(t-t_0)} + \int_{t_0}^{t} MN||x_s||_r e^{-\gamma(t-s)}ds \tag{12}$$

for $t_0 \le t < \beta_1$.
     Now define

$$v(t) = \begin{cases} M||\phi||_r & \text{for } t_0\text{-}r \le t \le t_0 \\ Me^{-\gamma(t-t_0)}||\phi||_r + e^{-\gamma t}\int_{t_0}^{t} MNe^{\gamma s}||x_s||_r ds & \text{for } t_0 < t < \beta_1. \end{cases}$$

Then v is continuous and nonegative, and $||x(t)|| \leq v(t)$
for $t_0 - r \leq t < \beta$.   Moreover, for $t_0 \leq t < \beta_1$,

$$v'(t) = -\gamma v(t) + MN||x_t||_r \leq -\gamma v(t) + MN||v_t||_r.$$

Since $MN < \gamma$, it follows from Lemma D that

$$||x(t)|| \leq v(t) \leq M||\phi||_r e^{-\lambda(t-t_0)} \qquad \text{for} \quad t_0 \leq t < \beta_1, \qquad (13)$$

where λ is the positive solution of

$$\lambda = \gamma - MNe^{\lambda r}. \qquad (14)$$

But inequality (13) together with Theorem 26-C shows that
$\beta_1 = \infty$ provided

$$||\phi||_r < H/M.$$

Then, by familiar arguments, inequality (13) yields the uniform asymptotic stability of the trivial solution of Eq.
(2).   □

Example 1.   Let the unperturbed system be the linear ordinary
differential system

$$y'(t) = A_1(t)y(t) \qquad (15)$$

and let the perturbed system be

$$x'(t) = A_1(t)x(t) + \sum_{j=2}^{m} A_j(t)x(t-r_j(t)), \qquad (16)$$

where each A_j is a continuous matrix-valued function, and
each r_j is continuous with $0 \leq r_j(t) \leq r$ for all $t > \alpha$.
Let the trivial solution of (15) be uniformly asymptotically
stable so that for some $M \geq 1$ and $\gamma > 0$

$$||y(t;t_0,\xi)|| \leq M||\xi||e^{-\gamma(t-t_0)} \qquad \text{for all} \quad t \geq t_0 > \alpha$$
$$\text{and} \quad \xi \in \mathbb{R}^n.$$

Regarding (15) as a delay differential equation, this gives, instead of (4) and (7),

$$||y(t;t_0,\phi)|| \leq M||\phi||_r e^{-\gamma(t-t_0)} \quad \text{for} \quad t \geq t_0 > \alpha$$

where either $\phi \in \mathscr{L}$ or $\phi = \xi u$. Why?

It now follows from Theorem E that the trivial solution of (16) is uniformly asymptotically stable provided there exists a number N such that

$$\sum_{j=2}^{m}||A_j(t)|| \leq N < \gamma/M \quad \text{for all} \quad t > \alpha. \tag{17}$$

Corollary F. The assertion of Theorem E remains true if condition (11) is replaced by

$$||h(t,\psi)|| \leq w(||\psi||_r)||\psi||_r \quad \text{for all} \quad (t,\psi) \in (\alpha,\infty) \times \mathscr{C}_D, \tag{18}$$

where $w: [0,\infty) \to [0,\infty)$ satisfies $\lim_{\rho \to 0+} w(\rho) = 0$.

Proof. Let M and γ be as in (4). Then reduce H, if necessary, so that

$$\sup_{0<\rho<H} w(\rho) < \gamma/M.$$

Now (18) gives condition (11). $\square$

Example 2. The trivial solution of

$$x'(t) = ax(t) + bx(t-r) + cx(t)x(t-r/2)$$

is uniformly asymptotically stable provided $a < 0$ and $|b| < |a|$ or $a < 0$ and $-1/r < b \leq 0$. To show this one uses the results of Example 32-4 and Problem 32-4(c) for the linear part of the equation, and then applies Corollary F.

Problem 32-4(c) asserts that the trivial solution of $x'(t) = ax(t) + bx(t-r)$ can be uniformly asymptotically stable even though $|b| \geq |a|$, provided $a < 0$ and r is sufficiently small. It would be nice to have an analogous result for system (16). In other words we want a stability condition for (16) which, unlike (17), uses the smallness of r. The smallness of r might help to preserve stability.

Theorem E is incapable of providing results in which changes of the delays, between systems (1) and (2), play an essential role. This is because of the nature of condition (11). Our objective now is to relax the hypothesis of smallness of h so that perturbations of the delays can be treated. In the process, the theorem becomes more complicated.

Theorem G. Let the trivial solution of (1) be uniformly asymptotically stable and let M and γ be as in (4). Let constants $K > 0$ and $N \in (0, \gamma/M)$ exist such that

(i) $||F(t,\psi) - F(t,\tilde{\psi})|| \leq K||\psi - \tilde{\psi}||_r$ when (t,ψ), $(t,\tilde{\psi})$
     $\in (\alpha,\infty) \times \mathscr{C}_D$ (a global Lipschitz condition), and

(ii) $||h(t,\psi)|| \leq N \max\{||\psi||_r, ||\psi'||_r/K\}$ just for
     those $(t,\psi) \in (\alpha,\infty) \times \mathscr{C}_D$ with ψ continuously
     differentiable.

Then the trivial solution of (2) is uniformly asymptotically stable.

Remarks. The first theorem considering the effects on stability of changes in the delays appears to have been proved by Repin [1957]. A theorem qualitatively similar to Theorem G was proved by Grossman and Yorke [1972] for more general systems. But they required a smaller N than the relatively large value we permit here (any $N < \gamma/M$).

In the proof we will need to consider some functions defined on $[-2r,0]$ instead of $[-r,0]$. The following notation and convention will be used:

$$||\psi||_{2r} = \sup_{-2r \leq \sigma \leq 0} ||\psi(\sigma)|| \quad \text{for} \quad \psi \in C([-2r,0],\mathbb{R}^n),$$

and whenever $||x_t||_{2r}$ occurs we shall interpret x_t to be defined by

$$x_t(\sigma) = x(t+\sigma) \quad \text{for} \quad -2r \leq \sigma \leq 0.$$

Proof of Theorem G. We are going to apply variation of parameters for $t \geq t_0+r$, rather than for $t \geq t_0$ as in the proof of Theorem E.

Let any $\varepsilon \in (0,H)$ be given. Then choose $\delta = \varepsilon e^{-Kr}/M$ so that, by Theorem 25-E, whenever $t_0 > \alpha$ and $||\phi||_r < \delta$ the unique solution of (2) with $x_{t_0} = \phi$ satisfies

$$||x(t;t_0,\phi)|| \leq \varepsilon/M \quad \text{for} \quad t_0 \leq t \leq t_0+r.$$

Let (t_0,ϕ) be any point in $(\alpha,\infty) \times \mathscr{L}_D$ with $||\phi||_r < \delta$ and let x on $[t_0-r,\beta_1)$ be the unique noncontinuable solution of (2) through (t_0,ϕ). The choice of δ assures us that $\beta_1 > t_0+r$.

Now for $t_0 \leq t < \beta_1$

$$||x'(t)|| = ||F(t,x_t)|| \leq K||x_t||_r.$$

Hence, for $t_0+r \leq s < \beta_1$,

$$||x_s'(\sigma)|| \leq K||x_s||_{2r} \quad \text{for} \quad -r \leq \sigma \leq 0.$$

So, by hypothesis (ii),

$$||h(s,x_s)|| \leq N||x_s||_{2r} \quad \text{for} \quad t_0+r \leq s < \beta_1.$$

Apply variation of parameters as in the proof of Theorem E except that now t_0+r replaces t_0, x_{t_0+r} replaces ϕ, and $||\cdot||_{2r}$ replaces $||\cdot||_r$. One then gets, instead of (12),

$$||x(t)|| \leq Me^{-\gamma(t-t_0-r)}||x_{t_0+r}||_{2r}$$
$$+ \int_{t_0+r}^{t} Me^{-\gamma(t-s)}N||x_s||_{2r}ds, \tag{19}$$

for $t_0+r \leq t < \beta_1$.

It then follows, as did (13), that

$$||x(t)|| \leq M||x_{t_0+r}||_{2r}e^{-\lambda(t-t_0-r)} \quad \text{for} \quad t_0+r \leq t < \beta_1, \tag{20}$$

where λ is the positive solution of

$$\lambda = \gamma - MNe^{2\lambda r}.$$

Thus, by virtue of the choice of δ,

$$||x(t)|| < \begin{cases} \varepsilon/M & \text{for} \quad t_0-r \leq t \leq t_0+r \\ \varepsilon e^{-\lambda(t-t_0-r)} & \text{for} \quad t_0+r \leq t < \beta_1, \end{cases} \tag{21}$$

This together with Theorem 26-C shows that $\beta_1 = \infty$ and that the trivial solution of (2) is uniformly stable.

With no further work, (21) also yields the other requirement for uniform asymptotic stability: Choose (and fix) some $H_1 \in (0,H)$, and choose $\delta_1 = H_1e^{-Kr}/M$ (just as δ was chosen for ε). Let $(t_0,\phi) \in (\alpha,\infty) \times \mathscr{C}_D$ with $||\phi||_r < \delta_1$ and let $\eta > 0$ be given. Then $||x(t)|| < \eta$ for all $t \geq t_0+T$ provided $T = T(\eta) > 0$ satisfies

$$H_1e^{-\lambda(T-r)} < \eta. \quad \square$$

Example 3. Let us apply Theorem G to system (16) by now considering the unperturbed system to be

$$y'(t) = \sum_{j=1}^{m} A_j(t)y(t). \tag{22}$$

and

$$h(t,\psi) = \sum_{j=2}^{m} A_j(t)[y(t-r_j)-y(t)]. \tag{23}$$

Let the trivial solution of (22) be uniformly asymptotically stable. Then one might expect that for sufficiently small r the trivial solution of (16) would also be uniformly asymptotically stable. We shall show that this is indeed the case.

The solutions of Eq. (22) satisfy $||y(t;t_0,\xi)|| \leq M||\xi||e^{-\gamma(t-t_0)}$ for $t \geq t_0 > \alpha$ for some $M \geq 1$ and $\gamma > 0$. Thus, regarding (22) as a delay differential system,

$$||y(t;t_0,\phi)||_r \leq M||\phi||_r e^{-\gamma(t-t_0)} \quad \text{for} \quad t \geq t_0 > \alpha$$

and any $\phi \in \mathcal{C}$. We shall show that if each A_j is a __bounded__ continuous matrix-valued function on (α,∞), then the trivial solution of system (16) is uniformly asymptotically stable provided

$$r \sup_{t>\alpha} \sum_{j=2}^{m} ||A_j(t)|| \cdot \sup_{t>\alpha} \sum_{j=1}^{m} ||A_j(t)|| < \gamma/M. \tag{24}$$

Take $H = \infty$. Then condition (i) of Theorem G is satisfied with

$$K = \sup_{t>\alpha} \sum_{j=1}^{m} ||A_j(t)||.$$

For condition (ii) we observe that if $(t,\psi) \in (\alpha,\infty) \times \mathcal{C}$ and ψ is continuously differentiable, then

$$||h(t,\psi)|| \leq \sup_{t>\alpha} \sum_{j=2}^{m} ||A_j(t)|| \cdot ||\psi'||_r \cdot r.$$

This gives condition (ii) of Theorem G with

$$N = r \sup_{\substack{t > \alpha}} \sum_{j=2}^{m} ||A_j(t)|| \cdot K,$$

and, by (24), $N < \gamma/M$.

Here we have a case of small perturbation of the delay itself. The unperturbed delay in this example happens to be zero.

Condition (24) was obtained by Halanay [1966].

Problems

1. (a) Prove that if $M \geq 1$ and $\gamma > 0$, then conditions (4) and (5) are equivalent.

    (b) Prove that condition (4) implies the uniform asymptotic stability of the trivial solution of (1).

2. Apply the theorems or examples of this section to each of the following to find sufficient conditions for the uniform asymptotic stability of the trivial solution of the given equation. Find values for M and γ so that you can obtain specific conditions in the form of inequalities in each case. The linear unperturbed equation to be used in each case is given in parentheses.

    (a) $x'(t) = ax(t) + bx(t-r)$, where $a < 0$, $r \geq 0$.
         $(y'(t) = ay(t).)$ Compare your result with that of Example 32-4.

    (b) $x'(t) = ax(t) + bx(t-r)$, where $a + b < 0$, $r > 0$.
         $(y'(t) = (a + b)y(t).)$ Compare with the result of Problem 32-4(c).

(c) $x_1'(t) = \sqrt{\frac{k}{m}} x_2(t),$

$x_2'(t) = -\sqrt{\frac{k}{m}} x_1(t) - \frac{b}{m} x_2(t) - \frac{q}{m} x_2(t-r),$

where   m, b, k, q, and   r   are positive constants
with $b^2 < 4mk$.   Hint: Use the result of Problem
32-1.   This system arises from Eq. 21-(15), but it
will be difficult to do as well as Example 32-7.

$$(y'(t) = \begin{pmatrix} 0 & \sqrt{k/m} \\ -\sqrt{k/m} & -b/m \end{pmatrix} y(t).)$$

(d)   The system of (c) with $(b+q)^2 < 4mk$.

$$(y'(t) = \begin{pmatrix} 0 & \sqrt{k/m} \\ -\sqrt{k/m} & -(b+q)/m \end{pmatrix} y(t).)$$

3. Find sufficient conditions for the uniform asymptotic
stability of the trivial solution of

(a) $x'(t) = \tilde{a}(t)x(t) + \tilde{b}(t)x(g(t)),$

where $|\tilde{a}(t) - a| < \Delta a$, $|\tilde{b}(t) - b| < \Delta b$ and
$t-r \leq g(t) \leq t$ for all $t > \alpha$.   Assume that the
given functions   a, b, and   g   are continuous, and
assume $a + b < 0$.

(b) $mz''(t) + bz'(t) + qz'(t-r) + kz(t) = \varepsilon[z'(t-r)]^3,$
"Minorsky's equation".   See Minorsky [1948] and
Pinney [1958], Chapter 11.   All the given constants,
m, b, q, k, r, and ε are positive.

4. Instead of using Lemma D, one might try to apply Lemma
8-A to inequality (12) in the proof of Theorem E.   Show
that this can be done at the expense of imposing a
stronger restriction on   h, namely (11) with
$0 < N < \gamma e^{-\gamma r}/M.$

Chapter IX. <u>AUTONOMOUS</u> <u>ORDINARY</u> <u>DIFFERENTIAL</u> <u>SYSTEMS</u>

Differential equations which arise in real-world physical problems are often too complicated to solve exactly. And even if an exact solution is obtainable, the required calculations may be too complicated to be practical, or the resulting solution may be difficult to interpret.

Thus it is useful to be able to learn something about the qualitative behavior of the solutions of a differential system without actually solving.

The type of "qualitative" information which is most often sought is that which describes the behavior of solutions as $t \to +\infty$. One important aspect of this -- stability -- was the subject of the previous chapter.

The present chapter introduces some of the more elementary standard methods for getting further information about solutions as $t \to +\infty$ in the special case of autonomous ordinary differential systems. The emphasis will actually be on second order systems.

This chapter only scratches the surface of the known techniques. Further development can be found by looking up the names Poincaré, Bendixson, and Perron in a more advanced book on ordinary differential equations. Particularly nice presentations are given in the little books of Hurewicz [1943] and Petrovskiĭ [1964].

It should also be remarked that in many cases, when confronted by a new equation, one must invent his own methods (or tricks) for studying that particular problem.

35.   TRAJECTORIES AND CRITICAL POINTS

Systems of ordinary differential equations representing real-world physical problems very often have the special
form

$$x'(t) = f(x(t)). \qquad (1)$$

Here f is a given function mapping $D \to \mathbb{R}^n$ for some open
set $D \subset \mathbb{R}^n$.

When, as in (1), the right hand side of a differential
system does not depend explicitly on t, the system is said
to be autonomous. If (1) represents some physical process,
we would say that the physical laws governing that process do
not change with time.

It is generally assumed that all basic physical laws
should be represented by autonomous systems. In other words,
these basic laws do not change as time goes on. However,
there is also a place for the more general nonautonomous equations

$$x'(t) = f(t,x(t)).$$

Nonautonomous equations would be used whenever one wants to
include the effects of some external, nonconstant influence
on the system without analyzing that external influence itself. For example the linear motion of an object of mass m
under the influence of an external force $g(t)$ might be
governed by the nonautonomous equation

$$mx'' + bx' = g(t)$$

or the nonautonomous system

$$x_1' = x_2$$

$$x_2' = -\frac{b}{m}x_2 + \frac{g(t)}{m},$$

where b is a coefficient of resistance. On the other hand, suppose we know that the force acting on the mass arises from the fact that the mass is suspended on the end of a (linear) spring. Then we get an autonomous system

$$x_1' = x_2$$

$$x_2' = -\frac{b}{m}x_2 - \frac{k}{m}x_1.$$

Any nonautonomous system,

$$x'(t) = f(t,x(t)), \tag{2}$$

can actually be transformed into an autonomous system. But this is done at the expense of increasing the order of the system. Specifically, let the given nonautonomous system be of n'th order. Then introduce an additional component to x by defining $x_{n+1}(t) = t$, and consider the equivalent autonomous system

$$x_1'(t) = f_1(x_{n+1}(t),x_1(t),\ldots,x_n(t)),$$
$$\vdots$$
$$x_n'(t) = f_n(x_{n+1}(t),x_1(t),\ldots,x_n(t)),$$
$$x_{n+1}'(t) = 1. \tag{3}$$

The present chapter will be devoted primarily to the autonomous system (1). Since an autonomous system can always be considered on $J = (-\infty,\infty)$, we no longer need the notation (α,β) for J. Thus we will now denote the domain of a non-continuable solution by (α,β) instead of the more cumbersome

(α_1, β_1).

Standing Hypothesis. We shall assume throughout the remainder
of this chapter that f is locally Lipschitzian on D so
that the solution of system (1) with given initial data
$x(t_0) = x_0$ is unique.

The reader should observe that in this (autonomous)
case f being locally Lipschitzian implies that f is continuous on D.   Problem 1.

The key property of an autonomous system is that,
given any solution, one can always produce another solution
by a mere translation in t.   Cf. Lemma 28-E.

Recall also that, if the trivial solution of an autonomous system is stable, then it is uniformly stable. If it
is asymptotically stable, then it is uniformly asymptotically
stable.

For an autonomous system, much information about the
solutions can be represented graphically by a "sketch" in $\mathbb{R}^n$
instead of in $\mathbb{R}^{1+n}$. Given any solution x   on (α, β) one
can imagine graphing the set of values taken by the point
$x(t)$, i.e., the set

$$\{(x_1(t), \ldots, x_n(t)): \ \alpha < t < \beta\}. \tag{4}$$

The resulting graph will be called a trajectory. From (4) we
see that a trajectory is a set (often a curve) in $\mathbb{R}^n$ which
can be described parametrically considering t   as the parameter.

Example 1. Graph the solution in $\mathbb{R}^3$ and its trajectory in
$\mathbb{R}^2$ of

$$\left.\begin{array}{l} x_1' = -x_1 - 8x_2 \\ x_2' = 8x_1 - x_2 \end{array}\right\} \quad \text{with} \quad x(0) = \begin{pmatrix} 1 \\ 0 \end{pmatrix}.$$

By substitution, the reader can easily verify that the unique
solution is

$$x(t) = \begin{pmatrix} e^{-t} \cos 8t \\ e^{-t} \sin 8t \end{pmatrix} \quad \text{for} \quad t \in \mathbb{R}.$$

As an aid in graphing we observe that

$$x_1^2(t) + x_2^2(t) = e^{-2t}.$$

The graph of the solution for $t \geq 0$ is represented in
Figure 1 and the graph of the trajectory in Figure 2.

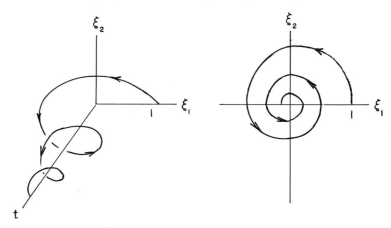

Figure 1   Figure 2

The solution curve, in Figure 1, spirals around the t
axis in corkscrew fashion, always becoming closer to that
axis as t increases. The graph of the trajectory, in
Figure 2, is just a two-dimensional spiral about the origin.
It represents the projection of the solution curve onto the
(ξ_1, ξ_2) plane.

The graph of a trajectory contains less information than the graph of a solution. One generally uses arrowheads, as in Figure 2, to indicate the direction of motion of a point along the trajectory as t increases. But we have given up the knowledge of the explicit location of a point at time t.

On the other hand, a single trajectory, such as that in Figure 2, actually represents infinitely many solution curves for an autonomous system of differential equations. Any time-translate of a solution is another solution. But both have the same trajectory.

The space $\mathbb{R}^n$ in which we graph the trajectories of a differential system is called _phase space_. If we had graphs or descriptions of _all_ trajectories for a differential system (instead of just one as in Figure 2) we would call the result a _phase portrait_ for the system.

Our examples will generally have n = 2 because it is then much easier to sketch the trajectories. In case n = 2 the phase space is often called the _Poincaré phase plane_. Since a great many practical problems lead to second order differential equations, which in turn become systems of two first order equations, the case n = 2 is of considerable importance.

An Aid to Sketching Trajectories in the Plane. If $x_1'(t) \neq 0$ at some t in (α,β), then the function x_1^{-1} is well defined in a neighborhood of $x_1(t)$. Thus if $\xi_1 = x_1(t)$ we have $t = x_1^{-1}(\xi_1)$, and we can represent the trajectory $\{(x_1(t),x_2(t)): \alpha < t < \beta\}$ by the equation

$$\xi_2 = x_2(x_1^{-1}(\xi_1)) \quad \text{as long as} \quad x_1' \neq 0. \tag{5}$$

Similarly the trajectory can be represented by

$$\xi_1 = x_1(x_2^{-1}(\xi_2)) \quad \text{as long as} \quad x_2' \neq 0. \tag{6}$$

Since x_1 and x_2 are solutions of (1), the conditions $x_1'(t) \neq 0$ and $x_2' \neq 0$ become respectively

$$f_1(x_1(t),x_2(t)) \neq 0, \quad \text{and} \quad f_2(x_1(t),x_2(t)) \neq 0.$$

If $f_1(\xi_1,\xi_2) \neq 0$, then along a trajectory as represented by (5) we find

$$\frac{d\xi_2}{d\xi_1} = \frac{x_2'(x_1^{-1}(\xi_1))}{x_1'(x_1^{-1}(\xi_1))} = \frac{f_2(\xi_1,\xi_2)}{f_1(\xi_1,\xi_2)}. \tag{7}$$

And, if $f_2(\xi_1,\xi_2) \neq 0$, we find similarly from (6) that along the trajectory

$$\frac{d\xi_1}{d\xi_2} = \frac{f_1(\xi_1,\xi_2)}{f_2(\xi_1,\xi_2)}. \tag{8}$$

Note that the slopes are computed in (7) and (8) without any need for solving the differential system.

Applied to Example 1, (7) and (8) give, along trajectories,

$$\frac{d\xi_2}{d\xi_1} = \frac{8\xi_1 - \xi_2}{-\xi_1 - 8\xi_2} \quad \text{whenever} \quad \xi_1 \neq -8\xi_2$$

and

$$\frac{d\xi_1}{d\xi_2} = \frac{-\xi_1 - 8\xi_2}{8\xi_1 - \xi_2} \quad \text{whenever} \quad \xi_2 \neq 8\xi_1.$$

The reader should verify that these give appropriate values for the slopes in Figure 2 at a couple of points, say at $(\xi_1,\xi_2) = (1,0)$ and at $(\xi_1,\xi_2) = (0,e^{-\pi/16})$.

A special situation arises at any point $(\eta_1,\dots,\eta_n)$ in phase space at which every f_i vanishes.

Definition. If, for some $\eta = (\eta_1, \ldots, \eta_n) \in D$,

$$f_i(\eta) = 0 \quad \text{for} \quad i = 1, \ldots, n,$$

then η is called a <u>critical point</u> for system (1).

If η is a critical point for system (1), then clearly $x(t) \equiv \eta$ for $-\infty < t < \infty$ defines a solution. Thus the point η itself is a trajectory.

The following theorem is an easy consequence of uniqueness. Its proof is left to the reader as Problem 2.

Theorem <u>A</u>. No two trajectories of system (1) ever intersect, unless they are identical. In particular, if η is a critical point for system (1), then no trajectory passes through η unless it is identically η.

Example 2. If x is a solution of

$$\left.\begin{array}{l} x_1' = x_1^2 - 17x_2 \\ x_2' = \sin x_1 + 1 - e^{x_2} \end{array}\right\} \quad \text{with} \quad x(0) = \begin{pmatrix} 0 \\ 2 \end{pmatrix}$$

then, since $(0,0)$ is a critical point for the system, $x(t)$ never becomes zero.

If x is a solution of (1) on $(-\infty, \infty)$, then it <u>is</u> possible that $x(t)$ approaches some critical point as a limit when $t \to \infty$ or $t \to -\infty$. In fact, the next theorem asserts that critical points are the only points in D which a solution can ultimately approach; and if such limiting behavior occurs, then it must occur as $t \to +\infty$ or $t \to -\infty$.

Theorem <u>B</u>. If x on (α, β) is a noncontinuable solution of system (1) with $\lim_{t \to \beta} x(t) = \eta$ (or $\lim_{t \to \alpha} x(t) = \eta$) for some $\eta \in D$, then $\beta = \infty$ (or $\alpha = -\infty$) and η is a critical point for system (1).

Proof. Let $\lim_{t \to \beta} x(t) = \eta \in D$. Then it follows from Theorem 24-C that $\beta = \infty$. Why? And, since f is continuous,

$$\lim_{t \to \infty} x'(t) = \lim_{t \to \infty} f(x(t)) = f(\eta).$$

Now suppose (for contradiction) that $f_i(\eta) \neq 0$ for some $i = 1, \ldots$, or n. If $f_i(\eta) > 0$, then for all sufficiently large t, say for $t \geq T$, we have $x_i'(t) > f_i(\eta)/2 > 0$. Thus

$$x_i(t) \geq x_i(T) + \tfrac{1}{2}f_i(\eta)(t-T) \qquad \text{for} \quad t \geq T.$$

But this contradicts the assumption that $x_i(t)$ has a limit as $t \to \infty$. Similarly one finds a contradiction to the supposition that $f_i(\eta) < 0$. It follows that $f(\eta) = 0$, i.e., η is a critical point.

The proof is entirely similar for the case $x(t) \to \eta$ as $t \to \alpha = -\infty$. $\square$

Example 1 illustrates the assertion of Theorem B, for $x(t) \to (0,0)$, the one and only critical point, as $t \to \infty$.

Another relatively simple result is the following criterion for the existence of periodic solutions. By a periodic solution of (1) we mean a solution, x, on $(-\infty, \infty)$ having the property that for some $T > 0$, $x(t+T) = x(t)$ for all t. The number T is called the period of x, or more correctly a period of x. It follows that any positive integral multiple of T is also a period of x. Note that a constant (critical point) solution provides a trivial example of a periodic solution, and in this case every $T > 0$ is a period.

Theorem C (Periodic Solutions). Let x on (α,β) be a non-continuable solution of (1) such that for some $t_0 \in (\alpha,\beta)$ and some $t_0+T \in (\alpha,\beta)$ with $T > 0$, $x(t_0+T) = x(t_0)$. Then $(\alpha,\beta) = (-\infty,\infty)$, and the solution is periodic with period T.

Proof. We know that the function y defined by

$$y(t) = x(t-T) \quad \text{for} \quad \alpha+T < t < \beta+T$$

is also a noncontinuable solution of (1). Moreover

$$y(t_0+T) = x(t_0) = x(t_0+T).$$

Now, since x and y are two solutions of (1) with the same value at t_0+T, it follows from uniqueness that $x(t) = y(t)$ as far as both are defined, i.e.,

$$x(t) = y(t) \quad \text{for} \quad \alpha+T < t < \beta.$$

But, since x and y are both noncontinuable solutions, it follows that $(\alpha,\beta) = (\alpha+T,\beta+T)$. Therefore $\alpha = -\infty$ and $\beta = +\infty$. Moreover $x(t+T) = y(t+T) = x(t)$ for all t. $\square$

Example 3. The equation

$$z'' + 2z^3 = 0 \tag{9}$$

can be thought of as governing the motion on a frictionless, horizontal, plane surface of a mass attached to a certain non-linear spring. Let us consider the solution of (9) subject to initial conditions

$$z(0) = z_0, \quad z'(0) = u_0, \tag{10}$$

where z_0 and u_0 are not both zero. An equivalent system is

$$x_1' = x_2,$$
$$x_2' = -2x_1^3,$$

$$(11)$$

with

$$x_1(0) = z_0, \qquad x_2(0) = u_0.$$

$$(12)$$

The existence of a unique noncontinuable solution of (9) and (10) or (11) and (12) follows from Theorem 24-C.   In an attempt to discover the domain of the solution, let us consider the "energy" function suggested by Eq. 24-(18),

$$v(t) \equiv \frac{1}{2} x_2^2(t) + \int_0^{x_1(t)} 2\xi^3 d\xi,$$

$$= \frac{1}{2} x_2^2(t) + \frac{1}{2} x_1^4(t).$$

$$(13)$$

Making use of (11), one finds

$$v'(t) = x_2(t)x_2'(t) + 2x_1^3(t)x_1'(t) = 0.$$

Thus $v(t)$ is a constant $= v(0)$, or

$$x_2^2(t) + x_1^4(t) = u_0^2 + z_0^4 \equiv K > 0.$$

$$(14)$$

This in turn gives $|x_1(t)| + |x_2(t)| \le K^{1/4} + K^{1/2}$, and it follows from Theorem 24-C that (9) and (10) or (11) and (12) have a unique solution on $(-\infty, \infty)$.

Equation (14) also shows that, in the phase plane, the trajectory of the solution must be some portion of the closed curve

$$\xi_1^4 + \xi_2^2 = K.$$

$$(15)$$

Now the only critical point of system (11) is $(0,0)$, and this does not lie on the curve. Hence the trajectory cannot be a single point on the curve (15). The motion along the trajectory can never reverse itself, since the velocity never

vanishes. Moreover, it is easily found directly from (11)
that $x_1(t)$ is increasing in the first quadrant. Thus it
follows that the trajectory is the entire curve (15), traced
out in the clockwise sense as indicated in Figure 3. Therefore, by Theorem C, the solutions are periodic.

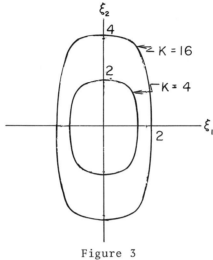

Figure 3

Theorem D. Any nonconstant periodic solution x of (1) has
a smallest period T. Moreover, the trajectory of x is a
closed curve in R^n described parametrically by
$\{x(t): 0 \leq t \leq T\}$ where $x(0) = x(T)$ but $x(t_1) \neq x(t_2)$
whenever $0 \leq t_1 < t_2 < T$.

   Proof. Suppose (for contradiction) that x has no
smallest period. Then for every given period $T > 0$ there
also exists a period $T_* \in (0,T)$. Thus there is a sequence
$\{T_k\}$ of periods such that

$$T_1 > T_2 > T_3 > \cdots > 0 \quad \text{and} \quad \lim_{k \to \infty} T_k = 0.$$

But, since x' exists, this means that at any t

$$x'(t) = \lim_{\Delta t \to 0} \frac{x(t+\Delta t) - x(t)}{\Delta t} = \lim_{k \to \infty} \frac{x(t+T_k) - x(t)}{T_k} = 0.$$

Thus   x   is a constant, contradicting our hypothesis.   We conclude that there does exist a smallest period $T > 0$.

The fact that the trajectory can be represented by $\{x(t): 0 \leq t \leq T\}$ is seen as follows.   For any t_1 in $\mathbb{R}$ there exists a unique integer   k   such that

$$t_1 = kT + t \quad \text{where} \quad 0 \leq t < T,$$

and $x(t_1) = x(t+kT) = x(t)$.   Of course we have, in particular, $x(0) = x(T)$.   And we cannot have $x(t_1) = x(t_2)$ for $0 \leq t_1 < t_2 < T$, or else, by Theorem C, x   would have a period $t_2 - t_1 < T$. $\square$

Problems

1. Verify that if $f: D \to \mathbb{R}^n$ is locally Lipschitzian, then f   is continuous.   Show that this is not true for a general (nonautonomous) $f: (\alpha, \beta) \times D \to \mathbb{R}^n$.

2. Prove Theorem A.

3. Let   A   be a constant $n \times n$ matrix.   Under what condition does the linear system $y' = Ay$ have only one critical point?   What is the critical point?

4. Use Eq. (14) to show that the (smallest) period of the solution of (9) and (10) is

$$T = 4 \int_0^{K^{1/4}} (K - \xi^4)^{-1/2} d\xi,$$

where $K = u_0^2 + z_0^4$.   Prove that this improper integral converges.

5. Determine the critical points and sketch some representa-
    tive trajectories in the phase plane for each of the
    following systems

    (a) $x_1' = x_2$, $x_2' = -4x_1$ (from $z'' + 4z = 0$).

    (b) $x_1' = x_2$, $x_2' = -K/x_1^2$ (from $z'' = -K/z^2$), $K > 0$.

    (c) $x_1' = -4x_1 + 2x_2$, $x_2' = x_1 - 5x_2$.

    (d) $x_1' = -x_1$, $x_2' = 0$.

6. Find all critical points for the scalar equation $(n = 1)$
    $$x' = kx - \frac{k}{p}x^2,$$
    and verify the assertions of Theorems A and B for this
    particular case. (You may use the solution of Eq. 21-(5).)

7. If an autonomous scalar equation, $x' = f(x)$, has no
    critical points in $D = (\gamma, \delta)$, show that either every
    solution is strictly increasing or every solution is
    strictly decreasing.

8. Let $g: \mathbb{R}^2 \to (0, \infty)$ be a given locally Lipschitzian func-
    tion. Then prove that the trajectories of the (locally
    Lipschitzian) system
    $$x_1' = f_1(x_1, x_2), \quad x_2' = f_2(x_1, x_2)$$
    are exactly the same as those of the system
    $$y_1' = g(y_1, y_2)f_1(y_1, y_2), \quad y_2' = g(y_1, y_2)f_2(y_1, y_2).$$
    (This result applies just as well to a system of n
    equations.)

9. Apply the idea of Problem 8 to find the phase portrait
    of the system
    $$x_1' = -x_1^3 - x_1x_2^2, \quad x_2' = -x_2^3 - x_1^2x_2.$$

36.   LINEAR SYSTEMS OF SECOND ORDER

Many practical applications lead to nonlinear differential systems of the form

$$x' = Ax + h(x),$$

where   h   represents, in some sense, a "small", nonlinear
perturbation of the linear system $y' = Ay$.   Sometimes in
such cases one can show that the behavior of the linear system, $y' = Ay$, gives a good approximation to the behavior of
the nonlinear system.   Thus it is of interest to study the
possible trajectories of linear autonomous systems.

The general linear autonomous system has the form

$$x' = Ax + b,$$

where   A   is a constant $n \times n$ matrix and   b   is a constant
n   vector.   We shall assume

$$\det A \neq 0.$$

Then it follows that A^{-1} exists so that our system can be
rewritten as

$$x' = A(x + A^{-1}b),$$

or as $y' = Ay$ where $y \equiv x + A^{-1}b$.

Thus, having assumed $\det A \neq 0$, we might as well con-
fine our attention to the homogeneous linear system with
constant coefficients,

$$x' = Ax. \tag{1}$$

We shall assume that the elements of   A   are all real.   And,
for simplicity, we shall concentrate on the case $n = 2$, so
that

$$A = \begin{pmatrix} a_{11} & a_{12} \\ a_{21} & a_{22} \end{pmatrix}.$$ (2)

The various types of qualitative behavior to be encountered in this section all have counterparts for general $n \geq 2$.

We shall discuss six typical examples illustrating the behavior of solutions under various assumptions on the eigenvalues and eigenvectors of A. Note that, since we are assuming det $A \neq 0$, 0 will not be an eigenvalue.

The presentation in this section was influenced by that of Simmons [1972].

Case 1. Distinct Real Eigenvalues of Like Sign. Consider system (1) under the assumption that the eigenvalues of A are both negative with $\lambda_2 < \lambda_1 < 0$. Then A has two linearly independent eigenvectors $\xi_{(1)}$ and $\xi_{(2)}$, and the general solution of (1) is

$$x(t) = c_1 e^{\lambda_1 t} \xi_{(1)} + c_2 e^{\lambda_2 t} \xi_{(2)}.$$

Two special types of trajectory result if $c_2 = 0$ or $c_1 = 0$, namely straight rays toward the origin (as $t \to \infty$) in the directions of $\pm \xi_{(1)}$ and $\pm \xi_{(2)}$ respectively. If neither c_1 nor c_2 is zero, the solution still tends to zero as $t \to \infty$ since both λ_1 and λ_2 are negative. But for large t the dominant contribution is always $c_1 e^{\lambda_1 t} \xi_{(1)}$, since $\lambda_1 > \lambda_2$.

The phase portrait is indicated in Figure 1 for the specific example

$$x' = \begin{pmatrix} -4 & 2 \\ 1 & -5 \end{pmatrix} x$$

which has the general solution

$$x(t) = c_1 e^{-3t} \begin{pmatrix} 2 \\ 1 \end{pmatrix} + c_2 e^{-6t} \begin{pmatrix} 1 \\ -1 \end{pmatrix}.$$

Further clues for sketching the trajectories are obtained from the expression for the slope along trajectories, cf. Eq. 35-(7),

$$\frac{d\xi_2}{d\xi_1} = \frac{\xi_1 - 5\xi_2}{-4\xi_1 + 2\xi_2} .$$

We see, for example, that $d\xi_2/d\xi_1 = 0$ when $\xi_2 = \xi_1/5$, $d\xi_2/d\xi_1 = \infty$ when $\xi_2 = 2\xi_1$, and $d\xi_2/d\xi_1 = 1$ when $\xi_2 = 5\xi_1/7$.

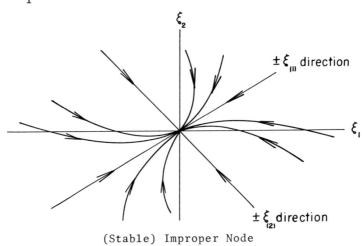

(Stable) Improper Node

with $\lambda_2 < \lambda_1 < 0$

Figure 1

The origin in this case is called a stable _improper_ _node_. In case the eigenvalues were positive and distinct, so that all nontrivial solutions would be unbounded as $t \to \infty$, the origin would be called an unstable improper node. Then all arrowheads in Figure 1 would be reversed. In that case we would have all solutions approaching $(0,0)$ as $t \to -\infty$.

Case 2.   Identical Real Eigenvalues and Two Linearly Independent Eigenvectors.   If system (1) has identical eigenvalues,
then   A   has two linearly independent eigenvectors if and
only if   A   is diagonal.   Why?   (Recall Problem 15-2.)   Thus
(1) becomes

$$x' = \begin{pmatrix} \lambda & 0 \\ 0 & \lambda \end{pmatrix} x,$$

and the general solution can be written as

$$x(t) = c_1 e^{\lambda t} \begin{pmatrix} 1 \\ 0 \end{pmatrix} + c_2 e^{\lambda t} \begin{pmatrix} 0 \\ 1 \end{pmatrix} = e^{\lambda t} \begin{pmatrix} c_1 \\ c_2 \end{pmatrix}.$$

If $\lambda < 0$ the phase portrait is easily found to be as indicated in Figure 2.   The trajectories are all rays directed
toward the origin.

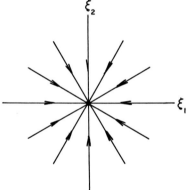

(Stable) Proper Node

$$\lambda_1 = \lambda_2 < 0$$

Figure 2

The origin is then called a stable proper node.   If
the common eigenvalue had been positive, the origin would
have been an unstable proper node and all trajectories would
have been directed away from the origin.

Case 3.   Identical Real Eigenvalues and Only One Linearly
Independent Eigenvector.   In this case the solution of (1) is
found by the method described in Section 15.   In particular,
the proof of Theorem 15-B shows that the general solution is

$$x(t) = c_1 e^{\lambda t} \xi + c_2 e^{\lambda t} (t\xi + \eta),$$

where λ is the eigenvalue, ξ is an eigenvector, and η is
a generalized eigenvector.   If $c_2 = 0$ the trajectory is
linear in the direction of $\pm\xi$.   If $c_2 \neq 0$ then ultimately
as $t \to \infty$ the dominant contribution to $x(t)$ is $c_2 t e^{\lambda t} \xi$.
Thus all solutions eventually tend toward the direction $\pm\xi$
as $t \to \infty$.

For $\lambda < 0$ all trajectories are directed toward the
origin as indicated in Figure 3.   The origin is then said to
be a stable improper node (as in Case 1).   If λ were positive the trajectories would be directed out away from the
origin, and the origin would be an unstable improper node.

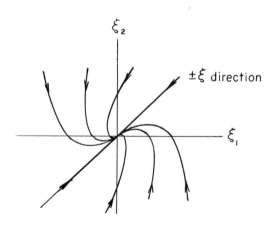

(Stable) Improper Node

$$\lambda_1 = \lambda_2 < 0$$

Figure 3

<u>Case 4</u>. <u>Distinct Real Eigenvalues of Opposite Sign</u>. If the eigenvalues of A are $\lambda_1 > 0$ and $\lambda_2 < 0$ then the general solution has the form

$$x(t) = c_1 e^{\lambda_1 t} \xi_{(1)} + c_2 e^{\lambda_2 t} \xi_{(2)}.$$

Only in the case $c_1 = 0$ does $x(t) \to 0$ as $t \to \infty$, and then the trajectory has a constant direction of $\pm \xi_{(2)}$. In all other cases the trajectory tends out away from the origin as $t \to \infty$, either in the direction $\pm \xi_{(1)}$ if $c_2 = 0$ or else along a curve which tends to the direction $\pm \xi_{(1)}$.

The behavior is illustrated in Figure 4 for the specific system

$$x' = \begin{pmatrix} -1 & -4 \\ -3 & -2 \end{pmatrix} x.$$

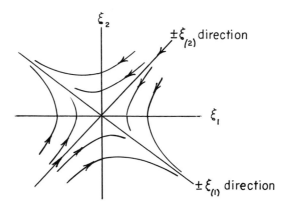

Saddle point

$$\lambda_2 < 0 < \lambda_1$$

Figure 4

The origin is then called a <u>saddle point</u>. The trivial solution is always unstable in the case of a saddle point.

Case 5.   Complex Conjugate Eigenvalues with Nonzero Real Part.
If $\lambda_1, \lambda_2 = \mu \pm i\omega$ where μ and ω are real, then the general solution of system (1) has the form

$$x(t) = c_1 e^{\mu t}(\cos \omega t + i \sin \omega t)\xi_{(1)} \tag{3}$$

$$+ c_2 e^{\mu t}(\cos \omega t - i \sin \omega t)\xi_{(2)},$$

where $\xi_{(1)}$ and $\xi_{(2)}$ are (complex) eigenvectors. The entire expression can be rewritten in terms of real parameters
only, as we have seen in Chapter IV.   If $\mu < 0$ then clearly
$x(t) \to 0$ as $t \to \infty$; and if $\mu > 0$ then any nontrivial solution is unbounded as $t \to \infty$ but tends to zero as $t \to -\infty$.

To see more specifically how $x(t)$ behaves as it tends
to 0 or becomes unbounded, it is useful to consider the
solution in polar coordinates.   We introduce

$$r(t) = [x_1^2(t) + x_2^2(t)]^{1/2}. \tag{4}$$

Then, for a nontrivial solution, $r(t) > 0$ and we define $\theta(t)$
by

$$x_1(t) = r(t) \cos \theta(t), \qquad x_2(t) = r(t) \sin \theta(t). \tag{5}$$

Admittedly this leaves $\theta(t)$ ambiguous in the sense that one
can add or subtract any integral multiple of 2π.   But that
will not matter so long as we choose $\theta(t)$ so that it varies
continuously with   t.

Instead of attempting to write the general solution in
polar coordinates, it is easier to transform the given differential system into polar coordinates.   From Eqs. (4) and
(5), one can easily verify that

$$rr' = x_1 x_1' + x_2 x_2' \tag{6}$$

and

$$r^2\theta' = x_1 x_2' - x_2 x_1'.\tag{7}$$

Problem 3. But x_1 and x_2 satisfy

$$x_1' = a_{11}x_1 + a_{12}x_2$$
$$x_2' = a_{21}x_1 + a_{22}x_2.\tag{1'}$$

Hence Eq. (7) becomes

$$\theta' = a_{21}\cos^2\theta + (a_{22}-a_{11})\cos\theta\sin\theta - a_{12}\sin^2\theta.\tag{8}$$

We are going to find that

$$(a_{22} - a_{11})^2 + 4a_{21}a_{12} < 0.\tag{9}$$

Then it will follow from (8) that θ' is either always posi-tive or always negative. Why? So the point $x(t)$ spirals around the origin in the phase plane as $t \to \infty$.

To establish (9) recall that the roots of the quadratic equation

$$\begin{vmatrix} a_{11}-\lambda & a_{12} \\ a_{21} & a_{22}-\lambda \end{vmatrix} = \lambda^2 - (a_{11}+a_{22})\lambda + a_{11}a_{22} - a_{12}a_{21} = 0\tag{10}$$

are complex. So it follows that the discriminant

$$(a_{11} + a_{22})^2 - 4(a_{11}a_{22} - a_{12}a_{21}) < 0.$$

But this is equivalent to (9).

Figure 5 presents a sketch of trajectories spiraling around the origin and progressing outward as $t \to \infty$ -- the case of $\mu > 0$. The origin is then called an unstable spiral point (or an unstable focus). If $\mu < 0$ all solutions tend to zero as $t \to \infty$ and the origin is then a stable spiral point (or a stable focus).

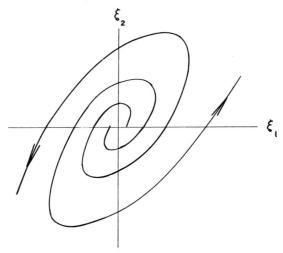

(Unstable) Spiral Point

$$\lambda_1, \lambda_2 = \mu \pm i\omega, \quad \mu > 0, \quad \omega \neq 0$$

Figure 5

Case 6. Conjugate Pure Imaginary Eigenvalues. If $\lambda_1, \lambda_2 = \pm i\omega$ where $\omega \neq 0$ is real, it follows by setting $\mu = 0$ in Eq. (3) that the solutions are periodic with period $2\pi/\omega$. We will find that the trajectories are actually ellipses centered at the origin.

From Eq. (10) and the fact that $\lambda = \pm i\omega$ it follows that

$$a_{22} = -a_{11} \quad \text{and} \quad a_{11}^2 + a_{12}a_{21} = -\omega^2. \tag{11}$$

By straightforward differentiation one then verifies that

$$-a_{21}x_1^2 + 2a_{11}x_1x_2 + a_{12}x_2^2 = \text{constant}. \tag{12}$$

This computation uses system (1') and the first of conditions (11). Problem 4. But the equation

$$A\xi_1^2 + B\xi_1\xi_2 + C\xi_2^2 = \text{constant.}$$

represents a conic section and, as shown in analytic geometry, the conic will be an ellipse (or degenerate) if $B^2 - 4AC < 0$. For Eq. (12) we have

$$B^2 - 4AC = 4a_{11}^2 + 4a_{12}a_{21} = -4\omega^2 < 0.$$

Hence the trajectories represented by Eq. (12) are ellipses.

In this case the origin is called a <u>center</u>.

Typical trajectories for the system

$$x' = \begin{pmatrix} 1 & 2 \\ -1 & -1 \end{pmatrix} x$$

are sketched in Figure 6. The reader should be able to easily verify from the differential system that points travel about

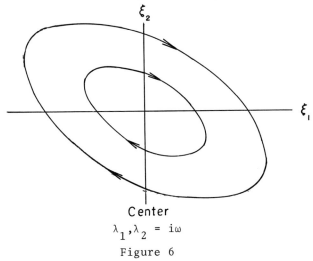

Center
$$\lambda_1, \lambda_2 = i\omega$$
Figure 6

the origin in the clockwise sense as t increases.

Problems

1. Sketch the phase portrait for the system

$$x' = \begin{pmatrix} -1 & -1 \\ 1 & -3 \end{pmatrix} x.$$

The origin is a stable improper node for this system.

2. Discuss the solutions of $x'' = kx$, where k is a positive constant. If this equation represented a real-world
physical phenomenon (which it apparently does not), what
behavior would you expect as $t \to \infty$? Why?

3. Verify Eqs. (6) and (7).

4. Verify Eq. (12).

5. Verify that all solutions of

$$x' = \begin{pmatrix} 1 & 2 \\ -1 & -1 \end{pmatrix} x,$$

the example used for Case 6, satisfy

$$x_1^2(t) + 2x_1(t)x_2(t) + 2x_2^2(t) = \text{constant} \geq 0.$$

6. In all cases where the origin has been described as a
stable node or spiral point, show that the trivial solution of (1) is actually uniformly asymptotically stable.

37. <u>CRITICAL</u> <u>POINTS</u> <u>OF</u> <u>QUASI-LINEAR</u> <u>SYSTEMS</u> <u>OF</u> <u>SECOND</u> <u>ORDER</u>

Consider again the autonomous system

$$x'(t) = f(x(t)),$$

where $f: D \to \mathbb{R}^2$ is locally Lipschitzian on some open set

$D \subset \mathbb{R}^2$.

A critical point $\eta \in D$ is said to be __isolated__ if there are no other critical points arbitrarily close to η. More precisely, η is an isolated critical point if, for some $\delta > 0$, the open ball

$$B_\delta(\eta) \equiv \{\xi \in \mathbb{R}^2 : (\xi_1 - \eta_1)^2 + (\xi_2 - \eta_2)^2 < \delta^2\}$$

contains no critical points other than η itself. We are going to consider the nature of trajectories in the vicinity of an isolated critical point. By a translation of the axes we can always move a critical point to the origin. Thus, without loss of generality, let us assume $\eta = (0,0)$.

Isolated critical points at $(0,0)$ for nonlinear systems will be classified using the same terminology introduced for the linear case. If as $t \to \infty$ each nontrivial solution tends to $(0,0)$ in the ξ_1, ξ_2-plane and has a tangent direction at $(0,0)$, then the origin is called a stable node. If this type of approach occurs as $t \to -\infty$ (instead of $+\infty$), the origin is an unstable node. If in some directions solutions $\to (0,0)$ as $t \to +\infty$ and in other directions solutions $\to (0,0)$ as $t \to -\infty$, the origin is a saddle point. If all trajectories spiral indefinitely around and towards the origin either as $t \to +\infty$ or as $t \to -\infty$ (and they are not closed), the origin is a spiral point. Finally if each trajectory near $(0,0)$ is a closed curve surrounding $(0,0)$, then the origin is called a center.

We shall restrict our attention here to the case when f can be expressed, near the origin, in terms of a nonsingular linear part plus a small perturbation. More specifically our differential system becomes

$$x' = Ax + h(x), \tag{1}$$

where

$$A = \begin{pmatrix} a_{11} & a_{12} \\ a_{21} & a_{22} \end{pmatrix} \quad \text{with} \quad \det A \neq 0 \tag{2}$$

and

$$h(\xi) = \begin{pmatrix} h_1(\xi) \\ h_2(\xi) \end{pmatrix}$$

with

$$\lim_{\xi \to 0} \frac{||h(\xi)||}{||\xi||^{1+p}} = 0 \quad \text{for some} \quad p > 0. \tag{3}$$

This is a slightly stricter condition than that in Corollary 33-D where we had $p = 0$. But note that if $f(\xi)$ can be represented by a Taylor's series near $(0,0)$, as is often the case, then $h(\xi)$ will represent just the terms of second and higher order. In that case condition (3) will be satisfied using any $p \in (0,1)$.

In most cases, if conditions (2) and (3) hold, the nature of the critical point $(0,0)$ of system (1) is determined by the associated <u>linear approximation</u>

$$y' = Ay. \tag{4}$$

For example, if for system (4) the origin is either a stable node or a stable spiral point, then the real parts of the eigenvalues of A must be negative. This implies that the trivial solution of (4) is uniformly asymptotically stable. But then it follows from Corollary 33-D that the trivial solution of (1) is also uniformly asymptotically stable. (This holds even if $p = 0$ in condition (3).) Thus, all trajectories for (1) near the origin tend to the origin as $t \to \infty$.

Regarding the type of critical point system (1) has at the origin we state the following result.

Theorem A. Let conditions (2) and (3) hold and let h be continuously differentiable on D. Then if the origin is a node, saddle point, or spiral point for system (4), it is respectively a node, saddle point, or spiral point for system (1).

The proof of this assertion and more detailed information about critical points can be found in the book by Petrovskiĭ [1964]. Here we shall give the proof for just the one easiest case.

Proof of Theorem A for the Case of a Stable Spiral Point. If the origin is a stable spiral point for (4) then, as noted above, the trivial solution of (1) is uniformly asymptotically stable.

To show that $(0,0)$ is a spiral point for system (1) let us rewrite a typical nontrivial solution in polar notation. As described in Eqs. 36-(4) and 36-(5), $r(t) > 0$ and

$$x_1(t) = r(t) \cos \theta(t), \qquad x_2(t) = r(t) \sin \theta(t).$$

Then, from 36-(7),

$$\theta' = \frac{x_1 x_2' - x_2 x_1'}{r^2}$$

$$= a_{21}\cos^2\theta + (a_{22}-a_{11})\cos \theta \sin \theta - a_{12}\sin^2\theta$$

$$+ \frac{1}{r} \cos \theta \, h_2(r \cos \theta, r \sin \theta)$$

$$- \frac{1}{r} \sin \theta \, h_1(r \cos \theta, r \sin \theta).$$

This can be rewritten as

$$\theta' = F(\theta) + H(r,\theta)$$

where

$$F(\theta) = a_{21}\cos^2\theta + (a_{22}-a_{11})\cos\theta \sin\theta - a_{12}\sin^2\theta$$

and

$$|H(r,\theta)| \leq \frac{1}{r}||h(r\cos\theta, r\sin\theta)||$$

$$\leq 2||h(r\cos\theta, r\sin\theta)||/[|r\cos\theta| + |r\sin\theta|].$$

But it follows from inequality 36-(9),

$$(a_{22} - a_{11})^2 + 4a_{21}a_{12} < 0,$$

that $F(\theta)$ is either always positive or always negative. More specifically, since F is continuous and periodic on $\mathbb{R}$, there exists $\varepsilon > 0$ such that for all θ either

$$F(\theta) > \varepsilon \quad \text{or} \quad F(\theta) < -\varepsilon.$$

But now, by the stability of the trivial solution, if $r(t)$ starts small it remains small. And if $r(t)$ remains sufficiently small condition (3) gives

$$|H(r,\theta)| < \varepsilon,$$

so that θ' is either always positive or always negative. Hence the trajectories spiral around $(0,0)$ as $t \to \infty$. $\square$

Note that in this special case it suffices to have condition (3) with $p = 0$. The fact that this does not suffice in all cases is shown by Problem 1.

Example 1. Consider the equations of the simple pendulum with friction,

$$x_1' = x_2$$

$$x_2' = -\frac{g}{\ell}\sin x_1 - \frac{b}{m}x_2, \tag{5}$$

where m, ℓ, g, and b are positive constants. Applying Taylor's theorem as in Example 33-2, we can rewrite system (5) as

$$x' = \begin{pmatrix} 0 & 1 \\ -g/\ell & -b/m \end{pmatrix} x + \begin{pmatrix} 0 \\ (g/\ell)(\cos \zeta)x_1^3/3! \end{pmatrix}.$$

where ζ is between 0 and $x_1(t)$. This is in the form of Eq. (1), and conditions (2) and (3) are satisfied. The eigenvalues of the matrix A are

$$\lambda = -\frac{b}{2m} \pm [(\frac{b}{2m})^2 - \frac{g}{\ell}]^{1/2}.$$

So it follows that for system (5) the origin is a stable spiral point if $b^2/m^2 < 4g/\ell$ and, using the unproved part of Theorem A, a stable node if $b^2/m^2 \geq 4g/\ell$.

Theorem A says nothing about the nature of the critical point for (1) when the origin is a center for system (4). In this case the nonlinear system may also have a center at $(0,0)$ or it may have a spiral point, as will be found in Problem 4.

Further details on the perturbation of a linear system near a critical point can be found in more advanced books on ordinary differential equations, e.g., Hurewicz [1943], Coddington and Levinson [1955], or Petrovskiĭ [1964].

Problems

1. Consider the system

$$x_1' = -x_1 - \frac{x_2}{\ell n \ (x_1^2+x_2^2)^{1/2}}, \quad x_2' = -x_2 + \frac{x_1}{\ell n \ (x_1^2+x_2^2)^{1/2}}.$$

Verify that this system is in the form of (1) and that

conditions (2) and (3) are satisfied but with $p = 0$.
Then show that the origin is a stable spiral point, even
though for the linear approximation it is a stable proper
node.

2. Assume $b^2/m^2 < 4g/\ell$ in system (5).

(a) Sketch a trajectory for system (5) near $(0,0)$.

(b) Note that system (5) has other critical points such
as $(\pi,0)$ and $(2\pi,0)$. What type of critical points
are these?

(c) What do you conjecture would be the behavior of a
solution of (5) with initial data $x(0) = (0,\omega_0)$
where ω_0 is large?

3. Classify, if possible, the type of critical point at
$(0,0)$ for the systems

(a) $x_1' = a_1 x_1 - b_1 x_1 x_2,$
$x_2' = -a_2 x_2 + b_2 x_1 x_2,$
where a_1, a_2, b_1, b_2 are positive constants.

(b) $x_1' = x_2 + x_1(1 - x_1^2 - x_2^2),$

$x_2' = -x_1 + x_2(1 - x_1^2 - x_2^2).$

(c) $x_1' = x_2 - \mu(x_1^3 - 3x_1)/3,$
$x_2' = -x_1,$
where μ is a positive constant.

4. Consider the system

$$x_1' = -x_2 - x_1(x_1^2 + x_2^2),$$

$$x_2' = x_1 - x_2(x_1^2 + x_2^2).$$

Verify that conditions (2) and (3) are satisfied. Then show, nevertheless, that the origin is a spiral point, even though for the linear approximation (0,0) is a center.

38.   GLOBAL BEHAVIOR FOR SOME NONLINEAR EXAMPLES

In the last section we discussed, mostly without proof, the behavior of trajectories of an almost linear system near an isolated critical point. But these results are of limited value since they only apply to trajectories "sufficiently close" to the critical point. And in practice it is generally difficult to decide how close is sufficiently close.

A considerable theory has been developed for the behavior "in the large" of two-dimensional autonomous systems. But instead of presenting this theory, we shall settle for a discussion of four examples. Examples 1, 2, and 4 arise in famous applied problems, and Example 3 is a popular but artificial illustrative system. To pursue this topic further, the reader can look up "Poincaré-Bendixson theory" in any one of a multitude of more advanced books on ordinary differential equations, e.g., those listed at the end of Section 37.

Each of the examples to be analyzed illustrates a somewhat different method designed for the type of problem at hand.

Example 1, The Undamped Pendulum. The equation of motion for a simple frictionless pendulum moving under the sole external influence of gravity is

$$\theta'' + \frac{g}{\ell} \sin \theta = 0. \tag{1}$$

Problem 24-9 showed that (1) together with any given initial data of the form

$$\theta(0) = \theta_0, \quad \theta'(0) = \omega_0, \tag{2}$$

has a unique solution on the entire axis $(-\infty, \infty)$.

To apply the theorems of this chapter, we consider the first-order system equivalent to (1),

$$x_1' = x_2 \tag{3}$$

$$x_2' = -\frac{g}{\ell} \sin x_1$$

with

$$x_1(0) = \theta_0, \quad x_2(0) = \omega_0, \tag{4}$$

where $x_1(t) \equiv \theta(t)$ and $x_2(t) \equiv \theta'(t)$. System (3) has critical points at $(n\pi, 0)$ for $n = 0, \pm 1, \pm 2, \ldots$

If we now consider, as in Example 24-1,

$$v(t) \equiv \frac{1}{2}x_2^2(t) + \frac{g}{\ell}[1 - \cos x_1(t)], \tag{5}$$

we find

$$v'(t) = x_2(t)x_2'(t) + \frac{g}{\ell}[\sin x_1(t)]x_1'(t) = 0.$$

Thus, for all t, $v(t) = v(0)$, i.e.,

$$\frac{1}{2}x_2^2(t) + \frac{g}{\ell}[1 - \cos x_1(t)] = \frac{1}{2}\omega_0^2 + \frac{g}{\ell}[1 - \cos \theta_0] \equiv K. \tag{6}$$

An equation such as (6) obtained from a given differential system by integration is called a "first integral" for the system.

Let us now assume that

$$-\pi < \theta_0 < \pi \quad \text{and} \quad K = \frac{1}{2}\omega_0^2 + \frac{g}{\ell}[1 - \cos \theta_0] < 2\frac{g}{\ell}. \tag{7}$$

The second condition says that the "energy" is not too great.

Then, from (6) and (7) it follows that for all t,
$\cos x_1(t) > -1$ and $x_2^2(t) < 4g/\ell$. Using the continuity of
x_1, we conclude from this that

$$-\pi < x_1(t) < \pi \quad \text{and} \quad |x_2(t)| < 2(g/\ell)^{1/2} \quad \text{for all t. (8)}$$

Note that the only critical point satisfying (8) is (0,0).

We shall show that $x_1(t)$ takes the value zero repeatedly and that on some of these occasions $x_2(t)$ repeats its
value. The fact that the solution is periodic will then
follow from Theorem 35-C.

Suppose (for contradiction) that there is some $t_1 > 0$
such that

$$\theta(t) = x_1(t) > 0 \quad \text{for all } t > t_1.$$

Then $\theta''(t) = x_2'(t) < 0$ for all $t > t_1$. If θ' ever becomes ≤ 0 we will find $\theta(t) = 0$ a finite time later. Thus
the only possibility is that

$$\theta'(t) = x_2(t) > 0 \quad \text{for all } t > t_1.$$

Thus $x_1(t) = \theta(t)$ is monotone and, by (8), bounded. Hence
$x_1(t)$ must approach a limit as $t \to \infty$. But then it follows
from (6) that $x_2(t)$ must also approach a limit, and by
Theorem 35-B we conclude that $(x_1(t), x_2(t)) \to (0,0)$ as
$t \to \infty$. Also from Eq. (6) one finds

$$\frac{1}{2}x_2^2(t) + \frac{g}{\ell}[1 - \cos x_1(t)] = \lim_{t \to \infty} v(t) = 0,$$

which contradicts the fact that $x_2(t) > 0$.

Similarly one shows that it is also impossible to have
$\theta(t) = x_1(t) < 0$ for all $t >$ some t_1.

Thus $x_1(t)$ becomes zero infinitely often and, from

Eq. (6), each time $x_1(t) = 0$, $x_2(t)$ must take one of the
two values,

$$\pm[\omega_0^2 + \frac{2g}{\ell} (1 - \cos \theta_0)]^{1/2} = \pm\omega_1 .$$

Thus, infinitely often $(x_1(t),x_2(t))$ arrives at a certain
point, either $(0,\omega_1)$ or $(0,-\omega_1)$. Theorem 35-C then assures
that the solution is periodic.

        To graph the trajectories for this example we can use
Eq. (6) (the conservation of energy equation).   Trajectories
are sketched in Figure 1 for four different values of K.
Two of these $(K = \frac{g}{2\ell}$ and $K = \frac{3g}{2\ell})$ satisfy the condition
$K < 2g/\ell$ imposed in (7).   Thus these curves represent
periodic solutions.

        The other two curves in Figure 1 show what happens if
$K \geq 2g/\ell$.   In case $K = 2g/\ell$ the trajectory either <u>is</u> a
critical point or it <u>approaches</u> a critical point as $t \to \infty$.
The critical points in question, $((2k+1)\pi,0)$ where k is an
integer, all correspond to the improbable case of the pendulum at rest in the vertical upward position.

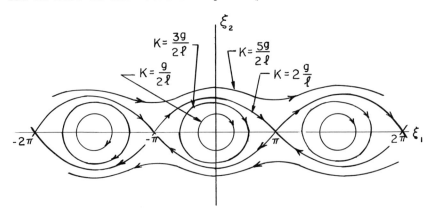

Figure 1

For $K = \dfrac{5g}{2\ell}$ (or any $K > 2g/\ell$) the pendulum has so
much energy that it swings around over the top repeatedly,
i.e., takes values such as π, 3π, 5π, ... In this case
the motion is really periodic again, but our phase plane
trajectory does not properly show this since the points
(ξ_1,ξ_2) and $(\xi_1+2\pi,\xi_2)$ appear as different points.

Example 2, The Lotka-Volterra Equations. The pair of ordinary differential equations

$$x_1' = a_1 x_1 - b_1 x_1 x_2$$

$$x_2' = -a_2 x_2 + b_2 x_1 x_2,$$

(9)

studied by Lotka [1920], [1925] and Volterra [1928], [1931]
was introduced in Section 21. The unknowns $x_1(t)$ and $x_2(t)$
are considered to be the populations of a species of prey and
a species of predators respectively. The coefficients, a_1,
a_2, b_1, and b_2, are positive constants.

Volterra considered system (9) in an attempt to explain
the fluctuations in observed data on the catches of two
species of fish in the Adriatic. The populations of a species
of small fish and a species of larger fish which feed on the
small ones appeared to fluctuate periodically, but out of
phase, as suggested by Figure 2. Presumably after the large
fish (the predators) kill off too many small fish (the prey)
the large fish themselves begin to die out due to food shortage. This in turn gives the small-fish population a chance
to recover. But then the large-fish population begins to
boom again because of the abundance of food. And so on.

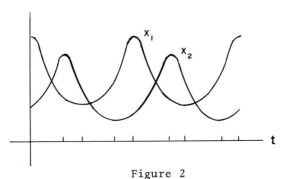

Figure 2

Volterra succeeded in proving that the system (9) does have solutions of this type.

The critical points of system (9) are $(\xi_1, \xi_2) = (0,0)$ and

$$(\xi_1, \xi_2) = (a_2/b_2, \ a_1/b_1). \tag{10}$$

Because of the application we have in mind, it is only relevant to consider the cases when $x_1(t)$ and $x_2(t)$ are non-negative. Thus we begin by considering trajectories near the critical point (10).

Introducing new unknowns

$$X_1 \equiv x_1 - a_2/b_2 \quad \text{and} \quad X_2 \equiv x_2 - a_1/b_1$$

we can translate the critical point in question to the origin. System (9) then becomes

$$X_1' = -(a_2 b_1/b_2) X_2 - b_1 X_1 X_2$$

$$X_2' = \ (a_1 b_2/b_1) X_1 + b_2 X_1 X_2, \tag{11}$$

and the associated linear system,

$$Y_1' = -(a_2 b_1 / b_2) Y_2$$

$$Y_2' = (a_1 b_2 / b_1) Y_1$$

has a center at $(0,0)$. This does not prove that $(0,0)$ is a center for system (11), but we are going to find that it is.

It was proved by Lotka and Volterra that, in fact, all trajectories of the original system (9) which pass through any noncritical point in the first quadrant are closed curves about the critical point (10).

Returning to the original notation of system (9), we note that along a trajectory

$$\frac{d\xi_2}{d\xi_1} = \frac{-a_2 \xi_2 + b_2 \xi_1 \xi_2}{a_1 \xi_1 - b_1 \xi_1 \xi_2} .$$

This is only valid when $\xi_1 \neq 0$ and $\xi_2 \neq a_1 / b_1$, but let us proceed to solve formally by separation of variables. Thus we write

$$\frac{a_1 - b_1 \xi_2}{\xi_2} \frac{d\xi_2}{d\xi_1} = \frac{-a_2 + b_2 \xi_1}{\xi_1}$$

or

$$a_1 \ln \xi_2 - b_1 \xi_2 + C = -a_2 \ln \xi_1 + b_2 \xi_1. \qquad (12)$$

Although its derivation was questionable, the validity of Eq. (12) when $\xi_1 = x_1(t)$ and $\xi_2 = x_2(t)$ can now be verified directly by differentiation with respect to t. Thus (12) provides a first integral for system (9).

The question is, What type of curves in the phase plane does Eq. (12) represent? The following argument is adapted from Plaat [1971].

Let us introduce polar coordinates, $r > 0$ and θ, based at the critical point by defining

$$r \cos \theta \equiv \xi_1 - a_2/b_2, \qquad r \sin \theta \equiv \xi_2 - a_1/b_1.$$

Then consider

$$G(\xi_1,\xi_2) \equiv b_2\xi_1 + b_1\xi_2 - a_2 \ln \xi_1 - a_1 \ln \xi_2$$

$$= a_2 + b_2 r \cos \theta + a_1 + b_1 r \sin \theta$$

$$- a_2 \ln (a_2/b_2 + r \cos \theta) - a_1 \ln (a_1/b_1 + r \sin \theta)$$

$$\equiv H(r,\theta).$$

Now Eq. (12) can be rewritten as

$$H(r,\theta) = C. \tag{13}$$

We are going to show that $H(r,\theta)$ is strictly increasing to $+\infty$ as (ξ_1,ξ_2) moves away from $(a_2/b_2, a_1/b_1)$ along any ray (θ fixed). Then it will follow that for each $C >$ $H(0,0)$ and each θ the equation $H(r,\theta) = C$ has exactly one solution $r > 0$. Hence the curve described by (13) or, equivalently, (12) is a simple closed curve about the critical point $(\xi_1,\xi_2) = (a_2/b_2, a_1/b_1)$. Moreover, the resulting curve is smooth since the partial derivatives of G,

$$\frac{\partial G}{\partial \xi} = b_2 - a_2/\xi_1 \qquad \text{and} \qquad \frac{\partial G}{\partial \xi} = b_1 - a_1/\xi_1,$$

never vanish simultaneously, except at the critical point itself.

The proof that $H(r,\theta)$ increases strictly to ∞ as r increases (with (ξ_1,ξ_2) remaining in the first quadrant) is simply a matter of considering

$$\frac{\partial H}{\partial r}(r,\theta) = \frac{b_2 r \cos^2 \theta}{r \cos \theta + a_2/b_2} + \frac{b_1 r \sin^2 \theta}{r \sin \theta + a_1/b_1}.$$

Clearly $\partial H/\partial r$ is positive. And, for $0 \leq \theta \leq \pi/2$,

$\partial H/\partial r \to b_2 \cos \theta + b_1 \sin \theta > 0$ as $r \to \infty$. While for $\pi/2 < \theta < 2\pi$ it follows directly from the definition of G that $H(r,\theta) = G(\xi_1, \xi_2) \to \infty$ as (ξ_1, ξ_2) approaches one of the coordinate axes.

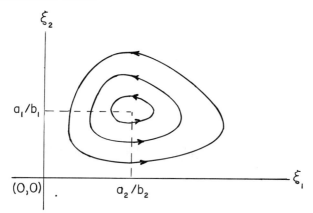

Figure 3

The closed trajectories are indicated in Figure 3, the direction of motion around each trajectory being easily determined from system (9) itself. Each of these trajectories represents a family of periodic solutions differing only by a time translation. A time plot of such a periodic solution could be represented by graphing $x_1(t)$ and $x_2(t)$ against t, and it would look something like Figure 2.

Example 3. Consider the system presented in Problem 37-3(b).

$$x_1' = x_2 + x_1(1 - x_1^2 - x_2^2)$$
$$x_2' = -x_1 + x_2(1 - x_1^2 - x_2^2).$$

(14)

Once again the use of polar coordinates will be helpful. We take $r(t) > 0$, $x_1(t) = r(t) \cos \theta(t)$, and $x_2(t) = r(t) \sin \theta(t)$. Then, with the aid of Eqs. 36-(6) and 36-(7), one finds

$$r' = r(1 - r^2), \qquad \theta' = -1. \tag{15}$$

The reader is asked to find the general solution of system (15),

$$r(t) = (1 + ce^{-2t})^{-1/2}, \qquad \theta(t) = -t + t_1, \tag{16}$$

where c and t_1 are arbitrary real numbers. Problem 1.
Note that if $c < 0$, $r(t)$ is only defined for $t > \frac{1}{2} \ln (-c)$.
The general solution of (14) is, therefore,

$$x_1(t) = \frac{\cos (-t + t_1)}{(1 + ce^{-2t})^{1/2}}, \qquad x_2(t) = \frac{\sin (-t + t_1)}{(1 + ce^{-2t})^{1/2}}. \tag{17}$$

Actually the trajectories are most easily analyzed from (16).
We see that every nontrivial solution winds around the origin
clockwise, making one turn every 2π units of time.   In case
$c = 0$, $r(t) \equiv 1$ so the trajectory is closed.   And this is
the only closed trajectory for system (14).   Thus if $c = 0$
the solutions are periodic with period 2π.   If $c > 0$ the
trajectories spiral outward, as $t \to \infty$, toward the unique
closed trajectory.   And if $c < 0$ the trajectories spiral
inward toward the unique closed trajectory.   See Figure 4.

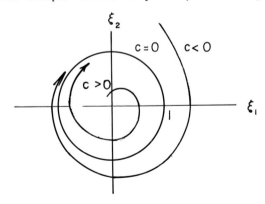

Figure 4

In this example the rate of rotation about the origin is the same for all trajectories. Hence it follows that every non-trivial solution of (14) tends toward some periodic solution as $t \to \infty$.

Definition. If, for some autonomous system, every trajectory which starts sufficiently close to some closed trajectory, C_0, tends to C_0 as $t \to \infty$, C_0 is called a stable limit cycle.

The artificial Example 3 was particularly easy to treat since we could find a simple exact expression for the stable limit cycle. The following example involves a physically motivated differential system and is more diffi-cult to analyze.

Example 4, The van der Pol Equation. Problem 24-11 called for a proof that every solution of the van der Pol equation,

$$z'' + \mu(z^2 - 1)z' + z = 0, \qquad \mu > 0, \tag{18}$$

can be continued to $+\infty$. Let us now try to analyze the be-havior of such solutions.

Instead of working with the usual position and velocity variables, it is customary to follow the procedure used by A. Liénard in 1928. Let $x_1 = z$ and

$$x_2 = z' + \int_0^z \mu(\xi^2 - 1)d\xi = z' + \mu(z^3/3 - z).$$

Then, introducing

$$F(\xi) = \mu(\xi^3 - 3\xi)/3, \tag{19}$$

Eq. (18) is equivalent to the system

$$x_1' = x_2 - F(x_1)$$
$$x_2' = -x_1. \tag{20}$$

The only critical point of system (20) is the origin and it is an unstable spiral point, assuming $\mu < 2$. We are going to show that system (20) has a unique closed trajectory, C_0, enclosing the origin and that every nontrivial trajectory tends to C_0 as $t \to \infty$. This will imply that every nontrivial solution behaves more and more like a periodic solution as $t \to \infty$.

Figure 5 will be the basis for our arguments to establish the existence of a unique closed trajectory. The first thing plotted there was the graph of $\xi_2 = F(\xi_1)$.

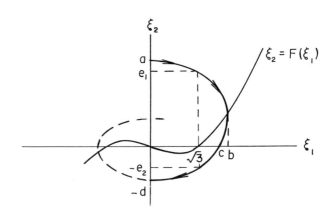

Figure 5

It follows from the first equation of (20) that a point on a trajectory must move to the right (as t increases) if the point is above the curve $\xi_2 = F(\xi_1)$, and to the left if the point is below that curve. Moreover, the rate at which the point moves right or left increases as the point

gets farther away from the curve $\xi_2 = F(\xi_1)$.   Similarly, a
point moves downward if it is to the right of the ξ_2 axis,
and upward if it is to the left.   From these observations,
convince yourself that a solution starting from any point in
the right half plane -- for example, from $(b,F(b))$ for some
$b > 0$ -- must eventually cross the negative ξ_2 axis as t
increases; and (going backwards) it must eventually cross the
positive ξ_2 axis as t decreases.

It also follows that whenever a trajectory crosses the
curve $\xi_2 = F(\xi_1)$ it crosses vertically and when it crosses
the ξ_2 axis it crosses horizontally.

Note also that Eqs. (20) are symmetric with respect to
reflection through the origin.   That is, the replacement of
(x_1,x_2) with $(-x_1,-x_2)$ leaves the equations unchanged.
Hence any information about a trajectory in the right half
plane immediately gives corresponding information about a
trajectory in the left half plane by reflection though $(0,0)$.
Indeed this is one of the advantages in our definition of x_1
and x_2.

Let us follow the progress of a point on a trajectory,
C, starting from $(0,a)$ on the positive ξ_2 axis.   As t
increases, the point moves to the right and downward until it
crosses the curve $\xi_2 = F(\xi_1)$ at $\xi_1 = b$.   Then it moves to
the left and downwards until it meets the positive ξ_1 axis
at $(c,0)$ and eventually the negative ξ_2 axis at $(0,-d)$.
Now if it should happen that $d = a$, the trajectory would continue to the left of the ξ_2 axis as a curve symmetric,
through $(0,0)$, to that already obtained.   Hence C would
again meet the positive ξ_2 axis at $(0,d) = (0,a)$, forming

a closed curve. All solutions corresponding to such a trajectory are periodic.

In general, however, $d \neq a$ as in Figure 5. Then the
trajectory will never meet itself. Why? Thus the corresponding solutions will not be periodic.

For the same reason, any increase in the value of a
or b or d implies an increase for all of them. Thus we
can just as well characterize the trajectory C by the value
of b instead of a.

To show the existence of a closed trajectory we must
show that an appropriate choice of b leads to d = a. Let

$$I(b) \equiv \frac{1}{2} d^2 - \frac{1}{2} a^2.$$

Then we must show that $I(b) = 0$ for some $b > 0$. But

$$I(b) = \int_{t_a}^{t_d} \frac{d}{dt}[\frac{1}{2}x_1^2(t) + \frac{1}{2}x_2^2(t)]dt = \int_{t_a}^{t_d} F(x_1(t))x_2'(t)dt,$$

where t_a and t_d are the instants at which some particular
solution with trajectory C passes through $(0,a)$ and
$(0,-d)$ respectively. This gives

$$I(b) = \int_a^{-d} F(\xi_1)d\xi_2, \tag{21}$$

where the integration is along C.

Note that ξ_2 is decreasing. So if $b \leq \sqrt{3}$, then
$F(x_1(t)) \leq 0$ for $t_a < t < t_d$ and $I(b) > 0$.

If $b > \sqrt{3}$, split the integral into three parts

$$I(b) = I_1 + I_2 + I_3,$$

where, taking all integrals along C, and introducing the

points $(\sqrt{3},e_1)$ and $(\sqrt{3},-e_2)$ indicated on Figure 5,

$$I_1 = \int_a^{e_1} F(\xi_1)d\xi_2, \quad I_2 = \int_{e_1}^{-e_2} F(\xi_1)d\xi_2, \quad I_3 = \int_{-e_2}^{-d} F(\xi_1)d\xi_2.$$

$$(22)$$

Then one finds $I_1 > 0$, $I_2 < 0$, and $I_3 > 0$. With the aid of (20) we can replace $d\xi_2$ in I_1 and I_3 with

$$\frac{d\xi_2}{d\xi_1}d\xi_1 = \frac{-\xi_1}{\xi_2 - F(\xi_1)}d\xi_1.$$

Thus

$$I_1 = \int_0^{\sqrt{3}} \frac{-F(\xi_1)\xi_1}{\xi_2 - F(\xi_1)}d\xi_1 \qquad (\xi_2 > 0)$$

$$(23)$$

$$I_3 = \int_{\sqrt{3}}^0 \frac{-F(\xi_1)\xi_1}{\xi_2 - F(\xi_1)}d\xi_1 \qquad (\xi_2 < 0).$$

Since different trajectories cannot intersect each other, it follows that any increase in b causes an outward expansion of C. Thus, considering I_1 in (23), an increase in b increases the value of ξ_2 at each ξ_1 in $[0,\sqrt{3})$. Hence I_1 decreases. By a similar argument, an increase in b decreases I_3. To determine the effect on I_2 of an increase in b, note that along with b, the values of e_1 and e_2 increase, and so does the value of ξ_1 at each ξ_2. Thus the positive quantity $-I_2$ increases, i.e., I_2 decreases.

In summary, I(b) decreases as b increases; and clearly I(b) depends continuously on b. We have already noted that if $b \leq \sqrt{3}$, then I(b) > 0. To show that I(b) takes the value zero for some unique b > 0 it will suffice to prove that I(b) → -∞ as b → ∞, or more particularly that $-I_2$ → +∞ as b → ∞.

Note that, integrating along C, Eq. (22) gives

$$-I_2 = \int_{-e_2}^{e_1} F(\xi_1)d\xi_2 > \int_0^{F(b)} F(\xi_1)d\xi_2 > F(c)F(b) \to \infty$$

as b → ∞.

Thus there is a unique $b = b_0 > 0$ such that $I(b_0) = 0$, i.e., $d_0 = a_0$. The resulting closed trajectory, C_0, represents infinitely many periodic solutions which differ only by a time translation.

To prove that C_0 is a stable limit cycle we shall show that every nontrivial trajectory spirals either in or out tending to C_0 as $t \to \infty$.

Observe that for $b > b_0$, $I(b) < 0$ so that $d < a$. It follows by the nonintersecting property of trajectories and by the symmetry through $(0,0)$ that when $b > b_0$, the trajectory is outside C_0 and is spiraling inward as t increases.

Similarly, if $b < b_0$ the trajectory is inside C_0 and is spiraling outward as t increases.

Consider any trajectory with $b > b_0$, and let the successive points of intersection with the positive ξ_2 axis be $(0,a)$, $(0,a_1)$, $(0,a_2)$, ... The sequence $\{a_i\}_{i=1}^{\infty}$ must be decreasing and bounded below by a_0 (since the trajectory never intersects itself). Hence

$$\lim_{i \to \infty} a_i = a*$$

exists where $a* \geq a_0$. We shall show that $a* = a_0$. A particular solution having the given trajectory is $x(\cdot;0,(0,a))$. Let $t_1, t_2, \ldots$ be the increasing sequence of instants such that

$$x(t_k;0,(0,a)) = (0,a_k).$$

Consider also the solution

$$x = x(\cdot;0,(0,a*)).$$

Then, since the right hand side of (20) is globally Lipschitzian on any bounded set, we can invoke continuous dependence (e.g., Theorem 8-B or 25-E), to show that for each
$t > 0$

$$\lim_{k \to \infty} x(t;0,x(t_k;0,(0,a))) = \lim_{k \to \infty} x(t;0,(0,a_k)) = x(t;0,(0,a^*)).$$

But, since system (20) is autonomous, this becomes

$$\lim_{k \to \infty} x(t+t_k;0,(0,a)) = x(t;0,(0,a^*)).$$

In other words, every point on the trajectory defined by

$$\{x(t;0,(0,a^*)): t \geq 0\} \tag{24}$$

is the limit of some sequence of points of the trajectory
defined by

$$\{x(t;0,(0,a)): t \geq 0\}. \tag{25}$$

It follows that the trajectory (24) must be closed, which implies $a^* = a_0$. For otherwise trajectory of (24) would return to the positive ξ_2 axis at some point $(0,a^{**})$ where
$a^{**} < a^*$, and then $(0,a^{**})$ could not be approached by points
of (25).

Since there is only one closed trajectory, the trajectory of (24) must be C_0; and the trajectory of (25) must become arbitrarily close to C_0 as $t \to \infty$.

A similar argument shows that all trajectories inside
C_0 spiral out toward C_0 as $t \to \infty$.

Finally, we state two general theorems which relate
to the preceding examples. The interested reader can find
proofs in the books listed on page 429.

Both theorems pertain to the autonomous system

$$x' = f(x), \tag{26}$$

where $f: D \to \mathbb{R}^2$ is locally Lipschitzian on an open set $D \subset \mathbb{R}^2$.

Theorem A (Poincaré).   Each closed trajectory (representing periodic solutions) surrounds at least one critical point of f.

Confirm this for the periodic solutions found in Examples 1, 2, 3, and 4.   See Figures 1, 3, 4, and 5.

Theorem B (Poincaré-Bendixson).   If a solution x   remains inside a closed bounded set $A \subset D$ for all $t \geq t_0$ and   f   has no critical points in   A,   then   x   either _is_ a periodic solution or it _spirals_ (in or out) _toward_ a periodic solution.

Note that, in view of Theorem A, the closed set in Theorem B must surround a "hole" which contains at least one critical point.   For instance, in Example 3 one could take $A = \{\xi \in R^2: a^2 \leq \xi_1^2 + \xi_2^2 \leq b^2\}$ where $0 < a < 1 < b$.

Problems

1.   Show that Eqs. (16) give the general solution of system (15).

2.   Show that the analysis given for Example 4 is also valid for the equation

$$z'' + f(z)z' + z = 0$$

   where   f   is a given continuous function such that $f(-\xi) = f(\xi)$ for all $\xi \in \mathbb{R}$, and if $F(\xi) \equiv \int_0^\xi f(s)ds$ then $F(\xi) = 0$ for one and only one positive number ξ.

APPENDICES

1. NOTATION FOR SETS, FUNCTIONS, AND DERIVATIVES

Some notation used throughout the text is assembled here for ready reference.

Sets

A set can be described either by listing its elements (or members or points), regardless of order, or by stating some properties which characterize the elements. For example, the set of all integers from one through four can be written

$$X = \{1,2,3,4\},$$
$$X = \{x: \quad x \quad \text{an integer, } 1 \le x \le 4\},$$
$$X = \{x \quad \text{an integer: } \quad 0 < x^3 < 117\},$$

or any number of other ways.

The symbol $\mathbb{R}$ stands for the set of all real numbers and $\mathbb{C}$ stands for the set of all complex numbers. For intervals of real numbers we use the notation

$$[\alpha,\beta] = \{x \text{ real:} \quad \alpha \le x \le \beta\}, \quad \text{a closed interval,}$$
$$(\alpha,\beta) = \{x \text{ real:} \quad \alpha < x < \beta\}, \quad \text{an open interval,}$$
$$\left.\begin{matrix} [\alpha,\beta) = \{x \text{ real:} \quad \alpha \le x < \beta\}, \\ (\alpha,\beta] = \{x \text{ real:} \quad \alpha < x \le \beta\}, \end{matrix}\right\} \quad \text{half-open intervals.}$$

We use the standard symbols $\in$, $\subset$, $\cup$, and $\cap$ which are interpreted as follows:

$x \in A$ is read "x is a member (or element) of A",

or "x belongs to A", or "x in A",

A $\subset$ B   is read   "A   is a subset of   B"   or   "A   is
      contained in   B",

A $\cup$ B   is read   "A   union   B"   or   "the union of   A   and
      B", and represents the set of all points belonging
      to   A   or   B   or both,

A $\cap$ B   is read   "A   intersect   B" or "the intersection
      of   A   and   B", and represents the set of all points
      belonging to both   A   and   B.

Two sets, A   and   B, are equal   (A = B)   if and only
if   A $\subset$ B   and   B $\subset$ A.   To negate either of the symbols $\in$
or $\subset$ we write $\notin$ or $\not\subset$ respectively.

Thus for example

$$5 \notin \{x: \quad x \quad \text{an integer,} \quad 0 < x^3 < 117\},$$
$$\{5,7\} \subset \{7,3,5\},$$
$$\{x \in \mathbb{R}: \quad x^2 = 1\} \cup \{i,-i\} = \{z \in \mathbb{C}: \quad z^4 - 1 = 0\},$$
$$\{x \in \mathbb{R}: \quad x^2 + 4x + 3 \leq 0\} \cap \{x \in \mathbb{R}: \quad x^2 > 2\} = [-3,-\sqrt{2}),$$

If we have a family of sets A_ν where ν ranges over some
"index set" Γ, then we denote the union and intersection of
all sets in the family by

$$\bigcup_{\nu \in \Gamma} A_\nu \quad \text{and} \quad \bigcap_{\nu \in \Gamma} A_\nu \quad \text{respectively.}$$

If X_i is a set for each   i = 1,2,...,n   then the
product of these sets,

$$X_1 \times X_2 \times \ldots \times X_n \quad \text{or} \quad \underset{i=1}{\overset{n}{\times}} X_i,$$

is the set of all ordered   n-tuples:

$$\{(x_1,\ldots,x_n): \quad x_i \in X_i \quad \text{for} \quad i = 1,\ldots,n\}.$$

We shall sometimes denote an ordered n tuple $(x_1, \ldots, x_n)$
simply by x. Thus, for example, we have for the "open rectangle":

$$(\alpha, \beta) \times (\gamma, \delta) = \{x = (x_1, x_2): \alpha < x_1 < \beta, \gamma < x_2 < \delta\}.$$

This last equation points up a minor defect in our
notation. The same symbol (,) can mean an open interval
or an ordered pair. Some authors avoid this difficulty by
writing an open interval as $]\alpha, \beta[$ instead of (α, β). However, we shall stick with the parentheses notation for open
intervals and assume that the meaning will always be clear
from the context.

In the special case where each $X_i = X$ we write
simply X^n, instead of $X_{i=1}^n X_i$, for the set of ordered n-
tuples of members of X.

In $\mathbb{R}^n$, an <u>open ball</u> of radius $\delta > 0$ with center at a
point $y = (y_1, \ldots, y_n)$ is defined by

$$B_\delta(y) = \{x = (x_1, \ldots, x_n) \in \mathbb{R}^n: \sum_{i=1}^n |x_i - y_i| < \delta\}.$$

<u>Definition</u>. A set $G \subset \mathbb{R}^n$ is said to be an <u>open set</u> if for
every $y \in G$ there exists a number $\delta > 0$ such that
$B_\delta(y) \subset G$.

It is easy to verify that, if $-\infty < \alpha < \beta < \infty$, the
"open interval" (α, β) is an open set in $\mathbb{R}$, but the inter-
vals $[\alpha, \beta]$, $[\alpha, \beta)$ and $(\alpha, \beta]$ are not open. And in $\mathbb{R}^n$ it
can be shown that an "open rectangle" $X_{i=1}^n (\alpha_i, \beta_i)$ is an
open set, and that an open ball itself is an open set.

<u>Definition</u>. A set $A \subset \mathbb{R}^n$ is said to be a <u>closed set</u> if

every convergent sequence of members of A has its limit in
A. That is, if $x_{(1)}, x_{(2)}, \ldots \in A$ and $\lim_{k \to \infty} x_{(k)} = \xi$,
then $\xi \in A$.

In the simplest case, when $n = 1$ and $-\infty < \alpha < \beta < \infty$,
we find that the "closed interval" $[\alpha, \beta]$ is a closed set, but
the intervals (α, β), $[\alpha, \beta)$, and $(\alpha, \beta]$ are not closed. One
can also show that, if $-\infty < \alpha_i < \beta_i < \infty$ for each i, the
"closed rectangle" $X_{i=1}^{n} [\alpha_i, \beta_i]$ is a closed set in $\mathbb{R}^n$. An
open ball $B_\delta(y)$ is not a closed set in $\mathbb{R}^n$.

Functions

If X and Y are two non-empty sets and if a rule f
is specified which assigns to each x in X a unique element
of Y, called f(x), then f is called a _function_ (or a _map-ping_ or a _transformation_). This will often be abbreviated

$$f: X \to Y.$$

We say that F maps X into Y and we refer to X as the
domain of F.

Note that, strictly speaking, _f stands for the func-tion_, while _f(x) is an element of Y_. But in many books,
including this one, the author occasionally refers to the
function itself as f(x).

The sets X and Y can be other than sets of real or
complex numbers. In such cases one often calls the function
a transformation or mapping. For example, let X =
$C^1([0,1], \mathbb{R})$ be the set of all (real-valued) continuously
differentiable functions on [0,1], with the understanding
that at 0 and 1 the derivatives are "one sided". Let
$Y = C([0,1], \mathbb{R})$ be the set of continuous functions mapping

$[0,1] \to \mathbb{R}$. (Thus $X \subset Y$.) We can define a transformation $D: X \to Y$ by $Dx(t) = x'(t)$, or $Dx = x'$.

If $X = C([0,1],\mathbb{R})$, we can define a mapping $T: \mathbb{R} \times X \to X$ by $(T(a,x))(t) = a + \int_0^t x(s)ds$.

Another special case is that of a transformation from a set $X \not\subset \mathbb{R}$ or $\mathbb{C}$ into $\mathbb{R}$ or $\mathbb{C}$. Such transformations are often called functionals. For example a functional $F: C([0,1],\mathbb{R}) \to \mathbb{R}$ can be defined by

$$F(x) = \int_0^1 x(t)dt.$$

Derivatives

For the derivative of a function $x: (\alpha,\beta) \to \mathbb{R}$, where (α,β) is an interval in $\mathbb{R}$, we shall generally use the notation x'. That is,

$$x'(t) = \lim_{h \to 0} \frac{x(t+h) - x(t)}{h} \quad \text{for} \quad \alpha < t < \beta.$$

On occasion it will be convenient to use Dx or dx/dt instead of x'.

The second derivative will be denoted by x'', or D^2x, or d^2x/dt^2. For the k'th derivative we will write $x^{(k)}$ or D^kx, or sometimes d^kx/dt^k.

If $D = \times_{i=1}^n (\alpha_i,\beta_i)$ is an open "rectangle" in $\mathbb{R}^n$, and v is a function (or mapping) from $D \to \mathbb{R}$, then we shall denote the first partial derivatives of v by D_iv for $i = 1,\ldots,n$ where

$$D_iv(x_1,\ldots,x_n) = \lim_{h \to 0} \frac{v(x_1,\ldots,x_i+h,\ldots,x_n) - v(x_1,\ldots,x_i,\ldots,x_n)}{h},$$

i.e., the partial derivative with respect to the i'th independent variable. We shall sometimes write $\frac{\partial v}{\partial x_i}$ instead of

$D_i v$.

Note that with the D_i notation we will have

$$D_2 v(t,x_1,x_2) = \frac{\partial v}{\partial x_1} (t,x_1,x_2).$$

In other words D_i indicates differentiation with respect to the i'th independent variable--and this need not be the variable called x_i.

2. SOME THEOREMS FROM CALCULUS

We shall state (mostly without proof) several important theorems from calculus. The proofs can be found in any number of books on advanced calculus, and sometimes in introductory calculus texts. The references given below are to Landau [1934] and Fulks [1969].

Theorem A (Mean Value Theorem). Let $[\alpha,\beta]$ be a closed, bounded interval, i.e., $-\infty < \alpha < \beta < \infty$. Let $f: [\alpha,\beta] \to \mathbb{R}$ be continuous and let f' exist on (α,β). Then there exists a number $\theta \in (\alpha,\beta)$ such that

$$f(\beta) - f(\alpha) = f'(\theta)(\beta - \alpha).$$

(Landau, Theorem 159; or Fulks, Theorem 3.10c.)

Theorem B (Taylor's Theorem, or the Generalized Mean Value Theorem). Let $[\alpha,\beta]$ be a closed bounded interval. Let $f: [\alpha,\beta] \to \mathbb{R}$ be such that, for some integer $m \geq 1$, $f^{(m-1)}$ is continuous on $[\alpha,\beta]$ and $f^{(m)}$ exists on (α,β). Then for each t_0 and t in $[\alpha,\beta]$ there exists θ between t_0 and t such that

$$f(t) = \sum_{k=0}^{m-1} \frac{f^{(k)}(t_0)}{k!}(t-t_0)^k + \frac{f^{(m)}(\theta)}{m!}(t-t_0)^m \tag{1}$$

(Landau, Theorems 177 and 178; or Fulks, Theorem 6.3c.)

Note. In case $t_0 = \alpha$ or $t_0 = \beta$, we interpret $f'(t_0), \ldots, f^{(m-1)}(t_0)$ as appropriate one-sided derivatives. Observe that Theorem A is a special case of Theorem B.

Definition. Let $J = (\alpha, \beta) \subset \mathbb{R}$ and let $f: J \to \mathbb{R}$. We shall say f is analytic on J if f has derivatives of all orders and for each t_0 in J there exists $\delta > 0$ such that $(t_0-\delta, t_0+\delta) \subset J$ and

$$f(t) = \sum_{k=0}^{\infty} \frac{f^{(k)}(t_0)}{k!}(t-t_0)^k \quad \text{for all} \quad t \quad \text{in} \quad (t_0-\delta, t_0+\delta). \tag{2}$$

Comparing Eqs. (1) and (2) we see that f is analytic on J if it has derivatives of all orders, and at each t_0 in J we have

$$\lim_{m \to \infty} \frac{f^{(m)}(\theta)}{m!}(t-t_0)^m = 0 \quad \text{for each} \quad t \quad \text{in} \quad (t_0-\delta, t_0+\delta),$$

where $\theta = \theta(t_0, t, m)$ is defined by Eq. (1). In this case, the infinite power series in (2) is called a Taylor's series.

Remark. It should be emphasized that the existence (and continuity) of all derivatives of f is not sufficient for analyticity. For example, if we define $f: \mathbb{R} \to \mathbb{R}$ by

$$f(t) = \begin{cases} 0 & \text{for} \quad t \leq 0 \\ e^{-1/t^2} & \text{for} \quad t > 0, \end{cases}$$

then it can be shown that f has derivatives of all orders for all t and, in fact, $f^{(j)}(0) = 0$ for $j = 1, 2, \ldots$. Thus f cannot be analytic on any open interval I which includes 0, since Eq. (2) would dictate $f(t) \equiv 0$ for

$-\delta < t < \delta$ for some $\delta > 0$. (Cf. Landau, Theorem 185; or Fulks, Section 15.3.) A function of similar form appears in the text in Example 22-5.

Theorem C (<u>Chain Rule for a Function of n Variables</u>). Let $D = \times_{j=1}^{n} (\alpha_j, \beta_j)$ be an open rectangle in $\mathbb{R}^n$, and let f be a real-valued function on D having continuous first partial derivatives. Let J be an open interval in $\mathbb{R}$ and let $g_j: J \to (\alpha_j, \beta_j)$ be differentiable at some t_0 for each $j = 1, \ldots, n$. If

$$F(t) \equiv f(g_1(t), \ldots, g_n(t)) \quad \text{for} \quad t \quad \text{in} \quad J,$$

then F is differentiable at t_0 and

$$F'(t_0) = \sum_{j=1}^{n} (D_j f)(g_1(t_0), \ldots, g_n(t_0)) g_j'(t_0). \tag{3}$$

(See Landau, Theorems 302 and 303 where $n = 2$; or Fulks, Theorems 9.2c and 9.4a.)

Remark. The chain rule theorem when f is a function of a single real variable (the case $n = 1$) has weaker hypotheses. Instead of requiring f to have a continuous derivative, it suffices in that case to require that the derivative of f exist, at the point in question. (Landau, Theorem 101; or Fulks, Theorem 3.9b.) To see that this is <u>not</u> <u>sufficient</u> if $n > 1$, consider the following example with $n = 2$. Let f be defined by

$$f(\xi_1, \xi_2) = \begin{cases} \dfrac{\xi_1 \xi_2^2}{\xi_1^2 + \xi_2^2} & \text{for} \quad (\xi_1, \xi_2) \neq (0,0) \\[2ex] 0 & \text{for} \quad (\xi_1, \xi_2) = (0,0), \end{cases}$$

and let $g_1(t) = g_2(t) = t$. Thus $F(t) = t/2$. Note that D_1f and D_2f exist everywhere, and yet

$$\tfrac{1}{2} = F'(0) \neq D_1f(0,0) \cdot g_1'(0) + D_2f(0,0) \cdot g_2'(0) = 0.$$

The function f <u>is</u> continuous at $(0,0)$, although this is not quite obvious.

An even more bizarre example is defined by

$$f(\xi_1,\xi_2) = \begin{cases} \dfrac{\xi_1\xi_2^2}{\xi_1^2 + \xi_2^4} & \text{for} \quad (\xi_1,\xi_2) \neq (0,0) \\[2ex] 0 & \text{for} \quad (\xi_1,\xi_2) = (0,0). \end{cases}$$

This function has partial derivatives, D_1f and D_2f everywhere. And yet f is not even continuous since, for $t \neq 0$, $f(t^2,t) = \tfrac{1}{2} \neq f(0,0)$ as $t \to 0$.

<u>Definitions</u>. The <u>line segment</u> connecting points ξ and $\tilde{\xi}$ in $\mathbb{R}^n$ is the set of points $\zeta(s) = (1-s)\xi + s\tilde{\xi}$ for $0 \leq s \leq 1$.

A set $A \subset \mathbb{R}^n$ is said to be <u>convex</u> if for each pair of points, ξ and $\tilde{\xi}$ in A, the line segment connecting ξ and $\tilde{\xi}$ is entirely in A.

<u>Theorem D</u> (<u>Mean Value Theorem for a Function of</u> n <u>Variables</u>). Let A be a convex set in $\mathbb{R}^n$, and let $f: A \to \mathbb{R}$ be continuous. Let f have continuous first partial derivatives (except possibly at boundary points of A). If $\xi = (\xi_1,\ldots,\xi_n)$ and $\tilde{\xi} = (\tilde{\xi}_1,\ldots,\tilde{\xi}_n)$ are any two points in A, then there exists a point $(\theta_1,\ldots,\theta_n)$ on the line segment connecting ξ and $\tilde{\xi}$ such that

$$f(\tilde{\xi}) - f(\xi) = \sum_{j=1}^{n} D_j f(\theta_1, \ldots, \theta_n)(\tilde{\xi}_j - \xi_j). \tag{4}$$

(This theorem is essentially in Landau, Theorem 304 for $n = 2$; and in Fulks, Theorem 9.5a. However we shall give a proof here as an application of Theorems A and C. Note that, as in Theorem C, we require continuity of the first partial derivatives of f; not mere existence as in Theorem A.)

Proof. Let $h_j = \tilde{\xi}_j - \xi_j$ for each $j = 1, \ldots, n$, and let us define a new function $F: [0,1] \to \mathbb{R}$ by setting

$$F(t) = f(\xi_1 + th_1, \ldots, \xi_n + th_n) \quad \text{for} \quad 0 \le t \le 1.$$

Then F, being a composition of continuous functions, is itself continuous on $[0,1]$. The existence of F' on $(0,1)$ follows from Theorem C, and (3) gives

$$F'(t) = \sum_{j=1}^{n} D_j f(\xi_1 + th_1, \ldots, \xi_n + th_n) \cdot h_j \quad \text{for} \quad 0 < t < 1.$$

From Theorem A we then conclude that there exists $\theta \in (0,1)$ such that

$$F(1) - F(0) = F'(\theta).$$

But this is (4) with $(\theta_1, \ldots, \theta_n) = (\xi_1 + \theta h_1, \ldots, \xi_n + \theta h_n)$. $\square$

Theorem E (An Implicit Function Theorem). Let $D = (\alpha_1, \beta_1) \times (\alpha_2, \beta_2)$, and let $f: D \to \mathbb{R}$. Let f, $D_1 f$, and $D_2 f$ be continuous on D with $D_2 f(t, \xi) \ne 0$ for all $(t, \xi) \in D$. Let

$$f(t_0, x_0) = 0 \quad \text{for some} \quad (t_0, x_0) \in D.$$

Then it follows that for some $q > 0$ there exists a unique function $x: (t_0 - q, t_0 + q) \to \mathbb{R}$ such that

$$(t, x(t)) \in D \quad \text{and} \quad f(t, x(t)) = 0$$

for $t_0 - q < t < t_0 + q$. Moreover, x is continuously differentiable and

$$x'(t) = - \frac{(D_1 f)(t, x(t))}{(D_2 f)(t, x(t))} . \tag{5}$$

(This is a weak form of Landau's Theorems 314 and 315. Corresponding results in higher dimensions are found in Fulks,
Theorem 10.6a and Section 9.4, Examples 1 and 2.)

If f is a function of two (or more) real variables
we denote mixed second partial derivatives as follows:

$$D_2 D_1 f = D_2 (D_1 f).$$

In general one should expect that $D_2 D_1 f \neq D_1 D_2 f$. For
example, if $f: \mathbb{R}^2 \to \mathbb{R}$ is defined by

$$f(\xi_1, \xi_2) = \begin{cases} \dfrac{\xi_1 \xi_2 (\xi_1^2 - \xi_2^2)}{\xi_1^2 + \xi_2^2} & \text{for } (\xi_1, \xi_2) \neq (0,0) \\ \\ 0 & \text{for } (\xi_1, \xi_2) = (0,0), \end{cases}$$

then a straightforward computation shows that $D_1 D_2 f(0,0) = -1$
while $D_2 D_1 f(0,0) = 1$. (Landau, p. 211; or Fulks, p. 239.)

But under the mild conditions of the next theorem it
turns out that the order of differentiation is interchange-
able.

Theorem F (Interchanging Order of Partial Differentiation).
Let f map some open set in $\mathbb{R}^2$ into $\mathbb{R}$. If f, $D_1 f$, $D_2 f$
and $D_2 D_1 f$ are continuous at (ξ_1, ξ_2), then $D_1 D_2 f(\xi_1, \xi_2)$
exists and

$$D_1 D_2 f(\xi_1, \xi_2) = D_2 D_1 f(\xi_1, \xi_2).$$

(See Landau, Theorem 299; or Fulks, Theorem 9.1a.)

Theorem $\underline{G}$ (<u>Bolzano-Weierstrass Theorem</u>). Let E be a bounded infinite subset of $\mathbb{R}$. (To say that E is bounded means $E \subset (\alpha, \beta)$, where $-\infty < \alpha < \beta < \infty$.) Then there exists at least one point $t_0 \in \mathbb{R}$ and at least one sequence $\{t_k\}$ of members of E such that $\lim_{k \to \infty} t_k = t_0$. (Landau, Theorem 129; or Fulks, Theorem 3.5a.)

Theorem $\underline{H}$ (<u>Heine-Borel Theorem</u> in $\mathbb{R}^n$). Let A be any closed, bounded set in $\mathbb{R}^n$. Let $\{G_\nu : \nu \in \Gamma\}$ be any collection of open sets such that

$$A \subset \bigcup_{\nu \in \Gamma} G_\nu.$$

(We say the collection of sets $\{G_\nu : \nu \in \Gamma\}$ is an <u>open cover-ing</u> of A.) Then there exists a finite subcollection of these open sets, say $\{G_{\nu_1}, \ldots, G_{\nu_p}\}$ such that

$$A \subset \bigcup_{j=1}^{p} G_{\nu_j}$$

This theorem can be rephrased as follows. If A is a closed bounded set in $\mathbb{R}^n$, then every open covering of A has a finite subcollection which is also an open covering of A. (Fulks, Theorem 8.2f.)

<u>Definition</u>. Let S be any set in $\mathbb{R}^n$ and let $f: S \to \mathbb{R}$. We say that f is <u>uniformly continuous</u> on S if for each $\varepsilon > 0$ there exists $\delta > 0$ such that

$$|f(\tilde{\xi}_1, \ldots, \tilde{\xi}_n) - f(\xi_1, \ldots, \xi_n)| < \varepsilon$$

whenever $(\xi_1, \ldots, \xi_n)$ and $(\tilde{\xi}_1, \ldots, \tilde{\xi}_n) \in S$ with

$$||\tilde{\xi} - \xi|| \equiv \sum_{j=1}^{n} |\tilde{\xi}_j - \xi_j| < \delta.$$

(Other books may use different "norms" for measuring distances in R^n. See Eqs. 5-(8a) and 5-(8b) of this text. But the resulting definitions of continuity and uniform continuity, and of closed and open sets, can be shown to be equivalent.)

Theorem I (Continuous Function on a Closed Bounded Set in R^n). Let A be a closed bounded set in R^n, and let f: A → R be continuous. Then f is bounded and uniformly continuous on A. (The special case n = 1 is in Landau, Theorems 145 and 154. The general case, n ≥ 1, is in Fulks, Theorems 8.6b and 8.6d.)

Theorem J (Cauchy Convergence Criterion). A sequence $\{t_k\}$ = $\{t_1, t_2, \dots\}$ of real numbers converges if and only if for every $\varepsilon > 0$ there exists K such that

$$j, k \geq K \quad \text{implies} \quad |t_j - t_k| < \varepsilon.$$

(Equivalent to Landau, Theorems 203 and 206; or Fulks, Theorem 3.6a.)

One part of this theorem is easy to prove. We present that proof in order to review the definition of convergence.

Proof of the easy part of Theorem J. Let the sequence converge to t, i.e., $\lim_{k \to \infty} t_k = t$. Then, for every $\varepsilon > 0$ there exists K such that

$$k \geq K \quad \text{implies} \quad |t - t_k| < \frac{\varepsilon}{2}.$$

Thus if j and k are both ≥ K we have

$$|t_j - t_k| \leq |t_j - t| + |t - t_k| < \frac{\varepsilon}{2} + \frac{\varepsilon}{2} = \varepsilon. \quad \square$$

We shall often consider the limit of a <u>sequence of</u>

functions, and then the notion of "uniform convergence" will be important.

Definition. Let S be any set and let g, g_1, g_2 ... be functions mapping $S \to \mathbb{R}$. Then we say $\{g_\ell\}$ converges uniformly to g on S if for every $\varepsilon > 0$ there exists K (independent of t) such that

$\ell \geq K$ implies $|g(t) - g_\ell(t)| < \varepsilon$ for all $t \in S$.

To appreciate the value of uniform convergence, let us first consider an example in which the convergence is not uniform: The sequence $\{g_\ell\}$ defined by $g_\ell(t) = t^\ell$ for $0 \leq t \leq 1$ ($\ell = 1,2,...$) converges to the function g defined by

$$g(t) = \begin{cases} 0 & \text{for } 0 \leq t < 1 \\ 1 & \text{for } t = 1 \end{cases}$$

It is easy to see that each g_ℓ is continuous while the limit, g, is discontinuous. Sketch the graphs of g_1, g_2, and g_3. With a little work one can show that the convergence of $\{g_\ell\}$ to g is not uniform.

The following important theorem shows that if the convergence had been uniform the limit function could not have been discontinuous. We give the proof in full.

Theorem K (Uniformly Convergent Sequence of Continuous Functions). Let J be an interval, and for each $\ell = 1,2,...$ let g_ℓ be a continuous function mapping $J \to \mathbb{R}$. Let $\{g_\ell\}$ converge uniformly to another function $g: J \to \mathbb{R}$. Then
   (i) g is continuous on J, and
   (ii) if $J = [\alpha, \beta]$ is closed and bounded, then

$$\int_\alpha^\beta g(t)dt = \lim_{\ell \to \infty} \int_\alpha^\beta g_\ell(t)dt.$$

(Part (i) is essentially Landau's Theorem 231; or Fulks'
Theorem 14.3a. Part (ii) is a very weak version of Landau's
Theorem 412; or Fulks' Theorem 14.3c.)

Proof of Theorem K. (i) Let any t_0 in J and any
$\varepsilon > 0$ be given. Choose ℓ, in accordance with the definition
of uniform convergence, sufficiently large so that

$$|g(t) - g_\ell(t)| < \varepsilon/3 \quad \text{for all} \quad t \in J.$$

Then, using this value of ℓ, choose $\delta > 0$ so small that

$t \in J$ with $|t-t_0| < \delta$ implies $|g_\ell(t) - g_\ell(t_0)| < \varepsilon/3$.

It follows that if $t \in J$ with $|t-t_0| < \delta$, then

$$|g(t) - g(t_0)| \leq |g(t) - g_\ell(t)| + |g_\ell(t) - g_\ell(t_0)|$$

$$+ |g_\ell(t_0) - g(t)|$$

$$< \frac{\varepsilon}{3} + \frac{\varepsilon}{3} + \frac{\varepsilon}{3} = \varepsilon.$$

(ii) Now let $J = [\alpha,\beta]$ as indicated. The existence of all
the integrals mentioned follows from the continuity of g,
$g_1, g_2, \ldots$. Let any $\varepsilon > 0$ be given. Choose K, in accordance with the definition of uniform convergence, sufficiently large so that

$\ell \geq K$ implies $|g(t) - g_\ell(t)| < \frac{\varepsilon}{\beta - \alpha}$ for all $t \in [\alpha,\beta]$.

Then, for $\ell \geq K$,

$$\left| \int_\alpha^\beta g(t)dt - \int_\alpha^\beta g_\ell(t)dt \right| \leq \int_\alpha^\beta |g(t) - g_\ell(t)|dt < \varepsilon. \quad \square$$

Remark. The concept of uniform convergence and the results of Theorem K are easily rephrased for infinite series of functions (in place of sequences). One merely applies the above definition and theorem to the sequences of "partial sums" of the series. Thus a series $\sum_{j=1}^{\infty} u_j(t)$ is treated by applying the above to the sequence $\{g_\ell\}$ defined by $g_\ell(t) = \sum_{j=1}^{\ell} u_j(t)$.

Corollary L (Uniformly Convergent Series of Continuous Functions). Let J be an interval, and for each $j = 1,2,\ldots$ let u_j be a continuous function mapping $J \to \mathbb{R}$. Let $\sum_{j=1}^{\infty} u_j$ converge uniformly to another function $g: J \to \mathbb{R}$.
Then

(i)   g   is continuous on   J, and

(ii) if $J = [\alpha,\beta]$ is closed and bounded, then

$$\int_\alpha^\beta g(t)dt = \sum_{j=1}^{\infty} \int_\alpha^\beta u_j(t)dt.$$

Theorem M (Interchanging Order of Iterated Integrals). Let a function v be defined and integrable as a double integral over the rectangle $[t_0,t_1] \times [s_0,s_1]$, and assume

$$\int_{s_0}^{s_1} v(t,s)ds \quad \text{exists for each}\quad t \quad \text{in}\quad [t_0,t_1],\text{ and}$$

$$\int_{t_0}^{t_1} v(t,s)dt \quad \text{exists for each}\quad s \quad \text{in}\quad [s_0,s_1].$$

Then

$$\int_{s_0}^{s_1} \int_{t_0}^{t_1} v(t,s)dtds = \int_{t_0}^{t_1} \int_{s_0}^{s_1} v(t,s)dsdt.$$

(Fulks, Corollary 11.3b.)

Corollary $\underline{N}$. Let v be defined and integrable as a double integral over the triangle

$$\Delta = \{(t,s): t_0 \le t \le t_1, t_0 \le s \le t\} \quad \text{if} \quad t_0 < t_1$$

(or $\Delta = \{(t,s): t_1 \le t \le t_0, t \le s \le t_0\}$ if $t_1 < t_0$).

Assume

$$\int_{t_0}^{t} v(t,s)ds \quad \text{exists for each} \quad t \quad \text{in} \quad [t_0,t_1] \quad (\text{or } [t_1,t_0])$$

and

$$\int_{s}^{t_1} v(t,s)dt \quad \text{exists for each} \quad s \quad \text{in} \quad [t_0,t_1] \quad (\text{or } [t_1,t_0]).$$

Then

$$\int_{t_0}^{t_1} \int_{t_0}^{t} v(u,s)dsdt = \int_{t_0}^{t_1} \int_{s}^{t_1} v(t,s)dtds. \tag{6}$$

Proof. Extend v to the entire rectangle $[t_0,t_1] \times [t_0,t_1]$ (or $[t_1,t_0] \times [t_1,t_0]$) by setting $v(t,s) = 0$ for $(t,s) \notin \Delta$. Then apply Theorem M. The reader should sketch the triangle Δ in the two cases, $t_0 < t_1$ and $t_1 < t_0$, as an aid in verifying Eq. (6). $\square$

REFERENCES

Arrow, K. J., Harris, T., and Marschak, J.
[1951] Optimal inventory policy, Econometrica 19 (1951),
        250-272. MR 13-368.

Babbage, C.
[1816] An essay towards the calculus of functions Part II,
        Philos. Trans. Roy. Soc. London 106 (1816) 179-256.
        Also same Transactions 105 (1815), 389-423 and 107
        (1817) 197-216.

Banks, H. T.
[1969] Representations for solutions of linear functional
        differential equations, J. Differential Equations 5
        (1969), 399-409. MR 38 #3547.

Barba, G.
[1930] Sopra l'equazione funzionale f(x)f'(x)=f[f(x)]
        collegata ad un problema geometrico, Atti Accad. Naz.
        Lincei Rend. Cl. Sci. Fis. Mat. Natur. (6) 11 (1930),
        655-658 and 735-740.

Bellman, R. and Cooke, K. L.
[1963] Differential-Difference Equations, Academic Press,
        New York, 1963. MR 26 #5259.

Boas, R. P.
[1960] A Primer of Real Functions, John Wiley, New York,
        1960. MR 22 #9550.

Boffi, V. and Scozzafava, R.
[1967] A first-order linear differential-difference equa-
        tion with N delays, J. Math. Anal. Appl. 17 (1967),
        577-589. MR 35 #6950.

Brayton, R. K.
[1967] Nonlinear oscillations in a distributed network,
        Quart. Appl. Math. 24 (1967), 289-301.

Callender, A., Hartree, D. R., and Porter, A.
[1936] Time-lag in a control system, Philos. Trans. Roy. Soc.
        London A235 (1936), 415-444.

Chow, S. N.
[1974] Existence of periodic solutions of autonomous func-
        tional differential equations, J. Differential Equa-
        tions 15 (1974), 350-378. MR 49 #779.

Coddington, E. A. and Levinson, N.
[1955] Theory of Ordinary Differential Equations, McGraw-Hill,
        New York, 1955. MR 16-1022.

Cooke, K. L.
  [1963] Differential-difference equations, Internat. Sympos.
          Nonlinear Differential Equations and Nonlinear Mech-
          anics, 155-171. Academic Press, New York, 1963.
          MR 26 #4003.

  [1967] Functional-differential equations: some models and
          perturbation problems, Differential Equations and
          Dynamical Systems, 167-183. Academic Press, New York,
          1967. MR 36 #5461.

Cooke, K. L. and Yorke, J. A.
  [1972] Equations modelling population growth, economic
          growth, and gonorrhea epidemiology, Ordinary Dif-
          ferential Equations, Academic Press, New York, 1972,
          35-53. See also Math. Biosci. 16 (1973), 75-101.
          MR 47 #1478.

Driver, R. D.
  [1962] Existence and stability of solutions of a delay-
          differential system, Arch. Rational Mech. Anal. 10
          (1962) 401-426. MR 25 #5260.

  [1963] A two-body problem of classical electrodynamics: the
          one-dimensional case. Ann. Physics 21 (1963), 122-
          142. MR 27 #1096.

  [1969] A "backwards" two-body problem of classical relativ-
          istic electrodynamics, Phys. Rev. 178 (1969), 2051-
          2057. MR 39 #1770.

  [1972] Some harmless delays, Delay and Functional Differen-
          tial Equations and their Applications, 103-119.
          Academic Press, New York, 1972.

  [1976a] Point data problems for functional differential equa-
          tions, Dynamical Systems, Vol. 2, An Internat. Sym-
          posium, 115-121. Academic Press, New York, 1976.

  [1976b] Linear differential systems with small delays, J.
          Differential Equations 21 (1976), 148-166.

Driver, R. D. and Norris, M. J.
  [1967] Note on uniqueness for a one-dimensional two-body
          problem of classical electrodynamics, Ann. Physics
          42 (1967), 347-351. MR 35 #3962.

Driver, R. D., Sasser, D. W., and Slater, M. L.
  [1973] The equation x'(t) = ax(t)+bx(t-τ) with "small" de-
          lay, Amer. Math. Monthly 80 (1973), 990-995. MR 48
          #4449.

Èl'sgol'c, L. È. (Эльсгольц, Л. Э.)
  [1955] Qualitative Methods in Mathematical Analysis, Ameri-
          can Math. Soc., Providence, 1964. (Translated from
          Russian original, 1955.) MR 30 #289.

[1964]   Introduction to the Theory of Differential Equations
         with Deviating Arguments, Holden-Day Inc., San
         Francisco, 1966 (translated from Russian original,
         1964). MR 30 #290. A revised edition by L. E.
         El'sgol'c and S. B. Norkin (1971) has also been trans-
         lated. Academic Press, New York, 1973. MR 50 #5134.

Ergen, W. K.
[1954]   Kinetics of the circulating-fuel nuclear reactor, J.
         Appl. Phys. 25 (1954), 702-711.

Fox, L., Mayers, D. F., Ockendon, J. R., and Tayler, A. B.
[1971]   On a functional differential equation, J. Inst. Math.
         Appl. 8 (1971), 271-307. MR 46 #488.

Fulks, W.
[1969]   Advanced Calculus: An Introduction to Analysis 2nd
         ed., John Wiley, New York, 1969.

Goodwin, R. M.
[1951]   The nonlinear accelerator and the persistence of
         business cycles, Econometrica 19 (1951), 1-17.

Gorjačenko, V. D. (Горяченко, В. Д.)
[1971]   Methods in the Theory of Stability in Nuclear Reactor
         Dynamics (Russian), Atomizdat, Moscow 1971.

Grodins, F. S., Buell, J., and Bart, A. J.
[1967]   Mathematical analysis and digital simulation of the
         respiratory control system, J. Appl. Physiology 22
         (1967), 260-276.

Gronwall, T. H.
[1918]   Note on the derivatives with respect to a parameter
         of the solutions of a system of differential equa-
         tions. Ann. of Math. 20 (1918) 292-296.

Grossman, S. E. and Yorke, J. A.
[1972]   Asymptotic behavior and exponential stability crit-
         eria for differential delay equations, J. Differen-
         tial Equations 12 (1972), 236-255. MR 49 #11006.

Gumowski, I.
[1962]   Calcul de la réponse transitoire à une onde carrée
         d'un amplificateur à transistors non linéarire. C. R.
         Acad. Sci. Paris 255 (1962), 2736-2738.

Gurtin, M. E. and Sternberg, E.
[1962]   On the linear theory of viscoelasticity, Arch. Ra-
         tional Mech. Anal. 11 (1962), 291-356. MR 26 #4565.

Halanay, A.
[1966]   Differential Equations: Stability, Oscillations,
         Time Lags, Academic Press, New York, 1966. MR 35
         #6938.

Hale, J.
[1971]   Functional Differential Equations, Springer-Verlag,
         New York, 1971.

Hartman, P.
[1964] Ordinary Differential Equations, John Wiley, New York,
        1964. MR 30 #1270.

Hoppensteadt, F. and Waltman, P.
[1970] A problem in the theory of epidemics, Math. Biosci.
        9 (1970), 71-91. MR 44 #7083. Part II, Math. Biosci.
        12 (1971), 133-145. MR 46 #1373.

Hsing, D. K.
[1977] An existence and uniqueness theorem for the one-
        dimensional backward two-body problem of electro-
        dynamics, Phys. Rev., to appear.

Hurewicz, W.
[1943] Lectures on Ordinary Differential Equations, M.I.T.
        Press, Cambridge, 1958 (from original mimeographed
        notes, 1943). MR 19-855.

Jarník, J. and Kurzweil, J.
[1975] Ryabov's special solutions of functional differential
        equations, Boll. Un. Mat. Ital. (4) 11 (1975), no. 3,
        suppl. 198-208.

Johnson, R. A.
[1972] Functional equations, approximations, and dynamic
        response of systems with variable time delay, IEEE
        Trans. Automatic Control AC-17 (1972), 398-401.

Jones, G. S.
[1962] The existence of periodic solutions of f'(x) =
        -αf(x-1)[1+f(x)], J. Math. Anal. Appl. 5 (1962),
        435-450. MR 25 #5234.

Kakutani, S. and Markus, L.
[1958] On the non-linear difference-differential equation
        y'(t)=[A-By(t-τ)]y(t), Contributions to the Theory of
        Nonlinear Oscillations 4 (1958), 1-18. MR 21 #755.

Kalecki, M.
[1935] A macrodynamic theory of business cycles, Econometrica
        3 (1935), 327-344.

Kamke, E.
[1944] Differentialgleichungen. Lösungsmethoden und Lösungen.
        Band I. 3rd ed. Akademische Verlagsgesellschaft,
        Leipzig, 1944. MR 9-33. See Chapter 10.

Kaplan, J. L. and Yorke, J. A.
[1975] On the stability of a periodic solution of a dif-
        ferential delay equation, SIAM J. Math. Anal. 6 (1975),
        268-282. MR 50 #13812.

Krasovskiĭ, N. N. (Красовский, Н. Н.)
[1956] On the application of the second method of A. M.
        Lyapunov to equations with time delays (Russian),
        Prikl. Mat. Meh. 20 (1956), 315-327. MR 18-128.

[1959]   Stability of Motion, Stanford Univ. Press, Stanford,
         1963 (translated from Russian original, 1959). MR
         21 #5047 and 26 #5258.

Lakshmikantham, V. and Leela, S.
[1969]   Differential and Integral Inequalities Vol. II,
         Academic Press, New York, 1969.

Landau, E.
[1934]   Differential and Integral Calculus, Chelsea Publish-
         ing Co., N.Y., 1951 (translated from German original,
         1934).

Levin, J. J. and Nohel, J. A.
[1960]   On a system of integro-differential equations occur-
         ring in reactor dynamics, J. Math. Mech. 9 (1960),
         347-368. MR 22 #8301. Part II, Arch. Rational Mech.
         Anal. 11 (1962), 210-243. MR 26 #5390.

London, W. P. and Yorke, J. A.
[1973]   Recurrent outbreaks of measles, chickenpox and mumps,
         I and II, Amer. J. Epidemiology 98 (1973), 453-482.

Lotka, A. J.
[1920]   Analytical note on certain rhythmic relations in
         organic systems, Proc. Nat. Acad. Sci. U.S.A. 6
         (1920), 410-415.

[1925]   Elements of Mathematical Biology, Dover Publications,
         N.Y., 1956 (originally published in 1925 as Elements
         of Physical Biology). MR 20 #782.

Lotka, A. J. and Sharpe, F. R.
[1923]   Contributions to the analysis of malaria epidemio-
         logy, Am. J. Hygiene 3, January Supplement (1923),
         1-121.

Lyapunov, A. M. (Ляпунов, А. М.)
[1892]   Problème Général de la Stabilité du Mouvement,
         Princeton Univ. Press, Princeton, N. J., 1947.
         (translated from Russian original, 1892). MR 9-34.

Manitius, A.
[1974]   Mathematical models of hereditary systems, Report
         CRM-462, Centre de Recherches Mathématiques,
         Université de Montréal, Montréal, Québec, 1974.

Martynjuk, D. I. (Мартынюк, Д. И.)
[1971]   Lectures on the Theory of Stability of Solutions of
         Systems with Aftereffect (Russian), Akad. Nauk
         Ukrain. SSR Inst. Mat., Kiev, 1971.

Massera, J. L.
[1949]   On Liapunoff's conditions of stability, Ann. of
         Math. 50 (1949), 705-721. MR 11-721.

Melzak, Z. A.
   [1961]   Some mathematical problems in retrograde nerve degen-
            eration. J. Math. Anal. Appl. 2 (1961), 264-272.
            MR 24 #B420.

Minorsky, N.
   [1947]   Experiments with activated tanks, Trans. ASME 69
            (1947), 735-747.

   [1948]   Self-excited mechanical oscillations, J. Appl. Phys.
            19 (1948), 332-338. MR 9-511.

   [1962]   Nonlinear Oscillations, D. Van Nostrand Co., Inc.,
            Princeton, 1962. See Chapter 21. MR 25 #1339.

Mitropol'skiĭ, Ju. A. and Martynjuk, D. I. (Митропольский,
Ю. А., Мартынюк, Д. И.)
   [1969]   Lectures on the Theory of Oscillations of Systems
            with Delay (Russian), Akad. Nauk Ukrain. SSR Inst.
            Mat., Kiev, 1969. MR 42 #629.

Myškis, A. D. (Мышкис, А. Д.)
   [1949]   General theory of differential equations with retarded
            argument, in Amer. Math. Soc. Translations, Series 1,
            Vol. 4 (1962), 207-267 (translated from Russian ori-
            ginal, 1949). Supplementary Bibliographical Mater-
            ials, Uspehi Mat. Nauk 5 (2) (1950), 148-154. MR
            11-365, MR 13-752, and MR 14-285.

   [1951]   Linear Differential Equations with retarded argu-
            ment (Russian), Second edition, Izdat. "Nauka",
            Moscow 1972 (revised and enlarged version of the
            1951 original). MR 14-52, MR 17-497, and MR 50 #5135.

Myškis, A. D. and El'sgol'c, L. È. (Мышкис, А. Д., Эльсгольц)
   [1967]   The status and problems of the theory of differential
            equations with deviating argument, Russian Math.
            Surveys 22 (1967), 19-57 (translated from Russian
            original, 1967). MR 35 #504.

Norkin, S. B. (Норкин, С. Б.)
   [1965]   Differential Equations of the Second Order with
            Retarded Argument, Translations of Mathematical Mono-
            graphs, Vol. 31. American Math. Soc., Providence,
            1972. (from Russian original, 1965). MR 33 #7656.

Oğuztöreli, M. N.
   [1966]   Time-Lag Control Systems, Academic Press, New York,
            1966. MR 36 #484.

Peano, G.
   [1885]   Sull'integrabilità delle equazioni differenziali di
            primo ordine, Atti R. Accad. Torino 21 (1885/1886),
            677-685.

Petrovskiĭ, I. G. (Петровский, И. Г.)
  [1964]   Ordinary Differential Equations, Prentice-Hall,
           Englewood Cliffs, 1966 (translated from Russian 5th
           edition, 1964). MR 30 #5005 and 33 #1518.

Pinney, E.
  [1958]   Ordinary Difference-Differential Equations, Univ. of
           California Press, Berkeley, 1958. MR 20 #4065.

Pitt, H. R.
  [1944]   On a class of integro-differential equations, Proc.
           Cambridge Philos. Soc. 40 (1944), 199-211. MR 6-273.

  [1947]   On a class of linear integro-differential equations,
           Proc. Cambridge Philos. Soc. 43 (1947), 153-163,
           MR 9-40.

Plaat, O.
  [1971]   Ordinary Differential Equations, Holden-Day, San
           Francisco, 1971.

Placzek, G.
  [1946]   On the theory of the slowing down of neutrons in
           heavy substances, Phys. Rev. 69 (1946), 423-438.
           MR 8-30.

Poisson, S. D.
  [1806]   Sur les équations aux différences mêlées, Journal
           de l'Ecole Polytechnique, Paris, (1) 6, cahier 13
           (1806), 126-147.

Popov, V. M.
  [1971]   Pointwise degeneracy of linear time-invariant, delay-
           differential equations, J. Differential Equations 11
           (1972), 541-561 (expanded version of Tech. Rept.
           R-71-03, Dept. of Electrical Engineering, Univ. of
           Maryland, College Park, 1971). MR 45 #5515.

Reid, W. T.
  [1930]   Properties of solutions of an infinite system of or-
           dinary linear differential equations of the first
           order with auxiliary boundary conditions, Trans.
           Amer. Math. Soc. 32 (1930), 284-318, see p. 296.

Repin, Ju. M. (Репин, Ю. М.)
  [1957]   On stability of solutions of equations with retarded
           argument (Russian), Prikl. Mat. Meh. 21 (1957), 253-
           261. MR 19-745.

Sherman, B.
  [1960]   The difference-differential equation of electron
           energy distribution in a gas, J. Math. Anal. Appl. 1
           (1960), 342-354. MR 23 #B1961.

Shimanov, S. N. (Шиманов, С. Н.)
  [1960]   On stability in the critical case of a zero root for
           systems with time lag, J. Appl. Math. Mech. 24, 653-
           668 (translated from Russian original, 1960). MR 22
           #9697.

Silberstein, L.
  [1940] On a hystero-differential equation arising in a pro-
          bability problem, Philos. Mag. 29 (1940), 75-84.
          MR 1-150.

Simmons, G. F.
  [1972] Differential Equations with Applications and Histori-
          cal Notes, McGraw-Hill, New York, 1972.

Stokes, A.
  [1962] A Floquet theory for functional differential equa-
          tions, Proc. Nat. Acad. Sci. U.S.A. 48 (1962), 1330-
          1334. MR 25 #5255.

Travis, S. P.
  [1975] A one-dimensional two-body problem of classical
          electrodynamics, SIAM J. Appl. Math. 28 (1975), 611-
          632. MR 51 #2472.

de Visme, G. H.
  [1961] The density of prime numbers, Math. Gaz. 45 (1961),
          13-14. MR 23 #A872.

Volterra, V.
  [1909] Sulle equazioni integro-differenziali della teoria
          dell'elasticità, Atti della Reale Accademia dei
          Lincei 18 (1909), 295-301. Reprinted in Vito Volterra,
          Opera Matematiche; Memorie e Note Vol. 3, Accademia
          Nazionale dei Lincei (Rome) 1957, pp. 288-293.
          MR 19-827.

  [1928] Variations and fluctuations of the number of indivi-
          duals in animal species living together, in Animal
          Ecology by R. N. Chapman, pp. 409-448. McGraw-Hill,
          New York, 1931. (Translated from Italian original,
          1928.)

  [1931] Leçons sur la Théorie Mathématique de la Lutte Pour
          la Vie, Gauthier-Villars, Paris, 1931.

Wangersky, P. J. and Cunningham, W. J.
  [1957] Time lag in prey-predator population models, Ecology
          38 (1957), 136-139.

Wilson, E. B. and Burke, M. H.
  [1942] The epidemic curve, Proc. Nat. Acad. Sci. U.S.A. 28
          (1942), 361-366. MR 4-201.

Winston, E. and Yorke, J. A.
  [1969] Linear delay differential equations whose solutions
          become identically zero, Rev. Roumaine Math. Pures
          Appl. 14 (1969), 885-887. MR 40 #7603.

Wright, E. M.
  [1945] On a sequence defined by a non-linear recurrence for-
          mula, J. London Math. Soc. 20 (1945), 68-73. MR 7-431.

[1948]   Linear difference-differential equations, Proc.
         Cambridge Philos. Soc. 44 (1948), 179-185. MR 10-125.

[1949]   The linear difference differential equation with con-
         stant coefficients. Proc. Roy. Soc. Edinburgh, A, 62
         (1949), 387-393. MR 11-182.

[1955]   A non-linear difference-differential equation, J.
         Reine Angew. Math. 194 (1955), 66-87. MR 17-272.

[1961]   A functional equation in the heuristic theory of
         primes, Math. Gaz. 45 (1961), 15-16. MR 23 #A873.

Yorke, J. A.
[1969]   Noncontinuable solutions of differential-delay equa-
         tions, Proc. Amer. Math. Soc. 21 (1969), 648-652.
         MR 39 #1774.

[1971]   Selected topics in differential delay equations,
         Japan-United States Seminar on Ordinary Differential
         and Functional Equations, pp. 16-28. Lecture Notes
         in Mathematics No. 243, Springer-Verlag, Berlin, 1971.

Ždanov, V. I. (Жданов, В. И.)
[1975]   The one-dimensional symmetric two-body problem in
         classical electrodynamics (Russian), to appear.
         See IV Vsesojuznaja Konferencija po Teorii i
         Priloženijam Differencial. Uravneniǐ s Otklon.
         Argumentom, Abstracts of Reports, 91, 92. Izdat.
         "Naukova Dumka", Kiev, 1975.

Zverkin, A. M. (Зверкин, А. М.)
[1959]   Dependence of the stability of solutions of differ-
         ential equations with a delay on the choice of the
         initial instant (Russian), Vestnik Moskov. Univ. Ser.
         Mat. Meh. Astr. Fiz. Him. 1959, no. 5, 15-20.
         MR 22 #4858.

[1965]   Series expansion of solutions of linear differential-
         difference equations, Part I (Russian), Trudy Sem.
         Teor. Differencial Uravneniǐ s Otklon. Argumentom
         Univ. Družby Narodov Patrisa Lumumby 3 (1965), 3-38.
         MR 34 #1646. Part II, Same Trudy 5 (1967), 3-50.
         MR 36 #2923.

[1971]   Pointwise completeness of systems with delay,
         Differential Equations 9 (1973), 329-333 (translated
         from Russian original, 1973). Results announced at
         VII Naucnaja Konferencija Fakul'teta Fiziko-Mat. i
         Estestvennyx Nauk, Univ. Družby Narodov Patrisa
         Lumumby, Abstracts of Reports, pp. 24-27. 1971.
         MR 47 #5403.

Zverkin, A. M., Kamenskiĭ, G. A., Norkin, S. B., and
El'sgol'c, L. È. (Зверкин, А. М., Каменский, Г. А.,
Норкин, С. Б., Эльсгольц, Л. Э.)
  [1962] Differential equations with deviating argument.
         Russian Math. Surveys 17, No. 2, 61-146 (translated
         from Russian original, 1962). MR 25 #4203. Part II
         (Russian), Trudy Sem. Teor. Differencial. Uravneniĭ s
         Otklon. Argumentom Univ. Družby Narodov Patrisa
         Lumum 2(1963), 3-49. MR 31 #4963.

ANSWERS AND HINTS

Section 2

3. (a) $x(t) = (\sin t - \cos t)/2 + ce^{-t}$.

   (b) The multiplier of c is e^{-t}, which is never zero.

4. $x(t) = 4e^{t-2} - 1$, $J = (-\infty,\infty)$. Unique.

5. $x(t) = (x_0+1)e^{-\cos t} + \cos t_0 - 1$, $J = (-\infty,\infty)$. Unique.

6. $x(t) = t/2 - 1/4 + ce^{-2t}$, $J = (-\infty,\infty)$. c is arbitrary.

7. $i(t) = \begin{cases} (E/R)(1 - e^{-Rt/L}) + i_0 e^{-Rt/L} & \text{if } R \neq 0. \\ Et/L + i_0 & \text{if } R = 0. \end{cases}$

   In each case $J = (-\infty,\infty)$ and solution is unique.

8. (a) $x(t) = t/2 + 1 - 3/2t$, $J = (0,\infty)$. Unique.

   (b) $x(t) = t/2 + 1 - 3/2t$, $J = (-\infty, 0)$. Unique.

   (c) The domains of definition are different.

9. $x(t) = 5e^{-t^2} + e^{-t^2}\int_0^t e^{s^2} ds$, $J = (-\infty,\infty)$. Unique.

10. $x(t) = t \ln |t| + ct$, $J = (-\infty,0)$ or $(0,\infty)$. c arbitrary.

11. $x(t) = \frac{1}{2}(t-1) + (t-1)\ln\frac{1-t}{2} - 1$, $J = (-\infty,1)$. Unique.

12. $x(t) = \sin t + \cos t$, $J = (-\pi/2, \pi/2)$. Unique.

13. $x(t) = \begin{cases} \sin t + \cos t & \text{for } t \in (-\pi/2, \pi/2) \\ \sin t + c \cos t & \text{for } t \in (\pi/2,3\pi/2) \text{ for any } c. \end{cases}$

   No contradiction since $a(t) = \tan t$ undefined at $t = \frac{\pi}{2}$.

14. $a = 0.0246$. 93 years.   15. $2 \ln 50 = 7.8$ min.

Section 3

3. $x'(t) = M(t)/[h'(h^{-1}[\int_{t_0}^t M(s)ds])] = \dfrac{M(t)}{h'(x(t))} = -\dfrac{M(t)}{N(x(t))}$.

8. (c) $x(t) = 1/(8-t)$, $J = (-\infty,8)$. Unique.

   (d) $x(t) = (7+2t-t^2)^{1/2}$, $J = (1-2\sqrt{2}, 1+2\sqrt{2})$ --- a semi-circle. Unique.

   (e) $x(t) = \begin{cases} t^2/4 & \text{for } t \geq 0 \\ -t^2/4 & \text{for } t \leq 0. \end{cases}$ Not unique. Another

   solution is $x(t) \equiv 0$. Find still a third solution.

(f) $x(t) = \pm(c^2-t^2)^{1/2}$, $J = (-|c|,|c|)$ in each case --
   two semicircles for each c. $c \neq 0$ arbitrary.

(g) $x(t) = -1/t$, $J = (0,\infty)$. Unique.

(h) $x(t) = \frac{c}{t}e^{-1/t}$, $J = (-\infty,0)$ or $(0,\infty)$. c arbitrary.

Section 4

1. Unique only on $(-3,\infty)$.

3. (a) $D = (0,\infty) \times \mathbb{R}$.   (b) $D = \mathbb{R}^2$.

   (c) $D = (-\frac{\pi}{2}, \frac{\pi}{2}) \times \mathbb{R}$.   (d) $D = \mathbb{R}^2$.

   (e) $D = \mathbb{R} \times (0,\infty)$.   (f) $D = \mathbb{R} \times (-\infty,0)$.

   (g) $D = \mathbb{R}^2$.   (h) $D = \mathbb{R}^2$.

4. $(0,\infty)$.

6. The function f defined by $f(t,\xi) = \xi^{2/3}$ does not satisfy
   a Lipschitz condition on any rectangle containing $(t,0)$.
   For if it did, with Lipschitz constant K, we would have
   $|f(t,\xi) - f(t,0)| \leq K|\xi|$ or $|\xi^{2/3}| \leq K|\xi|$ for all
   sufficiently small $|\xi|$. This would imply $|\xi^{-1/3}| \leq K$
   when $|\xi|$ is small, which is nonsense.

7. Prove that $f(t,\xi) = 1 + \xi^{2/3}$ is not Lipschitzian as in
   Problem 6. But now the solution is unique.

9. (a) $p/q \geq 1$.   (b) If $p/q > 1$, the solution is valid on
   $(-\infty, \frac{q}{p-q} x_0^{(q-p)/q})$. If $p/q \leq 1$, the solution is valid
   (but not necessarily unique) on $(-\infty,\infty)$.

Section 5

6. Given any $(t_0,x_0) = (t_0,x_{01},x_{02}) \in D$, there is at most one
   solution of $x' = f(t,x)$ -- on any subinterval of $(-1,2)$
   -- with $x(t_0) = x_0$.

7. (b) $K = 8$.

Section 6

5. If $x(t_0) = b_0$ and $x'(t_0) = b_1$ are given, there is at
   most one solution

   (a) on the interval $(n\pi, n\pi+\pi)$ containing t_0 where
       $n = 0$, or ± 1, or $\pm 2,\ldots$ (assuming t_0 is not a mul-
       tiple of π).

   (b) any open interval containing t_0.

   (c) any open interval (containing t_0) on which $x(t) \neq 0$
       (assuming $b_0 \neq 0$).

   (d) any open interval (containing t_0) on which $x'(t) \neq 1$
       (assuming $b_1 \neq 1$).

   (e) any open interval (containing $t_0 = 2$) on which $x(t) > 0$.

Section 8

2. This problem illustrates a common mistake made by beginners.
   You cannot differentiate an inequality. For example
   $\sin t \leq 1$ certainly does not imply $\cos t \leq 0$.

3. $|x(t) - \tilde{x}(t)| = |x_0 - \tilde{x}_0| e^{-\int_{t_0}^{t} a(s)ds}$. This is smaller, in gen-
   eral, than the right hand side of inequality 8-(5).

5. Theorem B gives $|x(t) - \tilde{x}(t)| \leq (|x(t_0) - \tilde{x}(t_0)| +$
   $|x'(t_0) - \tilde{x}'(t_0)| + |x''(t_0) - \tilde{x}''(t_0)|)e^{2|t-t_0|}$. (A sharper
   estimate can be obtained by direct solution of $x''' - x'' = 0$.)

6. (a) Use Theorem 4-A with $D = (-\infty, 1) \times \mathbb{R}$.

   (b) If $x(0) = x_0 \neq 1$ (by any slight amount) the solution
       of the linear equation is found to be
       $$x(t) = 1 + (x_0 - 1)/(1-t) \quad \text{which} \quad \to \pm\infty \quad \text{as} \quad t \to 1-.$$

   (c) No contradiction since f is not Lipschitzian on D.

7. (a) The solution $x(t) \equiv 1$ is unique as before.

   (b) If $x(0) = x_0 \neq 1$ the solution is
       $$x(t) = 1 + (x_0 - 1)e^{(3/2)[1-(1-t)^{2/3}]}.$$

Here $|x(t)-1| \leq |x_0-1|e^{3/2}$ for all $t < 1$.

(c) Again f is not Lipschitzian on $D = (-\infty,1) \times \mathbb{R}$. But this time the solution does "depend continuously" on x_0. While the inequality in (b) does not follow from Theorem 8-B, it could be proved directly from Lemma 8-A with $k(s) = (1-s)^{-1/3}$.

Section 10

6. (a) $y(t) = c_1e^{3t} + c_2e^{-3t}$.

(b) $y(t) = c_1\cos 3t + c_2\sin 3t$.

(c) $y(t) = c_1e^{3t} + c_2te^{3t}$.

(d) $y(t) = c_1e^{-2t}\cos \sqrt{5}\ t + c_2e^{-2t}\sin \sqrt{5}\ t$.

(e) $y(t) = c_1e^{-t} + c_2te^{-t} + c_3t^2e^{-t}$.

(f) $y(t) = c_1e^{t} + c_2te^{t} + c_3\cos t + c_4\sin t$.

(g) $y(t) = c_1\cos t + c_2\sin t + c_3t \cos t + c_4t \sin t$.

7. (a) $y(t) = e^{3t} - e^{-3t}$.

(b) $y(t) = -\frac{1}{3} \cos 3t - 2 \sin 3t$.

(c) $y(t) = -6e^{-3}e^{3t} + 5e^{-3}te^{3t}$.

8. (a) $(D+tI)(D+t^{-1}I) = D^2 + (t+t^{-1})D + (1-t^{-2})I$.

(b) $x(t) = t + c_1t^{-1}e^{-t^2/2} + c_2t^{-1}$ for $t \neq 0$.

(c) $(D+t^{-1}I)(D+tI) = D^2 + (t+t^{-1})D + 2I$.

Section 11

1. (a), (c), (d), and (f) are linearly independent sets. Sets (b) and (e) are linearly dependent.

2. If v_1 satisfied an equation $y'' + a_1(t)y' + a_0(t)y = 0$ with a_1 and a_0 continuous on $(-1,1)$, then the fact that $v_1(0) = 0$ and $v_1'(0) = 0$ would imply $v_1(t) \equiv 0$. Why? Similarly for v_2.

4. $x(t) = t - 2 + c_1e^{-t} + c_2te^{-t}$.

7. $y(t) = c_1 t^{\lambda_1} + c_2 t^{\lambda_2}$ where $\lambda_1, \lambda_2 = \dfrac{1-b_1 \pm [(1-b_1)^2 - 4b_0]^{1/2}}{2}$.

9. <u>Hint.</u> See the answer to Problem 2.

Section 12

2. (a) $x(t) = c_1 e^{-2t} + c_2 t e^{-2t} + c_3 t^2 e^{-2t}$.

(b) $x(t) = \frac{1}{2} t \sin t + c_1 \cos t + c_2 \sin t$.

(c) $x(t) = -\frac{1}{2} \cos t + \frac{1}{2} \sin t + c e^{-t}$.

(d) $x(t) = -\frac{1}{2} - \frac{1}{20} \cos 2t + \frac{3}{20} \sin 2t - \frac{1}{2} e^t + c_1 e^{2t} + c_2 e^{-t}$.

(e) $x(t) = e^t + c_1 + c_2 e^{2t} + c_3 t e^{2t}$.

(f) $x(t) = \frac{1}{4} t^2 e^{2t} + c_1 + c_2 e^{2t} + c_3 t e^{2t}$.

(g) $x(t) = -\frac{1}{10} t^2 - \frac{1}{50} t - \frac{1}{500}$
$\qquad + c_1 e^{2t} + c_2 e^{-t} \cos 2t + c_3 e^{-t} \sin 2t$.

(h) $x(t) = -\frac{11}{500} t - \frac{1}{100} t^2 - \frac{1}{30} t^3$
$\qquad + c_1 + c_2 e^{(-1+\sqrt{41})t/2} + c_3 e^{(-1-\sqrt{41})t/2}$

(i) $x(t) = -\frac{7}{9} t e^{-t} - \frac{1}{6} t^2 e^{-t} - \frac{1}{3} t e^{2t} + c_1 e^{-t} + c_2 e^{2t}$.

(j) $x(t) = \frac{1}{2} t e^{-2t} \sin t + c_1 e^{-2t} \cos t + c_2 e^{-2t} \sin t$.

$\qquad x(t) \to 0$ as $t \to \infty$.

(k) $x(t) = -\frac{3}{40} \cos 2t - \frac{1}{40} \sin 2t + c_1 e^{-t} + c_2 e^{-2t}$.

(ℓ) $x(t) = \frac{1}{8} - \frac{1}{8} t \sin 2t + c_1 \cos 2t + c_2 \sin 2t$.

3. (a) $x(t) = 2e^{-2t} + 4t e^{-2t} - t^2 e^{-2t}$.

(b) $x(t) = \frac{1}{2} t \sin t - \frac{1}{2} \cos t - \frac{\pi}{4} \sin t$.

(c) $x(t) = -\frac{1}{2} \cos t + \frac{1}{2} \sin t + \frac{3}{2} e^{\pi - t}$.

4. (a) $\tilde{x}(t) = At \cos t + Bt \sin t + Ct^2 \cos t + Dt^2 \sin t$.

(b) $\tilde{x}(t) = A + Bt + Ct^3 e^{-3t} + Dt^4 e^{-3t} + Et^5 e^{-3t}$.

5. (e) $C = 160 \times 10^{-12}$ farads. Then in (18) the ω_1 term has
$\qquad$ amplitude $a/563$ and the ω_2 term has amplitude $a/2$.

Section 13

1. $W(y_1, y_2)(t) = t^{-3} \neq 0$.

3. (a) $x(t) = c_1 |t|^{1/2} + c_2 |t|^{-1/2}$ on $(0, \infty)$ or on $(-\infty, 0)$.

(b) $x(t) = c_1|t|^{-1/2} + c_2|t|^{-1/2} \ln|t|$ on $(0,\infty)$ or $(-\infty,0)$.

(c) $x(t) = c_1 + c_2 t^2 + c_3 t^2 \ln|t|$ on $(0,\infty)$ or $(-\infty,0)$.

(d) $x(t) = c_1 t + c_2 t \ln|t| + c_3 t (\ln|t|)^2$ on $(0,\infty)$ or

     $(-\infty,0)$.

(f) $x(t) = c_1 \cos(t^2/2) + c_2 \sin(t^2/2)$ on $(0,\infty)$ or

     $(-\infty,0)$.   How do you conclude that this solution is

     valid beyond $\sqrt{2\pi}$?

4. Compute $x(t) = w(\ln|t|)$, $x'(t) = t^{-1} w'(\ln|t|)$,

   $x''(t) = t^{-2}[w''(\ln|t|) - w'(\ln|t|)]$, $x'''(t) = t^{-3}[w'''(\ln|t|)$

   $- 3w''(\ln|t|) + 2w'(\ln|t|)]$. Then you will find that   w

   satisfies a linear homogeneous equation with constant

   coefficients.

   (a) Find $4w''(s) - w(s) = 0$. So $w(s) = c_1 e^{s/2} + c_2 e^{-s/2}$

     or $x(t) = c_1|t|^{1/2} + c_2|t|^{-1/2}$ on $(0,\infty)$ or $(-\infty,0)$.

5. (a) $x(t) = c_1 t^{-2} \cos(\ln|t|) + c_2 t^{-2} \sin(\ln|t|)$.

   (b) $x(t) = c_1 t^2 + c_2 t^{-1} \cos(\ln|t|^3) + c_3 t^{-1} \sin(\ln|t|^3)$.

6. (a) $x(t) = -1 + t + te^{-t} + c_1 e^{-t} + c_2 e^{-2t}$.

   (b) $x(t) = t - 1 + ce^{-t}$.   Can you recognize your work as

     being equivalent to the method of Section 2?

   (c) $x(t) = -\dfrac{t}{2} \cos t + c_1 \cos t + c_2 \sin t$.

   (d) $x(t) = \dfrac{1}{2}e^t + c_1 \cos t + c_2 \sin t$.

   (e) $x(t) = -(\cos t) \ln|\sec t + \tan t| + c_1 \cos t + c_2 \sin t$.

   (f) $x(t) = (\cos t) \ln|\cos t| + t \sin t + c_1 \cos t + c_2 \sin t$.

   (g) $x(t) = -1 + (\sin t) \ln|\sec t + \tan t| + c_1 \cos t$

                                      $+ c_2 \sin t$.

   (h) $x(t) = -te^{-t} + c_1 + c_2 e^{-t} + c_3 e^{-2t}$.

   (i) $x(t) = te^t + c_1 t + c_2 t^2$.

   (j) $x(t) = (\ln|t|)^2/2t + c_1 t^{-1} + c_2 t^{-1} \ln|t|$.

7. $y(t) = c_1 t + c_2 t^2 + c_3 [t^2 e^{2^{-1} t^{-2}} + t \int_1^t s^{-2} e^{2^{-1} s^{-2}} ds]$.

8. (a) $x(t) = \frac{1}{6} \int_{t_0}^t h(s)[e^{t-s} - e^{-5(t-s)}] ds + c_1 e^t + c_2 e^{-5t}$.

   (b) $x(t) = \frac{1}{8} \int_{t_0}^t h(s)[e^{-2(t-s)} - \cos 2(t-s) + \sin 2(t-s)] ds$
   $+ c_1 e^{-2t} + c_2 \cos 2t + c_3 \sin 2t$.

   (c) $x(t) = \int_{t_0}^t h(s)\sin(t-s)\,ds + c_1 \cos t + c_2 \sin t$.

   (d) $x(t) = \int_{t_0}^t h(s) e^{-2(t-s)} \sin(t-s)\,ds + c_1 e^{-2t} \cos t$
   $+ c_2 e^{-2t} \sin t$.

Section 14

5. (a) $y(t) = c_1 e^{5t} \begin{bmatrix} 1 \\ 1 \end{bmatrix} + c_2 e^{-2t} \begin{bmatrix} 4 \\ -3 \end{bmatrix}$.

   (b) $x(t) = \frac{1}{5} \begin{bmatrix} 2 \\ -3 \end{bmatrix} + \frac{1}{65} \begin{bmatrix} -22 \cos t - 6 \sin t \\ 4 \cos t + 7 \sin t \end{bmatrix} + y(t)$,

   where $y(t)$ is as in part (a).

   (c) $x(t) = \frac{1}{4} \begin{bmatrix} -2+4t \\ 5-2t \\ -2 \end{bmatrix} + c_1 e^{-t} \begin{bmatrix} 1 \\ -1 \\ 0 \end{bmatrix} + c_2 e^{-2t} \begin{bmatrix} 2 \\ -1 \\ -2 \end{bmatrix}$

   $+ c_3 e^{-3t} \begin{bmatrix} 1 \\ -1 \\ -2 \end{bmatrix}$.

6. (a) Put $c_1 = -1/7$ and $c_2 = 2/7$ in solution of 5(a).

   (b) Put $c_1 = 9/65$ and $c_2 = 1/5$ in solution of 5(b).

Section 15

5. (a) $x(t) = \begin{bmatrix} te^{4t}+e^{4t}-2t-2 \\ -2te^{4t}+e^{4t}-2t-2 \end{bmatrix} + c_1 e^t \begin{bmatrix} 1 \\ 1 \end{bmatrix} + c_2 e^{4t} \begin{bmatrix} 1 \\ -2 \end{bmatrix}$.

   (b) $y(t) = C_1 e^{it} \begin{bmatrix} 1+i \\ -1 \end{bmatrix} + C_2 e^{-it} \begin{bmatrix} 1-i \\ -1 \end{bmatrix}$

   $= c_1 \begin{bmatrix} \cos t - \sin t \\ -\cos t \end{bmatrix} + c_2 \begin{bmatrix} \cos t + \sin t \\ -\sin t \end{bmatrix}$

   where $c_1 = C_1 + C_2$ and $c_2 = iC_1 - iC_2$.

   (c) $x(t) = \frac{1}{4} \begin{bmatrix} 3t-3 \\ t \end{bmatrix} + \begin{bmatrix} -5e^{-t} \\ -2e^{-t} \end{bmatrix} + c_1 e^{-2t} \begin{bmatrix} 1 \\ 1 \end{bmatrix} + c_2 e^{-2t} \begin{bmatrix} 1+t \\ t \end{bmatrix}$.

(d) $y(t) = c_1 e^t \begin{bmatrix} 0 \\ 1 \\ 0 \end{bmatrix} + c_2 e^t \begin{bmatrix} 3 \\ 0 \\ 2 \end{bmatrix} + c_3 e^{2t} \begin{bmatrix} 1 \\ 4 \\ 1 \end{bmatrix}$.

(e) $y(t) = c_1 e^t \begin{bmatrix} 0 \\ 1 \\ 0 \end{bmatrix} + c_2 e^t \begin{bmatrix} 3 \\ -2t \\ 2 \end{bmatrix} + c_3 e^{2t} \begin{bmatrix} 1 \\ 3 \\ 1 \end{bmatrix}$.

(f) $y(t) = c_1 e^{-2t} \begin{bmatrix} 1+t+t^2 \\ t-t^2 \\ 2+t-t^2 \end{bmatrix} + c_2 e^{-2t} \begin{bmatrix} 1+2t \\ 1-2t \\ 1-2t \end{bmatrix} + c_3 e^{-2t} \begin{bmatrix} 1 \\ -1 \\ -1 \end{bmatrix}$.

(g) $y(t) = c_1 e^{-2t} \begin{bmatrix} 1+t \\ -t \\ -t \end{bmatrix} + c_2 e^{-2t} \begin{bmatrix} 1 \\ 0 \\ 1 \end{bmatrix} + c_3 e^{-2t} \begin{bmatrix} 2 \\ -1 \\ 0 \end{bmatrix}$.

(h) $y(t) = c_1 e^{-t} \begin{bmatrix} 1 \\ -1 \\ 0 \end{bmatrix} + C_2 e^{-t+it} \begin{bmatrix} 1 \\ 1+i \\ -1 \end{bmatrix} + C_3 e^{-t-it} \begin{bmatrix} 1 \\ 1-i \\ -1 \end{bmatrix}$

$= c_1 e^{-t} \begin{bmatrix} 1 \\ -1 \\ 0 \end{bmatrix} + c_2 e^{-t} \begin{bmatrix} \cos t \\ \cos t - \sin t \\ -\cos t \end{bmatrix}$

$+ c_3 e^{-t} \begin{bmatrix} \sin t \\ \sin t + \cos t \\ -\sin t \end{bmatrix}$.

6. In the answers to Problem 5:

   (a) set $c_1 = 2$, $c_2 = 1$.   (b) set $c_1 = 1$, $c_2 = 0$.

   (c) set $c_1 = 2$, $c_2 = 23/4$. (d) set $c_1 = -26$, $c_2 = -2$,

   $c_3 = 7$.

   (e) set $c_1 = -19$, $c_2 = -2$, $c_3 = 7$.

7. (a) $y(t) = c_1 e^{-t} \begin{bmatrix} 1 \\ -1 \end{bmatrix} + c_2 e^{-3t} \begin{bmatrix} 1 \\ -3 \end{bmatrix}$. $z(t) = y_1(t)$.

   (b) $y(t) = c_1 e^{-2t} \begin{bmatrix} 1 \\ -2 \end{bmatrix} + c_2 e^{-2t} \begin{bmatrix} t \\ 1-2t \end{bmatrix}$. $z(t) = y_1(t)$.

Section 17

3. Since $T(t,s) = T(t-s,0)$ in each case, we list only $T(t,0)$.

   (a) $T(t,0) = \dfrac{1}{3} \begin{bmatrix} 2e^t + e^{4t} & e^t - e^{4t} \\ 2e^t - 2e^{4t} & e^t + 2e^{4t} \end{bmatrix}$

   (b) $T(t,0) = \begin{bmatrix} \cos t + \sin t & 2 \sin t \\ -\sin t & \cos t - \sin t \end{bmatrix}$

(d) $T(t,0) = \begin{pmatrix} 3e^t - 2e^{2t} & 0 & -3e^t + 3e^{2t} \\ 8e^t - 8e^{2t} & e^t & -12e^t + 12e^{2t} \\ 2e^t - 2e^{2t} & 0 & -2e^t + 3e^{2t} \end{pmatrix}$.

(e) $T(t,0) = \begin{pmatrix} 3e^t - 2e^{2t} & 0 & -3e^t + 3e^{2t} \\ 6e^t - 2te^t - 6e^{2t} & e^t & -9e^t + 2te^t + 9e^{2t} \\ 2e^t - 2e^{2t} & 0 & -2e^t + 3e^{2t} \end{pmatrix}$.

(f) $T(t,0) = e^{-2t} \begin{pmatrix} 1+t & t - t^2/2 & t^2/2 \\ -t & 1 - 2t + t^2/2 & t - t^2/2 \\ -t & -2t + t^2/2 & 1 + t - t^2/2 \end{pmatrix}$.

4. (a) $x(t) = \begin{pmatrix} x_{01}\cos t + (x_{01} + 2x_{02})\sin t \\ x_{02}\cos t - (x_{01} + x_{02})\sin t \end{pmatrix} + \begin{pmatrix} 3\cos t + \sin t - 3 \\ -\cos t - 2\sin t + 1 \end{pmatrix}$

(b) $x(t) = \begin{pmatrix} x_{01}\cos t + (x_{01} + 2x_{02})\sin t \\ x_{02}\cos t - (x_{01} + x_{02})\sin t \end{pmatrix} + \begin{pmatrix} t\cos t + (1+t)\sin t \\ -t\sin t \end{pmatrix}$

(c) $x(t) = \begin{pmatrix} x_{01}\cos t + (x_{01} + 2x_{02})\sin t \\ x_{02}\cos t - (x_{01} + x_{02})\sin t \end{pmatrix}$

$\qquad + \begin{pmatrix} 2\sin t - 2\cos t \ln(\sec t + \tan t) \\ -1 + \cos t - \sin t + (\sin t + \cos t)\ln(\sec t + \tan t) \end{pmatrix}$

(d) $x(t) = \begin{pmatrix} x_{01}e^{-2t} + (x_{01} - x_{02})te^{-2t} \\ x_{02}e^{-2t} + (x_{01} - x_{02})te^{-2t} \end{pmatrix} + \begin{pmatrix} -(t^2/2)e^{-2t} \\ (t - t^2/2)e^{-2t} \end{pmatrix}$.

6. $x(t) = \begin{pmatrix} x_{01}\cos \frac{t^2}{2} + x_{01}\sin \frac{t^2}{2} \\ -x_{01}\sin \frac{t^2}{2} + x_{02}\cos \frac{t^2}{2} \end{pmatrix} + \begin{pmatrix} 2 - 2\cos \frac{t^2}{2} \\ 2\sin \frac{t^2}{2} \end{pmatrix}$.

Section 18

1. $||A|| = 4$, $||B|| = 5$, $||A+B|| = 8$, $||AB|| = 14$, $||BA|| = 20$.

6. $||I||_S = n$, which is less desirable than $||I|| = 1$.

7. (a) If $y_{(j)}$ is the solution of $y' = A(t)y$ with $y(s) = I_{(j)}$, then, by Theorem F, $||y_{(j)}(t)|| \le e^{K|t-s|}$ on J.

Thus $||T(t,s)|| \le \max_j ||y_{(j)}(t)|| \le e^{K|t-s|}$.

(b) $||x(t)|| \leq e^{K|t-t_0|}||x_0|| + |\int_{t_0}^{t} e^{K|t-s|}Mds$

$= ||x_0||e^{K|t-t_0|} + \frac{M}{K}(e^{K|t-t_0|} - 1)$.

(c) By the mean value theorem, for some $\theta \in (0,|t-t_0|)$,

$e^{K|t-t_0|} - e^0 = K|t-t_0|e^{K\theta}$.

8. $||A|| = \max_j ||AI_{(j)}|| \leq \sup \{||A\xi||: ||\xi|| = 1\}$

$\leq \sup \{\frac{||A\xi||}{||\xi||}: \xi \neq 0\} \leq ||A||$ by Theorem A(v).

Section 19

1. (a) $e^{tA} = T(t,0)$ in answer to Problem 17-3(a).

(b) $e^{tA} = T(t,0)$ in answer to Problem 17-3(b).

(c) $e^{tA} = T(t,0)$ in answer to Problem 17-3(d).

(d) $e^{tA} = e^{-2t}\begin{bmatrix} 1+t & 2t & -t \\ -t & 1-2t & t \\ -t & -2t & 1+t \end{bmatrix}$.

3. $e^A e^B = \begin{bmatrix} e & e \\ 0 & 1 \end{bmatrix}$, $e^{A+B} = \begin{bmatrix} e & e-1 \\ 0 & 1 \end{bmatrix}$, $e^B e^A = \begin{bmatrix} e & 1 \\ 0 & 1 \end{bmatrix}$.

6. $e^{\int_0^t A(s)ds} = \begin{bmatrix} e^t & t(e^t-1) \\ 0 & 1 \end{bmatrix}$.

8. Use the facts that $||A^k|| \leq ||A||^k$ and that $\Sigma_{k=0}^{\infty} ||A||^k$ converges to prove that $\Sigma_{k=0}^{\infty} A^k$ converges. Then since

$(\sum_{k=0}^{\ell} A^k)(I - A) = I - A^{\ell+1}$,

we find, letting $\ell \to \infty$, $(\Sigma_{k=0}^{\infty} A^k)(I - A) = I$. Similarly one shows that $(I - A)\Sigma_{k=0}^{\infty} A^k = I$.

9. $\sin A \equiv \sum_{m=0}^{\infty} \frac{(-1)^m}{(2m+1)!} A^{2m+1}$ (convergent for all A).

(a) $\sin A = \begin{bmatrix} \sin 1 & 0 \\ 0 & \sin 2 \end{bmatrix}$.

(b) $\sin A = \frac{1}{7}\begin{bmatrix} 3\sin 5 - 4\sin 2 & 4\sin 5 + 4\sin 2 \\ 3\sin 5 + 3\sin 2 & 4\sin 5 - 3\sin 2 \end{bmatrix}$.

Section 20

1. (a) $x_{(1)}(t) = 1 + ct - c\frac{t^2}{2}$,

   $x_{(2)}(t) = 1 + ct + \frac{1}{2!}(ct)^2 - \frac{1}{3!}c^2t^3$,

   $x_{(3)}(t) = 1 + ct + \frac{1}{2!}(ct)^2 + \frac{1}{3!}(ct)^3 - \frac{1}{4!}c^3t^4$,

   $x(t) = e^{ct} = 1 + ct + \frac{1}{2!}(ct)^2 + \frac{1}{3!}(ct)^3 + \cdots$

   (b) $x_{(1)}(t) = 2 + 3(t-1) + \frac{1}{2}(t-1)^2$,

   $x_{(2)}(t) = 2 + 3(t-1) + 2(t-1)^2 + \frac{1}{3!}(t-1)^3$,

   $x_{(3)}(t) = 2 + 3(t-1) + 2(t-1)^2 + \frac{4}{3!}(t-1)^3 + \frac{1}{4!}(t-1)^4$,

   $x(t) = 4e^{t-1} - t - 1$

   $\qquad = 2 + 3(t-1) + 4[\frac{1}{2!}(t-1)^2 + \frac{1}{3!}(t-1)^3 + \cdots]$.

   (c) $x_{(1)}(t) = \begin{bmatrix} 1 - t \\ t \end{bmatrix}$, $\qquad x_{(2)}(t) = \begin{bmatrix} 1 - t - \frac{1}{t}t^2 \\ t - \frac{5}{2}t^2 \end{bmatrix}$

   $x(t) = \begin{bmatrix} 2e^{-2t} - e^{-3t} \\ e^{-2t} - e^{-3t} \end{bmatrix} = \begin{bmatrix} 1 - t - \frac{1}{2}t^2 + \frac{11}{6}t^3 - \cdots \\ t - \frac{5}{2}t^2 + \frac{19}{6}t^3 - \cdots \end{bmatrix}$.

3. (a) $|x(t) - x_{(4)}(t)| \le e^{|ct|}|ct|^5/5!$ for all t in $\mathbb{R}$.

   (b) If $\alpha_0 < 1$, $\beta_0 > 1$, and $B = \max(2+\beta_0, |2+\alpha_0|)$, then

   $|x(t) - x_{(4)}(t)| \le Be^{|t-1|}|t-1|^5/5!$ for $\alpha_0 \le t \le \beta_0$.

   (c) $||x(t) - x_{(4)}(t)|| \le \frac{108}{5}e^{6|t|}|t|^5$ for all t in $\mathbb{R}$.

4. Letting $\Delta = \max\{\beta-t_0, t_0-\alpha\}$, inequality (10) gives (by a
   very simple induction)

   $$||x_{(\ell+1)}(t) - x_{(\ell)}(t)|| \le M(K\Delta)^\ell \quad \text{on} \quad J,$$

   where $0 < K\Delta < 1$. Using this, and the convergence of
   geometric series $\Sigma_{p=0}^{\infty} M(K\Delta)^p$ instead of (11), one shows
   the uniform convergence of the series $\{x_{(\ell)}\}$ on J.

Section 21

1. (a) $cr < 2-\sqrt{2}$.

(b) "No" is probably the correct answer. It would be very difficult to find this condition by the method of steps. By other methods the appropriate condition has been found to be $cr < 1/e$. Driver [1972].

(c) Let $y(t) = qa/c - x(t)$.

4. $x(t) = -\dfrac{b}{a^2} - \dfrac{bt}{a} + (1 + \dfrac{b}{a^2})e^{at}$ on $[0,1]$,

$x(t) = \dfrac{2b^2}{a^3} - \dfrac{b^2}{a^2} + \dfrac{b^2t}{a^2} + [1 + \dfrac{b}{a^2} - (b + \dfrac{b}{a} + \dfrac{b}{a^2} + \dfrac{b^2}{a^2})e^{-a}]e^{at}$

$\qquad - \dfrac{2b^2}{a^3}\, e^{a(t-1)} + (b + \dfrac{b^2}{a^2})te^{a(t-1)}$ on $[1,2]$.

5. (a) $a = \dfrac{e}{e-1}$, $b = \dfrac{-e}{e-1}$.

(b) $x(t) = x_0 + c(e^t - 1)$ for arbitrary constant c.

(c) The solution is still not unique, but this is not obvious. A much stronger non-uniqueness result will be proved in Section 22.

6. A unique solution exists for all $t \geq 0$.

7. A unique solution exists for <u>all</u> $t \geq 0$.

9. If $x(t) = \theta(t)$ on $[t_0-1,t_0]$, where θ is a given continuous function, then a unique solution of the delay differential equation exists for all $t \geq t_0$.

10. On $\mathbb{R}^3$, let f be continuous and let f satisfy a Lipschitz condition with respect to its second argument. Then the delay differential equation has at most one solution for $t \geq t_0$ such that $x(t) = \theta(t)$ on $[t_0-1,t_0]$, where θ is a given continuous function.

11. (a) On $[0,r]$ first solve the ordinary linear first-order equation $y'(t) = -a_2 y(t) + b_2\theta_1(t-r)\theta_2(t-r)$ with $y(0) = \theta_2(0)$. Then, having found y on $[0,r]$, solve the ordinary Bernoulli equation

$x'(t) + [b_1 y(t) - a_1]x(t) = -(a_1/P)x^2(t)$

on [0,r] with x(0) = θ_1(0). Proceed similarly on
[r,2r], [2r,3r],...

(b) No.

12. One answer is q = 1, k = π^2, ω = π.

13. x(t) = tan (t - 1 + $\pi/4$) on [1, 1 + $\pi/4$). The solution
cannot be continued past 1 + $\pi/4$. Why?

14. A unique solution exists for all t $\geq$ 1.

15. (b) See Example 30-5.

16. (a) <u>Hint</u>. First show that on [0,1]
$$x(t) + 1 = [\theta(0) + 1]e^{\int_0^t c\theta(s-1)ds} .$$

(b) If (ii) does not hold, show that for large t, x(t)
and x'(t) must have constant but opposite signs.
Thus L = $\lim_{t\to\infty}$ x(t) exists. Now argue from Eq. (7)
that L must be either 0 or -1, and we know L $\neq$ -1.

(c) If x(t) < -1 for t $\geq$ 0, then x'(t) $\leq$ 0 for t $\geq$ 1.
Now x(t) $\not\to$ L since L would have to be either 0 or
-1. Hence x(t) $\to$ -∞ as t $\to$ ∞.

For further details see Wright [1955].

17. (a) c = ln 2.   (b) Yes, c > 0. z(n) = $(\ln n)^{-1}$.

(c) Since z(n) > 0, x(t) > -1. So, from Problem 16(b),
either z(n)(ln n) $\to$ 1 or z(n)(ln n) - 1 oscillates.

Section 23

5. Not unique. Two solutions are x(t) = 0 and x(t) =
$(t/12)^3$.

6. $||x(t) - \tilde{x}(t)|| \leq \sup_{-r\leq s\leq 0} ||\theta(s) - \tilde{\theta}(s)||e^{(|a|+|b|)t}$.

7. Two among many possible answers:
$|\theta(0)| + |\theta'(t)| < \varepsilon e^{-K\beta}$ on [-r,0], where
K = max ($|k/m|$, $|b/m|$ + 1) + $|q/m|$, or

$|\theta(0)| + |m/k|^{1/2}|\theta'(t)| < \varepsilon e^{-K\beta}$ on $[-r,0]$, where

$K = |k/m|^{1/2} + |b/m| + |q/m|$.

8. Find $y'(t) = f_1(t,h^{-1}(y(t)))$ with $y(t_0) = 0$, and show that $y(t)$ is uniquely determined.

Section 24

1. (a) $x_{(1)}(t) = 1 + t$,

   $x_{(2)}(t) = 1 + t + t^2 + \frac{1}{3}t^3$,

   $x(t) = \frac{1}{1-t} = 1 + t + t^2 + t^3 + \dots$ (for $|t| < 1$).

   (b) $x_{(1)}(t) = 1 + \frac{1}{3}t^3$,

   $x_{(2)}(t) = 1 + t + \frac{1}{6}t^4 + \frac{1}{63}t^7$.

2. **Hint.** Verify (11) directly by induction.

3. If $a = \infty$ and $b > 0$, then $|x(t) - x_{(4)}(t)| \leq \frac{2}{15}(b+1)^6|t|^5$ for $|t| \leq b/(b+1)^2$.

4. $\Delta = 1/4$, $\Delta_1 = 3/16$.

5. Let $v_\ell = \sup \{||x_{(\ell+1)}(t) - x_{(\ell)}(t)||: \; t_0 - \Delta \leq t \leq t_0 + \Delta\}$ for $\ell = 0,1,\dots$ Then from (7) we find $v_{\ell+1} \leq K\Delta v_\ell$ for $\ell = 0,1,\dots$ By mathematical induction this gives $v_\ell \leq 2b(K\Delta)^\ell$ for $\ell = 0,1,\dots$ Now using this instead of (8), the uniform convergence of the sequences $\{x_{(\ell)k}\}$ follows by comparison with the geometric series $\sum_{\ell=0}^{\infty} 2b(K\Delta)^\ell = 2b/(1-K\Delta)$.

6. From inequality (7), now valid on $[\alpha_1,\beta_1]$, compute

   $e^{-c|t-t_0|}||x_{(\ell+2)}(t) - x_{(\ell+1)}(t)||$

   $\leq e^{-c|t-t_0|}|\int_{t_0}^{t} Ke^{c|s-t_0|}v_\ell ds|$

   $= e^{-c|t-t_0|}\frac{K}{c}[e^{c|t-t_0|} - 1]v_\ell \leq \frac{K}{c}v_\ell$.

   Assuming $K/c < 1$, prove the uniform convergence of $\{x_{(\ell)}\}$ on $[\alpha_1,\beta_1]$.

7. By Problem 6, the equation $x'(t) = \sin x(t) - \sin t$ with $x(1) = 1$ has a unique solution on $(-\infty,\infty)$.

11. Consider $v(t) = [x'(t)]^2/2 + x^2(t)/2$. Then $v'(t) = -\mu(x^2 - 1)x'^2$. Whenever $x^2 \geq 1$, $v' \leq 0$, and whenever $x^2 < 1$, $v' \leq \mu x'^2 \leq 2\mu v$. Thus in all cases $v'(t) \leq 2\mu v(t)$, for all t. This gives, for $t \geq t_0$, $v(t) \leq v(t_0)e^{2\mu(t-t_0)}$. Now apply Theorem C.

12. For the associated system, $||f(t,\xi)|| \leq \max\{2,t^2\}||\xi||$. So Corollary E applies.

13. Let x on (α_1,β_1) be the unique noncontinuable solution. Note that $x(0) = x_0 > 0$ and $x'(0) = u_0 > 0$ implies $x''(0) > 0$. Hence x' is increasing and x is increasing. Thus $x(t) \geq x_0$ and $x'(t) \geq u_0$ for $0 \leq t < \beta_1$. Defining $v(t) = x'^2/2 - x^4/4$, we find $v'(t) = 0$. So $[x'(t)]^2 = u_0^2 + [x(t)]^4/2 - x_0^4/2$ on $[0,\beta_1)$. By separation of variables, this gives

$$\int_{x_0}^{x(t)} \frac{d\xi}{[u_0^2 + (\xi^4 - x_0^4)/2]^{1/2}} = t \quad \text{for} \quad 0 \leq t < \beta_1.$$

Show that the integral on the left is bounded for all $x(t) > x_0$, so that $\beta_1 < \infty$.

14. (a) No solution.

(b) $x(t) = \sqrt{2t}$ for $t > 0$ (unique). No solution for $t < 0$.

(c) $x(t) = \sqrt{-2t}$ for $t < 0$ (unique). No solution for $t > 0$.

(d) $x(t) = t$ for $t > 0$ (unique). $x(t) = -t$ for $t < 0$.

(e) No solution.   (f) $x(t) = t$ (unique).

(g) $x(t) = ct$ (nonunique).

(h) $x(t) = ce^{-1/t}$ (nonunique).

(i) $x(t) = 0$ (unique).

(j) $x(t) = ct/(c+t)$ (nonunique).

(k) $x(t) = t, 0,$ or $-t$ (nonunique).

Section 25

1. $F(t,\psi) = \begin{bmatrix} [a_1 - b_1\psi_2(0) - \int_{-r}^{0} h_1(\sigma)\psi_2(\sigma)d\sigma]\psi_1(0) \\ [-a_2 + b_2\psi_1(0) + \int_{-r}^{0} h_2(\sigma)\psi_1(\sigma)d\sigma]\psi_2(0) \end{bmatrix}.$

2. If one defines $x_1(t) = x(t)$ and $x_2(t) = (m/k)^{1/2}x'(t)$,

then

$F(t,\psi) = \begin{bmatrix} \sqrt{\frac{k}{m}}\,\psi_2(0) \\ -\sqrt{\frac{k}{m}}\,\psi_1(0) - \frac{b}{m}\psi_2(0) - \frac{q}{m}\psi_2(-r) \end{bmatrix}$

If you defined x_1 or x_2 differently, the correct form
of F will be different. The definitions suggested above
have the advantage that both x_1 and x_2 have the same
"dimensions" -- those of length.

3. On $[0,\infty)$ the delay is unbounded. On $[0, 100)$ one can
take $r = 90$.

5. On $\mathscr{C}_D$ $||\cdot||_r$ fails to satisfy conditions (iii) and (iv)
of Lemma A, in general.

6. (b) Inequality (5) holds with $K = 10^6$.

7. $K = |c|(1 + 2H)$.

8. $\mathscr{B} \subset \mathscr{C}_A$.

Section 26

1. The upper bound for $||x_{(\ell+1)}(t) - x_{(\ell)}(t)||$ would then
become $2bK^\ell(t-t_0)^\ell/\ell!$ for $t_0 \le t \le t_0+\Delta$. Prove it,
and show that this would suffice.

4. Given any $t_0 \in \mathbb{R}$ and $\phi \in C([-r,0],\mathbb{R})$, there is a unique
solution x on $[t_0-r,\infty)$ with $x_{t_0} = \phi$.

5. Assume $\Sigma_{j=1}^{\infty} |a_j|$ converges. Then, given any $t_0 \in \mathbb{R}$ and $\phi \in C([-1,0],\mathbb{R})$, there is a unique solution x on $[t_0-1,\infty)$ with $x_{t_0} = \phi$. Carefully verify Condition (C) and a global Lipschitz condition.

7. Hint. Note that, since $g_j(t) < t$ and g_j is continuous, there exists some $\Delta_1 > 0$ such that $g_j(t) \le t_0$ for $t_0 \le t \le t_0+\Delta_1$, $j = 2,\ldots,m$.

8. Hint. See the answer to Problem 24-6.

Section 27

1. There are infinitely many correct answers for each part of this question. But the correct answers found most naturally are

(a) $\tilde{x}(t) = -\dfrac{7+4\sqrt{2}}{34} \cos t + \dfrac{11-\sqrt{2}}{34} \sin t$.

(b) $\tilde{x}(t) = -\dfrac{2\pi}{\pi^2+4} t \cos t + \dfrac{4}{\pi^2+4} t \sin t$.

(c) $\tilde{x}(t) = t - 2$.

(d) $\tilde{x}(t) = t^2/4 + t/8$.

(e) $\tilde{x}(t) = t^3/3 + t^2/3$.

Section 28

3. (a) If $\lambda = \mu+i\omega$ with $\mu \ge 0$, then $|\omega r| = |bre^{-\mu r} \sin \omega r| < \pi/2$ so that $\mu = -be^{-\mu r} \cos \omega r < 0$ -- a contradiction.

(b) Modify Example 3.

(c) Modify Example 4.

4. Generalize Example 22-1.

5. $x(t) = -2 \cos 2t + y(t)$ where $y(t) \to 0$ exponentially as $t \to \infty$.

Section 29

3. (a) has unbounded solutions because the equation $\lambda = e^{-\lambda r}$
   has a positive solution.

   (b) has only bounded solutions. See Problem 28-3(a).

   (c) has unbounded solutions if $h(t) = \sin t$. See Problem 27-1(b).

   (d) has only bounded solutions. See Example 28-4.

Section 30

3. For each of the three cases, $b^2 < 4mk$, $b^2 = 4mk$, and
   $b^2 > 4mk$, one would have to find an upper bound for the
   maximum possible value of $||x(t;t_0,x_0)||$ in terms of
   $||x_0||$.

Section 31

2. (a) $w(s) = ks^2/4$, $W(s) = ks^2/2$.

   (b) $w(s) = s^2$, $W(s) = (1+|a|r)^{1/2}s$.

3. If $x_0 > 0$ is given, one finds $x(1) \geq e^{-2+t_0^{-1}} x_0 \to \infty$
   as $t_0 \to 0^+$.

4. (a) Take $H \leq 3$ and $V(t,\xi) = \xi^2$.

   (d) Take $V(t,\xi) = \frac{1}{2} k_1 \xi_1^2 + \frac{1}{2} k_2(\xi_3-\xi_1)^2 + \frac{1}{2} m_1 \xi_2^2$
   $+ \frac{1}{2} m_2 \xi_4^2$. Some care is required to show that this V
   satisfies condition (a) of Theorem A.

Section 32

4. (b) Take $H < 3$ and $V(t,\xi) = \xi^2$.

5. (a) Hint. To verify condition (c) of Theorem C, let
   $cb/m = (1-\delta)\{4c(k/m)^{1/2}[kb/m - c(k/m)^{1/2}]\}^{1/2}$. Then
   $\delta > 0$. Use this in the expression for $dV(t,x(t))/dt$.

6. Use Theorem D.

Section 33

3. (a) Find $M = 1$ and $\gamma = 2$.   Then need $H < (\sqrt{41} - 1)/10$,

say $H = 1/2$.   Then δ_1 = H/M = 1/2   will suffice.

   (b) Find $M = 7/5$ and $\gamma = 1$.   Can take any $H < 40/7$,

say $H = 28/5$.   Then $\delta_1 = 4$.

   (c) Find $M < 5/4$ and $\gamma = 1$.   Can take $H = 1/5$ and

$\delta_1 = 4/25$.

Section 34

2. (a) $|b| < |a|$, as in Example 32-4.

   (b) $(|a|+|b|)|b|r < |a+b|$.   This is slightly more general

than Problem 32-4(c) since it permits some cases with

$a > 0$.   Cf. Figure 31-1.

   (c) $q < \frac{b}{4}(1 - \frac{b^2}{4mk})^{1/2}$, much less general than Examples

28-3 and 32-7.

   (d) $(\frac{b}{m} + \frac{q}{m} + \sqrt{\frac{k}{m}})qr < (b+q)[1 - \frac{(b+q)^2}{4mk}]^{1/2}/4$.   This is less

general than condition 28-(16) if q is large, but

more general if q is small.

3. (a) $(1+|b|r)(\Delta a+\Delta b) + (|a|+|b|)|b|r < |a+b|$.

   (b) $q < b$ (by Example 28-3 and Corollary F), or

$(b/m + q/m + \sqrt{k/m})r < \pi/2$ (by Example 28-4 and

Corollary F).

Section 35

3. det $A \neq 0$.   Critical point is   0.

5. (a) Critical point: $(0,0)$.

One finds

$v(t) = \frac{1}{2}x_2^2 + 2x_1^2$

$= constant.$

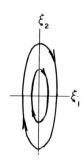

(b) No critical points. $\dfrac{x_2^2}{2} - \dfrac{K}{x_1} = \dfrac{u_0^2}{2} - \dfrac{K}{x_0} = v_0$.

Thus $x^2 = \pm\sqrt{2v_0 + \dfrac{2K}{x_1}}$.

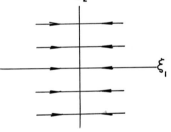

(c) Critical point: $(0,0)$. General solution:

$$x(t) = c_1 e^{-3t}\begin{pmatrix} 2 \\ 1 \end{pmatrix} + c_2 e^{-6t}\begin{pmatrix} 1 \\ -1 \end{pmatrix}.$$ Trajectories are

sketched in Figure 36-1.

(d) Critical points: $(0,\xi_2)$

for all ξ_2.

General solution:

$x_1(t) = c_1 e^{-t}$,

$x_2(t) = c_2$.

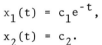

6. Critical points: 0 and P. See Figure 21-2.

9. Consider the system $x_1' = -x_1$, $x_2' = -x_2$. Even though

$g(\xi_1,\xi_2) = \xi_1^2 + \xi_2^2$ is zero at $(0,0)$, the new system has

the same critical point. $(0,0)$, as the given system.

The phase portrait is indicated in Figure 36-2.

Section 36

1. The only linearly independent
    eigenvector is $\begin{bmatrix} 1 \\ 1 \end{bmatrix}$.

    $\lambda = -2$.

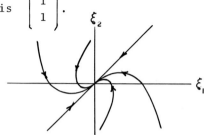

2. General solution: $x(t) = c_1 e^{\sqrt{k}\,t} + c_2 e^{-\sqrt{k}\,t}$. In the phase
    plane, $(0,0)$ is a saddle point. Unless the initial con-
    ditions are exactly such that $c_1 = 0$, a completely im-
    probable case, one will have $x(t) \to \infty$ like $c_1 e^{\sqrt{k}\,t}$ as
    $t \to \infty$.

Section 37

1. In polar coordinates, $r' = -r$ and $\theta' = (\ln r)^{-1}$ when
    $r > 0$.

2. (a) $(0,0)$ is a stable spiral point.

    (b) To study solutions near $(\pi,0)$, define $X_1(t) = x_1 - \pi$, $X_2(t) = x_2(t)$. Then
    $$X_1' = X_2, \quad X_2' = \frac{g}{\ell} \sin X_1 - \frac{b}{m} X_2,$$
    and one finds that $(\pi,0)$ is a saddle point. Simi-
    lar analysis shows that $(2\pi,0)$ is a stable spiral
    point.

    (c) Defining $v(t)$ as in Example 24-1, one finds
    $v'(t) \geq -(2b/m)v(t)$. From this, show that
    $x_2^2(t) + 4g/\ell \geq \omega_0^2 e^{-2bt/m}$ for $t \geq 0$. Thus if

$\omega_0 > 2\sqrt{g/\ell}$, then $x_2(t)$ remains positive as long as $\omega_0 e^{-bt/m} > 2\sqrt{g/\ell}$. Then eventually $(x_1(t), x_2(t))$ spirals in toward $(2n\pi, 0)$ for some positive integer n.

3. (a) Saddle point.

   (b) Unstable spiral point.

   (c) If $0 < \mu < 2$, an unstable spiral point.

   If $\mu = 2$, an unstable node.

   If $\mu > 2$, a saddle point.

4. In polar coordinates, $r' = -r^3$ and $\theta' = 1$ when $r > 0$.

INDEX

Abel's formula, 109, 150

Applications of delay differential equations, 226

references, 239

Asymptotic stability, 363

uniform, 364

Automobile shock absorbers, 91

Autonomous systems, 329, 349, 400

Backwards solution of a delay differential equation, 248

Barbašin-Krasovskiĭ Thm., 371

Basic initial problem, 255

Bendixson, 400, 431, 448

Bernoulli's equation, 38

Bessel's eq. of order $\frac{1}{2}$, 138

Bolzano-Weierstrass Thm., 460

Boundary value problems, 78

Bounded delays, 285

Capacitance, 119, 177, 178

Cauchy criterion, 461

Center, 423

Characteristic equation, 92, 163, 320

Closed interval, 449, 452

Closed set, 451

Complex exponentials, 67

Complex functions, 63

Continuation of solutions, 275, 304

Continuity Condition (C), 290

Continuous dependence on initial conditions, 74, 193, 264, 297, 350

Control system with delay, 234

Convergence estimate, 202

Critical damping, 91

Critical point, 407

isolated, 425

Damped oscillations, 90, 168

Damping coefficient, 90

Decoupling method, 152

Diagonalizing matrices, 151

Differential inequality, 33, 73

with delay, 389

Differential operators, 85

Eigenvalues, 147

Eigenvectors, 147

generalized, 156

Electrical circuits, 16, 18, 40, 77, 119, 177, 178

Electron two-body problem, 237

Energy, 280

Error estimate, 202, 220, 273

Errors, growth of, 74, 193, 265, 266, 297

Euler's equation, 108, 127, 129, 135

Existence of solutions, 212, 270, 299

Exponential decay, 244, 325, 363, 373, 377, 386

Exponential of a matrix, 197

First integral, 432, 437

First order o.d.e., 7, 27

Fixed point of a mapping, 274

Focus, 421

Frequency, natural, 119

Friction, 52, 119, 173

Functional diff. eq. 225, 288

Functions, 452

   of a matrix, 197

Fundamental matrix, 187

Generalized eigenvectors, 156

Global

   behavior of solutions, 431

   existence, 281

Gravitation, 58, 62

Gronwall's Lemma, 72

Growth of solutions, 168, 195, 267, 324, 336

Heine-Borel Theorem, 460

Homogeneous linear equations, 85, 140, 247, 314

Inductance, 17, 119, 178

Inequalities

   delay differential, 389

   Gronwall's, 72

   Halanay's, 390

Inequalities

   Reid's, 72

Infimum, 35

Infinitesimal upper bound, 354

Initial conditions, 3, 7, 41, 53

Initial function, 227

Integral equations, 212

Integral inequality, 72, 75

Integrating factor, 9, 208

Interchanging order of integration, 464, 465

Liénard, A., 441

Limit cycle, 441

Linear combination, 99, 162

Linear delay differential equations, 246, 313

   with constant coefficients, 319

   stability, 384

   variation of parameters, 331

Linear dependence, 101, 142

Linear independence, 101, 142

Linear operations, 43

Linear operators, 86, 99, 246

Linear ordinary diff. eqs.

   first order scalar, 7

   higher order scalar, 57, 84, 222

   systems, 49, 75, 139, 414

   variation of parameters, 125, 180

Linear vector space, 42

Lipschitz conditions

  for functionals, 292

  local, 259, 261, 293, 300

  for scalar functions, 36

  for vector functions, 46,
   258

Lotka-Volterra equations, 232,
  435

Lyapunov's direct method, 352

Lyapunov functions and
  functionals, 353

Matrix exponentials, 197

Matrix-valued function, 140

  continuous, 141, 192

  differentiable, 141, 192

  integrable, 141

Matrix sequences, 198

Matrix series, 199

Mean Value Theorem, 454

  function of n variables, 457

Method of steps, 228

Minorsky's equation, 399

Mixing of liquids, 13, 18,
  226, 240

Nodes

  improper, 416, 418

  proper, 417

Noncontinuable solutions, 275,
  304

Nonunique solutions, 24, 37,
  39, 80, 82, 241, 248, 250,
  267

Norm

  of a function, 291

  of a matrix, 190

  of a vector, 45

  weighted, 283

Open interval, 449, 451

Open set, 451

Order of a diff. eq., 6

Oscillations, 119, 168, 235,
  243, 244

Overdamping, 90

Peano, 72

Pendulum, 278, 362, 381, 383,
  428, 430, 431

Periodic solutions, 408

Perron, 379, 400

Phase plane, 405

Phase portrait, 405

Phase space, 405

Picard iteration, 212

Planetary motion, 58

Poincaré, 400, 405, 431, 448

Polar coordinates, 420, 427

Population growth, 229

Positive definite functional,
  354

Prey-predator models, 232, 435

Prime number theorem, 237

Quasi-bounded functionals, 305